T0350789

Modeling, Simulation, and Control of AI Robotics and Autonomous Systems

Tanupriya Choudhury
Graphic Era University, India

Anitha Mary X
Karunya Institute of Technology and Sciences, India

Subrata Chowdhury
Sreenivasa Institute of Technology and Management Studies, India

C. Karthik
Jyothi Engineering College, India

C. Suganthi Evangeline
Sri Eshwar College of Engineering, India

A volume in the Advances in Computational Intelligence and Robotics (ACIR) Book Series

Published in the United States of America by
IGI Global
Engineering Science Reference (an imprint of IGI Global)
701 E. Chocolate Avenue
Hershey PA, USA 17033
Tel: 717-533-8845
Fax: 717-533-8661
E-mail: cust@igi-global.com
Web site: http://www.igi-global.com

Library of Congress Cataloging-in-Publication Data

CIP DATA PROCESSING

ISBN 9798369319628(hc) | ISBN 9798369347119(sc) | eISBN 9798369319635

This book is published in the IGI Global book series Advances in Computational Intelligence and Robotics (ACIR) (ISSN: 2327-0411; eISSN: 2327-042X)

British Cataloguing in Publication Data
A Cataloguing in Publication record for this book is available from the British Library.

For electronic access to this publication, please contact: eresources@igi-global.com.

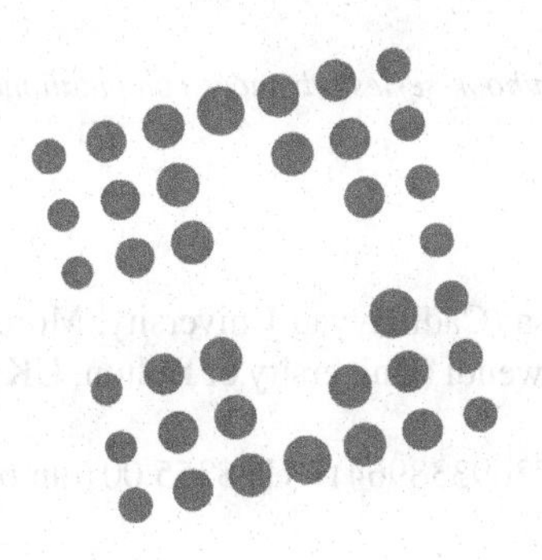

Advances in Computational Intelligence and Robotics (ACIR) Book Series

Ivan Giannoccaro
University of Salento, Italy

ISSN:2327-0411
EISSN:2327-042X

Mission

While intelligence is traditionally a term applied to humans and human cognition, technology has progressed in such a way to allow for the development of intelligent systems able to simulate many human traits. With this new era of simulated and artificial intelligence, much research is needed in order to continue to advance the field and also to evaluate the ethical and societal concerns of the existence of artificial life and machine learning.

The **Advances in Computational Intelligence and Robotics (ACIR) Book Series** encourages scholarly discourse on all topics pertaining to evolutionary computing, artificial life, computational intelligence, machine learning, and robotics. ACIR presents the latest research being conducted on diverse topics in intelligence technologies with the goal of advancing knowledge and applications in this rapidly evolving field.

Coverage

- Neural Networks
- Computational Logic
- Natural Language Processing
- Cognitive Informatics
- Synthetic Emotions
- Evolutionary Computing
- Adaptive and Complex Systems
- Brain Simulation
- Heuristics
- Fuzzy Systems

IGI Global is currently accepting manuscripts for publication within this series. To submit a proposal for a volume in this series, please contact our Acquisition Editors at Acquisitions@igi-global.com or visit: http://www.igi-global.com/publish/.

The Advances in Computational Intelligence and Robotics (ACIR) Book Series (ISSN 2327-0411) is published by IGI Global, 701 E. Chocolate Avenue, Hershey, PA 17033-1240, USA, www.igi-global.com. This series is composed of titles available for purchase individually; each title is edited to be contextually exclusive from any other title within the series. For pricing and ordering information please visit http://www.igi-global.com/book-series/advances-computational-intelligence-robotics/73674. Postmaster: Send all address changes to above address.

Titles in this Series

For a list of additional titles in this series, please visit: http://www.igi-global.com/book-series/advances-computational-intelligence-robotics/73674

AI and IoT for Proactive Disaster Management
Mariyam Ouaissa (Chouaib Doukkali University, Morocco) Mariya Ouaissa (Cadi Ayyad University, Morocco) Zakaria Boulouard (Hassan II University, Casablanca, Morocco) Celestine Iwendi (University of Bolton, UK) and Moez Krichen (Al-Baha University, Saudi Arabia)
Engineering Science Reference • copyright 2024 • 299pp • H/C (ISBN: 9798369338964) • US $355.00 (our price)

Utilizing AI and Machine Learning for Natural Disaster Management
D. Satishkumar (Nehru Institute of Technology, India) and M. Sivaraja (Nehru Institute of Technology, India)
Engineering Science Reference • copyright 2024 • 340pp • H/C (ISBN: 9798369333624) • US $315.00 (our price)

Shaping the Future of Automation With Cloud-Enhanced Robotics
Rathishchandra Ramachandra Gatti (Sahyadri College of Engineering and Management, India) and Chandra Singh (Sahyadri College of Engineering and Management, India)
Engineering Science Reference • copyright 2024 • 431pp • H/C (ISBN: 9798369319147) • US $345.00 (our price)

Bio-inspired Swarm Robotics and Control Algorithms, Mechanisms, and Strategies
Parijat Bhowmick (Indian Institute of Technology, Guwahati, India) Sima Das (Bengal College of Engineering and Technology, India) and Farshad Arvin (Durham University, UK)
Engineering Science Reference • copyright 2024 • 261pp • H/C (ISBN: 9798369312773) • US $315.00 (our price)

Comparative Analysis of Digital Consciousness and Human Consciousness Bridging the Divide in AI Discourse
Remya Lathabhavan (Indian Institute of Management, Bodh Gaya, India) and Nidhi Mishra (Indian Institute of Management, Bodh Gaya, India)
Engineering Science Reference • copyright 2024 • 355pp • H/C (ISBN: 9798369320150) • US $315.00 (our price)

Machine Learning Techniques and Industry Applications
Pramod Kumar Srivastava (Rajkiya Engineering College, Azamgarh, India) and Ashok Kumar Yadav (Rajkiya Engineering College, Azamgarh, India)
Engineering Science Reference • copyright 2024 • 307pp • H/C (ISBN: 9798369352717) • US $365.00 (our price)

Intelligent Decision Making Through Bio-Inspired Optimization
Ramkumar Jaganathan (Sri Krishna Arts and Science College, India) Shilpa Mehta (Auckland University of Technology, New Zealand) and Ram Krishan (Mata Sundri University Girls College, Mansa, India)
Information Science Reference • copyright 2024 • 275pp • H/C (ISBN: 9798369320730) • US $320.00 (our price)

701 East Chocolate Avenue, Hershey, PA 17033, USA
Tel: 717-533-8845 x100 • Fax: 717-533-8661
E-Mail: cust@igi-global.com • www.igi-global.com

Table of Contents

Preface xiv

Chapter 1
Autonomous System and AI 1
T. Preethiya, SRM Institute of Science and Technology, India
Priyanga Subbiah, SRM Institute of Science and Technology, India
T. Pandiarajan, Rajalakshmi Institute of Technology, India
Karthikeyan Subramanian, Birmingham City University, UAE
Prince Chelladurai, University College of Engineering, Villupuram, India
C. Selvalakshmi, Mangayarkarasi College of Engineering, India

Chapter 2
Explainable AI-Based Semantic Object Detection for Autonomous Vehicles 15
Elamathiyan A., Karpagam Academy of Higher Education, India
G. Dhivya, Karpagam Academy of Higher Education, India

Chapter 3
AI-Based In-Cabin Monitoring System for Autonomous Vehicles 33
Ajith K., Karpagam Academy of Higher Education, India
R. Sharmila, Karpagam Academy of Higher Education, India

Chapter 4
Real-Time Fault Detection and Condition Monitoring for Industrial Autonomous Vehicles 54
Thirusudhan M., Karpagam Academy of Higher Education, India
Sasikumar S., Karpagam Academy of Higher Education, India

Chapter 5
IoT in Day-to-Day Life: Vehicle Body Fatigue Analysis 73
Rajaram Vassudev Pai Kuchelkar, Liverpool John Moores University, India
Anupama Jawale, Narsee Monjee College of Commerce and Economics, India

Chapter 6
Design and Fabrication of a Softrobotic Gripper for Involving Underwater Vehicles in Seaweed Farming 95
Prabhakar Gunasekaran, Thiagarajar College of Engineering, India
Meenakshi S., Thiagarajar College of Engineering, India
Jainulafdeen A., K. Ramakrishnan College of Engineering, India
Ayyanar N., Thiagarajar College of Engineering, India
Rajalakshmi Murugesan, Thiagarajar College of Engineering, India

Chapter 7
Secure VANET Routing Protocols for Improved Vehicular Communication in Autonomous Systems 109
Abarna S., National Engineering College, India
Naskath J., National Engineering College, India
Rathi Pathi R., National Engineering College, India
Jeyalakshmi C., Mohamed Sathak Engineering College, India

Chapter 8
Autonomous Systems Revolutionizing Health Insurance Industry: Achieving Operational Excellence in Services 131
Anupa Stanly, Karunya Institute of Technology and Sciences, India
K. Aruna, Karunya Institute of Technology and Sciences, India

Chapter 9
Artificial Intelligence in Robotics 152
Preethiya T., SRM Institute of Science and Technology, India
Priyanga Subbiah, SRM Institute of Science and Technology, India
Pandiarajan T., Rajalakshmi Institute of Technology, India
Stephen Ojo, College of Engineering, Anderson University, USA
Vijayalakshmi S., SNS College of Engineering, India

Chapter 10
Synergistic Swarm: Multi-Robot Systems in Healthcare 166
Jaspreet Kaur, Chandigarh University, India

Chapter 11
Cogwheels of Care: Robotic Marvels in the Hospital Landscape 180
Jaspreet Kaur, Chandigarh University, India

Chapter 12
Bio-Inspired Nanorobots for Cancer Diagnosis and Therapy 196
Anshit Mukherjee, Abacus Institute of Engineering and Management, India
Gunjan Mukherjee, Brainware University, India

Chapter 13
Advanced Biomimetic Compound Continuum Robot for Minimally Invasive Surgical Applications213
Ranjit Barua, Omdayal Group of Institutions, India

Chapter 14
An Intelligent Robotic Fogger System for Predicting Dengue Outbreaks 232
D. Raveena Judie Dolly, Karunya Institute of Technology and Sciences, India
D. J Jagannath, Karunya Institute of Technology and Sciences, India
J. Dinesh Peter, Karunya Institute of Technology and Sciences, India

Compilation of References 245

Related References 267

About the Contributors 287

Index 294

Detailed Table of Contents

Preface xiv

Chapter 1
Autonomous System and AI 1
T. Preethiya, SRM Institute of Science and Technology, India
Priyanga Subbiah, SRM Institute of Science and Technology, India
T. Pandiarajan, Rajalakshmi Institute of Technology, India
Karthikeyan Subramanian, Birmingham City University, UAE
Prince Chelladurai, University College of Engineering, Villupuram, India
C. Selvalakshmi, Mangayarkarasi College of Engineering, India

Autonomous systems and AI have revolutionized industry automation and innovation. This abstract shows how this synergy changes our lives and work. AI has rapidly advanced autonomous systems like self-driving cars, drones, and robots. These systems can sense, make smart decisions, and adapt to changing environments with AI. They can navigate complex settings and interact with humans thanks to this superior sense Autonomous systems can make real-time judgments using AI methods like machine learning and deep learning. AI-driven decision-making assures adaptability and efficiency, whether a self-driving car chooses the safest route, or a drone optimizes its flying path. AI-assisted autonomous systems prioritize safety and reliability. These systems can detect anomalies and respond proactively to prevent accidents and failures through self-monitoring and diagnostics.

Chapter 2
Explainable AI-Based Semantic Object Detection for Autonomous Vehicles 15
Elamathiyan A., Karpagam Academy of Higher Education, India
G. Dhivya, Karpagam Academy of Higher Education, India

A goal is to facilitate the recognition and segmentation of the route driven by autonomous vehicles through the use of machine learning (ML) models. The pixel-wise road detection task, the semantic segmentation architectures underwent training and comparison. By using XAI, the authors are able to interpret and read the predictions generated by these abstract models. They generated arguments for the recommended segmentation model for autonomous vehicle road detection using a range of XAI approaches. Supervised learning is enabled by KNN, Decision Tree, and Random Forest, which are the current algorithms used for comparison. On the other hand, the newly built K means clustering function best when paired for image processing since they are good with images. To display the results of computations for the evaluation parameters of each algorithm, including accuracy, sensitivity, recall and precision, tables, and the necessary features from the evaluation matrices are utilised. The k-means clustering system for explainable AI-based semantic object detection in automated cars achieves 94.58% accuracy on both train and test sets.

Chapter 3
AI-Based In-Cabin Monitoring System for Autonomous Vehicles 33
Ajith K., Karpagam Academy of Higher Education, India
R. Sharmila, Karpagam Academy of Higher Education, India

Fully autonomous vehicles (FAVs) require internal monitoring in order to function without a human driver. The far-from-sufficient FAV, sufficient in-cabin monitoring is a prerequisite to ensure both people and vehicles. On public roadways, there are a lot of accidents that happen, most of them are the result of reckless driving. Modern driver monitoring systems evaluate driver behavior and, if necessary, highlight risky driving behaviours using special sensor technologies. The result accurately predicted bounding boxes and the real data show a considerable amount of overlap. Unlike most past efforts, the authors use a random forest to learn a template-based model. This way, forecast the object probability of a window in a sliding window technique and regress its aspect ratio using a single mode at the same time. Examined mobility services at increasing degrees autonomy, including the exercise caution and the best ways.

Chapter 4
Real-Time Fault Detection and Condition Monitoring for Industrial Autonomous Vehicles 54
Thirusudhan M., Karpagam Academy of Higher Education, India
Sasikumar S., Karpagam Academy of Higher Education, India

Real-time status monitoring and early defect diagnosis are becoming more and more necessary for modern industrial systems. Developing intelligent remote diagnostic technologies and integrating on-board and off-board diagnosis are two more unresolved research projects in the automotive industry. The automated transfer vehicle (ATV) equipment condition monitoring example in this chapter is part of a smart industrial use case. The suggested method successfully permits ATV fault scenarios to be seen in real time by expanding to a fleet of devices in an actual production plant. The application of a statistical threshold is the initial stage in creating a high-performance detection model for defect detection on stacked long short-term memory networks. For better computation speed, a second model trimming approach based on principal component analysis is suggested. The improved fault detection technique is ultimately applied by the airborne embedded computer platform with field-programmable gate arrays.

Chapter 5
IoT in Day-to-Day Life: Vehicle Body Fatigue Analysis 73
Rajaram Vassudev Pai Kuchelkar, Liverpool John Moores University, India
Anupama Jawale, Narsee Monjee College of Commerce and Economics, India

One of the analyses is fatigue analysis which helps determine the durability of the product for its intended life. The fatigue analysis requires an input data from vehicle-road interaction obtained as vertical acceleration. The input data is then used in finite element analysis software or in a test rig instrumentation to perform a fatigue analysis. The focus of this study is about collecting the input data using a data acquisition device. OEM of the automobiles use proprietary data acquisition device for testing their vehicles for durability. However, the automotive body building companies or vehicle body builders which are 1-Tier companies below OEM have a limited budget for such an analysis and therefore they principally lag in advance analysis approach such as a fatigue analysis for the design of vehicle body. Instead, to sustain durability of vehicle, the design is made heavier. This research study is an effort to provide an affordable setup for data acquisition by making use of IoT and web technologies which are cost effective.

Chapter 6
Design and Fabrication of a Softrobotic Gripper for Involving Underwater Vehicles in Seaweed Farming 95
Prabhakar Gunasekaran, Thiagarajar College of Engineering, India
Meenakshi S., Thiagarajar College of Engineering, India
Jainulafdeen A., K. Ramakrishnan College of Engineering, India
Ayyanar N., Thiagarajar College of Engineering, India
Rajalakshmi Murugesan, Thiagarajar College of Engineering, India

Seaweeds, crucial components of marine ecosystems, thrive in marine and coastal waters, notably in the Gulf of Mannar Biosphere. Despite their ecological importance, certain seaweed species pose a threat to coral reefs due to the release of hydrophobic allelochemicals. Fisherwomen, risking their lives in collecting seaweed seed stock from littoral zones, rocky shores, and deep-sea waters, encounter hazards such as sharp rocks, poisonous algae, cyanobacteria, and coral reefs. Societal research pinpoints risks, overexploitation, unorganized harvesting, and algal blooms as significant issues. The proposed solution involves integrating intelligent Soft Robotic Grippers into underwater vehicles for precise seaweed farming without harming coral reefs. Abacus FEA software aids in deformation analysis, guiding the gripper's design to safeguard fisherwomen and preserve coral reefs.

Chapter 7
Secure VANET Routing Protocols for Improved Vehicular Communication in Autonomous Systems 109
Abarna S., National Engineering College, India
Naskath J., National Engineering College, India
Rathi Pathi R., National Engineering College, India
Jeyalakshmi C., Mohamed Sathak Engineering College, India

Vehicular ad hoc networks (VANETs) represent an advanced iteration of mobile ad hoc networks (MANETs) designed specifically for internet communication within vehicles. VANETs aim to enable vehicle-to-vehicle (V2V) communication, enhancing safety and convenience for drivers and passengers. However, the open nature of ad hoc networks and the absence of a well-defined line of defense make security a crucial concern for VANETs. Prior to deploying mobile ad hoc networks in hostile or sensitive areas, it becomes imperative to establish robust security services. This study addresses the need for a trusted VANET routing protocol that incorporates a diverse range of security services. The proposed approach implements a secure routing protocol based on the Dijkstra Algorithm to identify the secure and shortest path. For ensuring secure routing, the protocol employs route request (RREQ) and route reply (RREP) mechanisms to identify trustworthy nodes. Additionally, message authentication is utilized to provide end-to-end, hop-to-hop, and entire-route authentication. To transmit messages securely, the Diffie-Hellman Key Exchange Protocol is employed for message encryption, ensuring safe delivery to the intended destination. To assess the performance of the suggested protocol, the authors conducted simulations using NS2. These simulation results demonstrate that the proposed routing protocol outperforms existing methods, affirming its effectiveness in VANET environments.

Chapter 8
Autonomous Systems Revolutionizing Health Insurance Industry: Achieving Operational Excellence in Services 131
Anupa Stanly, Karunya Institute of Technology and Sciences, India
K. Aruna, Karunya Institute of Technology and Sciences, India

Financial services, particularly the insurance service sector, are increasingly embracing technology. Autonomous systems, which include artificial intelligence (AI), machine learning, and automation, are driving a striking revolution in the health insurance industry. This research provides a comprehensive analysis of various facets, such as the use of autonomous systems, expediting claim processing, identifying, and preventing fraud, improving the customer experience, data-driven decision-making, and adherence to healthcare legislation. The study concludes by highlighting the sector's profound impact from autonomous systems and pointing to a promising future for health insurance that will be characterised by operational effectiveness and customer-centricity. This study sets out a thorough exploration of the dynamic world where autonomous systems are changing the laws governing health insurance. This investigation's main objectives are to achieve operational excellence and provide services that are utterly customer-centric.

Chapter 9
Artificial Intelligence in Robotics 152
Preethiya T., SRM Institute of Science and Technology, India
Priyanga Subbiah, SRM Institute of Science and Technology, India
Pandiarajan T., Rajalakshmi Institute of Technology, India
Stephen Ojo, College of Engineering, Anderson University, USA
Vijayalakshmi S., SNS College of Engineering, India

The advent of artificial intelligence (AI) has had a profound impact on the realm of robotics, fundamentally altering the capabilities of self-governing devices. This abstract examines the significant influence of artificial intelligence (AI) on the field of robotics, emphasizing notable progress and practical implementations. Artificial intelligence (AI)-powered robots demonstrate improved capabilities in perception, decision-making, and adaptability, which allows them to thrive in a wide range of jobs across several domains such as industry, healthcare, space exploration, and autonomous vehicles. Machine learning methodologies, such as deep learning and reinforcement learning, enable robots to acquire aptitudes, enhance their performance, and engage in intelligent interactions with their surroundings. The ethical considerations, safety measures, and societal repercussions pertaining to AI-driven robots are also examined and analyzed.

Chapter 10
Synergistic Swarm: Multi-Robot Systems in Healthcare.. 166
Jaspreet Kaur, Chandigarh University, India

The "synergistic swarm" investigates the incorporation of multi-robot systems in healthcare, introducing a fundamental change in patient care and medical operations. This abstract emphasises the collaborative synergy achieved by intelligently coordinating several robotic entities, resulting in improved efficiency, precision, and adaptability in healthcare environments. By utilising cutting-edge technology like artificial intelligence, robotics, and sensor networks, the system seeks to enhance many functions, including diagnostics and patient support, to their maximum efficiency. This chapter highlights the significant potential of combining different approaches to healthcare in order to improve the delivery of medical services. This could lead to more effective, patient-focused, and adaptable healthcare robotics in the changing healthcare industry.

Chapter 11
Cogwheels of Care: Robotic Marvels in the Hospital Landscape 180
Jaspreet Kaur, Chandigarh University, India

"Cogwheels of Care: Robotic Marvels in the Hospital Landscape" examines the incorporation of sophisticated robotics in contemporary healthcare. This abstract explores the profound influence of robots in hospitals, fundamentally changing patient care and enhancing medical capabilities. The chapter explores the ways in which these advanced robots enhance and simplify many jobs, ranging from surgical procedures to everyday activities, by maximising efficiency and accuracy. It emphasises the interdependent connection between technology and healthcare practitioners, focusing on the ethical considerations and societal consequences of this technological transformation. This research highlights the changing healthcare landscape, where the complex interaction between human expertise and robotic innovation is transforming the principles of compassionate and effective patient-centered care.

Chapter 12
Bio-Inspired Nanorobots for Cancer Diagnosis and Therapy 196
Anshit Mukherjee, Abacus Institute of Engineering and Management, India
Gunjan Mukherjee, Brainware University, India

Cancer is one of the most serious threats to human health and life. Despite the advances in conventional therapies, such as surgery, chemotherapy, radiotherapy, and immunotherapy, there are still many challenges and limitations in achieving effective and precise cancer treatment. Nanorobots, inspired by natural biological nanomachines, offer a promising alternative for cancer diagnosis and therapy. Nanorobots are nanoscale devices that can perform various tasks under the guidance of external stimuli, such as magnetic fields, light, ultrasound, or chemical gradients. Nanorobots can be designed to target specific cancer cells or tissues, deliver drugs or genes, sense tumor biomarkers, perform minimally invasive surgery, or combine multiple functions for comprehensive treatment. In this chapter, the authors review the recent progress and applications of bio-inspired nanorobots for cancer diagnosis and therapy, with a focus on magnetic field-driven nanorobots. They also discuss the challenges and future perspectives of nanorobots in clinical translation.

Chapter 13
Advanced Biomimetic Compound Continuum Robot for Minimally Invasive Surgical Applications213
Ranjit Barua, Omdayal Group of Institutions, India

the bio-inspired compound continuum robot represents a groundbreaking innovation in the realm of minimally invasive surgery (MIS). Drawing inspiration from the flexibility and adaptability observed in nature, this robotic system employs a novel approach to navigating complex anatomical structures with enhanced precision. Mimicking the serpentine motion of snakes, the robot utilizes a compound continuum structure composed of interconnected segments. This design allows for unparalleled maneuverability, enabling the robot to navigate through confined spaces and intricate pathways within the human body. By emulating the biomechanics of natural organisms, the robot can reach anatomical locations that traditional rigid instruments might struggle to access. In this chapter, the authors will discuss the advanced biomimetic compound continuum robot for minimally invasive surgical applications.

Chapter 14
An Intelligent Robotic Fogger System for Predicting Dengue Outbreaks 232
D. Raveena Judie Dolly, Karunya Institute of Technology and Sciences, India
D. J Jagannath, Karunya Institute of Technology and Sciences, India
J. Dinesh Peter, Karunya Institute of Technology and Sciences, India

Humans may encounter an arboviral illness through viruses transmitted by mosquitoes, commonly resulting in a fever known as breakbone fever. This term reflects the severity of muscle spasms and joint pains associated with the illness. While some cases are asymptomatic, others can be fatal. Dengue awareness often arises during seasonal changes. The integration of AI in dengue prediction becomes crucial for early diagnosis and treatment. Utilizing appropriate deep learning classifiers can aid in categorizing cases based on their severity. This article advocates for the implementation of an intelligent robotic fogger system in predicted areas. This approach employs interprofessional strategies to safeguard health workers and residents in regions prone to dengue outbreaks.

Compilation of References 245

Related References 267

About the Contributors 287

Index 294

Preface

Welcome to *Modeling, Simulation, and Control of AI Robotics and Autonomous Systems*, edited by Tanupriya Choudhury, Anitha Mary X, Subrata Chowdhury, C. Karthik, and C. Suganthi Evangeline.

Intelligent Robotics and Autonomous Systems (IRAS) represent a confluence of robotics, artificial intelligence (AI), and control systems aimed at crafting intelligent machines capable of autonomous task execution. This edited volume delves into the intricate realms of system modeling, simulation, and control, elucidating their paramount significance in advancing the frontiers of IRAS.

System modeling, the foundational pillar, entails the creation of mathematical constructs that delineate the intricate dynamics of robotic systems. It encompasses a meticulous portrayal of kinematics, dynamics, sensors, actuators, and their interplay, thereby offering invaluable insights into system behaviors across diverse scenarios.

Simulation emerges as a pivotal tool, affording engineers the capability to scrutinize and refine their designs virtually, obviating the need for resource-intensive real-world testing. Leveraging software frameworks like MATLAB/Simulink or ROS, researchers can orchestrate a myriad of simulated environments and sensor inputs, gauging the efficacy of their designs under varying conditions.

Control, propelled by AI methodologies, assumes a central role in endowing robots with autonomous decision-making prowess. From reinforcement learning algorithms facilitating experiential learning to intricate control schemes orchestrating complex maneuvers, the amalgamation of AI and control theory augments the autonomy and efficacy of robotic systems manifold.

The applications of these methodologies reverberate across multifarious domains, exemplified vividly in the realm of autonomous driving systems and industrial automation. These technologies herald a paradigm shift, revolutionizing industries and augmenting human capabilities.

Moreover, the resurgence of research in autonomous systems underscores a transformative era characterized by unprecedented technological advancements. From chess-playing algorithms to self-driving vehicles and manufacturing line robots, the landscape of autonomous systems burgeons with innovation and promise.

As we embark on this journey through the realms of robotics and autonomous systems, this edited volume endeavors to furnish readers with a panoramic vista of cutting-edge breakthroughs, theoretical frameworks, and computational paradigms. It is our fervent hope that this compendium serves as a beacon, illuminating pathways towards the realization of intelligent, autonomous systems poised to redefine the contours of human endeavor.

ORGANIZATION OF THE BOOK

Chapter 1 delves into the symbiotic relationship between Autonomous Systems and Artificial Intelligence (AI), which has revolutionized industry automation and innovation. It explores how this synergy has transformed various aspects of our lives and work, particularly focusing on advancements in self-driving cars, drones, and robots. The chapter elucidates how AI empowers autonomous systems to sense, make intelligent decisions, and adapt to dynamic environments. Furthermore, it delves into AI-driven decision-making processes, emphasizing safety, reliability, and proactive anomaly detection to prevent accidents and failures.

Chapter 2 focuses on Explainable AI-Based Semantic Object Detection for Autonomous Vehicles, detailing machine learning models' applications in route recognition and segmentation. It explores various XAI approaches to interpret model predictions and evaluates their performance in autonomous vehicle road detection, emphasizing accuracy and efficiency.

Chapter 3 presents an AI-Based In-Cabin Monitoring System for Autonomous Vehicles, highlighting the importance of internal monitoring for fully autonomous vehicles. It discusses driver behavior evaluation, risky driving behavior detection, and the implementation of a Random Forest model to enhance object detection and user privacy.

Chapter 4 addresses Real-Time Fault Detection and Condition Monitoring for Industrial Autonomous Vehicles, emphasizing the necessity of early defect diagnosis in modern industrial systems. It proposes intelligent diagnostic technologies and on-board/off-board diagnosis integration, exemplified through an Automated Transfer Vehicle (ATV) equipment condition monitoring application.

Chapter 5 explores IoT in Day-to-Day Life, focusing on Vehicle Body Fatigue Analysis. It introduces an affordable IoT-based data acquisition setup for fatigue analysis in vehicle body design, aiming to bridge the gap between automotive manufacturers and vehicle body builders in durability testing.

Chapter 6 discusses the Design and Fabrication of a Softrobotic Gripper for Involving Underwater Vehicles in Seaweed Farming, highlighting the ecological importance of seaweeds and the hazards associated with manual harvesting. The chapter proposes integrating Soft Robotic Grippers into underwater vehicles for precise and eco-friendly seaweed farming.

Chapter 7 introduces Secure VANET Routing Protocols for Improved Vehicular Communication in Autonomous Systems, emphasizing security concerns in Vehicular Ad Hoc Networks (VANETs). It proposes a secure VANET routing protocol based on the Dijkstra Algorithm, incorporating message authentication and encryption for secure communication.

Chapter 8 explores Autonomous Systems' Revolutionizing Health Insurance Industry, focusing on operational excellence in health insurance services. It examines the impact of autonomous systems, including AI, machine learning, and automation, in expediting claim processing, fraud detection, and enhancing customer experience.

Chapter 9 delves into Artificial Intelligence in Robotics, emphasizing AI's transformative influence on robotic capabilities. It explores practical implementations of AI-powered robots in various domains, highlighting machine learning methodologies' role in enhancing perception, decision-making, and adaptability in robots.

Chapter 10 investigates Synergistic Swarm: Multi-Robot Systems in Healthcare, emphasizing the collaborative synergy achieved by coordinating multiple robotic entities in healthcare settings. It explores the applications of AI, robotics, and sensor networks in enhancing diagnostics, patient support, and medical operations.

Chapter 11 examines Cogwheels of Care: Robotic Marvels in the Hospital Landscape, exploring the profound influence of advanced robotics on patient care and medical capabilities. It discusses the ethical considerations and societal implications of integrating robots into healthcare, emphasizing the transformative potential of this technological evolution.

Chapter 12 introduces Bio-Inspired Nanorobots for Cancer Diagnosis and Therapy, exploring nanorobots' promising applications in cancer treatment. It discusses the recent progress and challenges in designing nanorobots for targeted drug delivery, sensing tumor biomarkers, and performing minimally invasive surgery, with a focus on magnetic field-driven nanorobots.

Chapter 13 presents Advanced Biomimetic Compound Continuum Robot for Minimally Invasive Surgical Applications, detailing a groundbreaking robotic system inspired by natural biomechanics. It explores the compound continuum robot's flexibility and adaptability in navigating complex anatomical structures, particularly in minimally invasive surgery.

Chapter 14 discusses An Intelligent Robotic Fogger System for Predicting Dengue Outbreaks, focusing on utilizing AI-driven robotic systems for disease prevention. It proposes implementing an intelligent robotic fogger system in predicted dengue outbreak areas, emphasizing interprofessional strategies to safeguard public health.

IN CONCLUSION

As we draw the curtains on this compendium, *Modeling, Simulation,* and *Control of AI Robotics and Autonomous Systems,* we reflect on the myriad facets of innovation, ingenuity, and interdisciplinary collaboration encapsulated within its pages. The journey through the realms of intelligent robotics and autonomous systems has been nothing short of exhilarating, unveiling a tapestry of cutting-edge research, transformative methodologies, and visionary perspectives.

From the revolutionary fusion of autonomous systems and AI, redefining industrial automation paradigms, to the advent of edge computing and machine learning heralding a new era in aerial autonomy, each chapter resonates with the collective pursuit of excellence and advancement. Proposals for intelligent bio-inspired autonomous underwater vehicles, explainable AI-based semantic object detection for autonomous vehicles, and real-time fault detection for industrial autonomous vehicles epitomize the relentless quest for innovation and progress.

Moreover, the exploration of human-robot interaction, synergistic swarm multi-robot systems in healthcare, and the integration of sophisticated robotics in the hospital landscape underscores the transformative potential of robotics in reshaping healthcare delivery and patient outcomes. The chapters on secure VANET routing protocols and autonomous systems revolutionizing the health insurance industry underscore the pivotal role of technology in enhancing safety, efficiency, and accessibility across diverse domains.

As we navigate the intricate landscapes of artificial intelligence, machine learning, and robotics, it becomes increasingly evident that our collective endeavors are not merely confined to the realms of academia or industry but resonate deeply with societal aspirations and challenges. The deployment of bio-inspired nanorobots for cancer diagnosis and therapy and the implementation of intelligent robotic fogger systems for predicting dengue outbreaks underscore the profound impact of robotics in addressing pressing global health challenges.

In conclusion, *Modeling, Simulation, and Control of AI Robotics and Autonomous Systems* serves as a testament to the indomitable spirit of human ingenuity and innovation. It is our fervent hope that this compendium not only serves as a comprehensive reference for researchers, engineers, and enthusiasts but also inspires future generations to push the boundaries of what is possible in the dynamic and ever-evolving field of robotics and autonomous systems.

Tanupriya Choudhury
Graphic Era University, India

Anitha Mary X
Karunya Institute of Technology and Sciences, India

Subrata Chowdhury
Sreenivasa Institute of Technology and Management Studies, India

C. Karthik
Jyothi Engineering College, India

C. Suganthi Evangeline
Sri Eshwar College of Engineering, India

Chapter 1
Autonomous System and AI

T. Preethiya
https://orcid.org/0000-0003-3504-1884
SRM Institute of Science and Technology, India

Priyanga Subbiah
https://orcid.org/0000-0002-2395-7492
SRM Institute of Science and Technology, India

T. Pandiarajan
https://orcid.org/0009-0007-7808-8961
Rajalakshmi Institute of Technology, India

Karthikeyan Subramanian
Birmingham City University, UAE

Prince Chelladurai
University College of Engineering, Villupuram, India

C. Selvalakshmi
Mangayarkarasi College of Engineering, India

ABSTRACT

Autonomous systems and AI have revolutionized industry automation and innovation. This abstract shows how this synergy changes our lives and work. AI has rapidly advanced autonomous systems like self-driving cars, drones, and robots. These systems can sense, make smart decisions, and adapt to changing environments with AI. They can navigate complex settings and interact with humans thanks to this superior sense Autonomous systems can make real-time judgments using AI methods like machine learning and deep learning. AI-driven decision-making assures adaptability and efficiency, whether a self-driving car chooses the safest route, or a drone optimizes its flying path. AI-assisted autonomous systems prioritize safety and reliability. These systems can detect anomalies and respond proactively to prevent accidents and failures through self-monitoring and diagnostics.

DOI: 10.4018/979-8-3693-1962-8.ch001

INTRODUCTION

Systems that can operate independently of direct human assistance are known as autonomous systems. They have special traits that permit them autonomy, such as the ability to make judgments, perceive their surroundings through sense and perceptions, process data to make educated decisions, and execute actions or behaviors based on those conclusions. Systems with varying degrees of autonomy can do simple rule-based operations or more sophisticated decision-making powered by Artificial Intelligence (AI) algorithm.

Overview of Autonomous Systems

Autonomous systems come in many forms and are widely used in industries such as robotics and transportation. Self-driving cars, which use sensor data and AI algorithms to navigate roadways and make driving judgments, are a prominent example of the transportation industry. Additionally, a variety of industries, including delivery, disaster relief, and surveillance, have found use for unmanned aerial vehicles, or drones. Another important factor is robotics: service robots help humans with anything from housework to healthcare, while industrial robots streamline production processes (Aguirre & Rodriguez, 2017). Beyond these, there are more autonomous systems in industries such as agriculture and space exploration, each tailored to certain tasks and conditions.

Sensing devices for gathering environmental data and perception algorithms for interpreting it are essential parts of autonomous systems. To evaluate data and come to the right conclusions, these systems use AI-driven decision-making methods like machine learning and reinforcement learning. These systems can map out actions and govern their movements thanks to planning and control techniques. The last phase, known as actuator, is when choices are actually put into action. Examples of this include modifying a vehicle's course or moving a robot.

Although autonomous systems have a lot of potential, there are also a variety of obstacles and restrictions. Keeping these systems safe and dependable remains the top priority, particularly for vital uses like driverless cars and medical robotics (Tong et al., 2019). A number of ethical issues come up, including privacy concerns, decision-making procedures, and human-AI system interactions. Establishing frameworks for regulating new technologies and guaranteeing their appropriate deployment and use also requires addressing legal and regulatory obstacles.

Importance and Impact of AI in Autonomous Systems

The incorporation of Artificial Intelligence into Autonomous Systems represents a significant breakthrough, transforming their potential and influence across various sectors. The cornerstone that allows these systems to operate independently and intelligently is artificial intelligence. It is important because it enhances decision-making processes and enables systems to dynamically adjust to a variety of changing conditions. Through the utilization of AI techniques such as deep learning and machine learning, autonomous systems are capable of processing large volumes of data in real-time and deriving actionable insights. This combination increases the efficiency, accuracy, and adaptability of these systems by enabling them to maneuver through complicated environments, make snap decisions, and continuously learn from their experiences. AI in autonomous systems is having a broad impact on various industries. For example, self-driving cars are revolutionizing transportation, robotic automation is streamlining manufacturing, and surgical robots and diagnostic AI are improving healthcare. Additionally, AI-driven autonomy opens up new avenues for innovation, pushing industries to rethink procedures and discover uncharted territory in terms of technological incorporation while leading to progress in a variety of fields. To fully realize the transformative potential of AI within autonomous systems, however, and to address societal concerns and ensure responsible deployment and usage, strong regulatory frameworks are required, along with ethical considerations, safety assurances, and other issues.

UNDERSTANDING AUTONOMOUS SYSTEMS

Autonomous Systems exhibit distinctive traits, chiefly autonomy in decision-making, perception of their surroundings through sensors, sophisticated data analysis, and execution of actions based on these assessments. This autonomy can range from basic rule-based operations to complex, AI-driven decision-making processes. These systems often comprise a network of sensors—such as cameras, LIDAR, radar, or other environmental detectors—enabling them to gather data crucial for their operations (Cui et al., 2019; Hodge et al., 2021). Subsequently, these data inputs undergo processing and interpretation using AI algorithms, including machine learning and neural networks, to generate informed decisions and actions.

Autonomous systems comprise a broad range of technologies intended to function autonomously, across various domains, without continual human supervision. Autonomous vehicles are a standout system among the others. The future of mobility will be drastically altered by self-driving cars, which integrate a sophisticated network of sensors, artificial intelligence algorithms, and mapping technologies to navigate roads and make decisions in real time. In addition, unmanned aerial vehicles, also known as drones (Mehta et al., 2021), have become incredibly useful instruments in a variety of fields, including disaster relief, logistics, surveillance, and agriculture. These vehicles use AI and sensors to carry out tasks on their own. Another important category that is applicable to both the industrial and service sectors is robotics. Industrial robots increase productivity by performing repetitive, precise tasks, which optimizes manufacturing processes. On the other hand, service robots help people in a variety of contexts, such as healthcare and housework. demonstrating how flexible and useful autonomous technology can be to enhance daily tasks. Beyond their use on land, autonomous underwater and aerial vehicles also contribute to atmospheric research, space exploration, underwater exploration, and research and maintenance, all of which further the frontiers of scientific knowledge. These various kinds of autonomous systems, each customized to particular environments and tasks, use sensor technology, AI algorithms, and actuation mechanisms to operate independently. They support a range of industries and are revolutionizing the execution and management of tasks in multiple domains.

Autonomous Vehicles

Autonomous vehicles, which integrate cutting-edge technology to navigate and operate without constant human intervention, represent a transformative innovation in transportation that will revolutionize mobility. Leading this category are self-driving cars, which are outfitted with a complex web of sensors, cameras, radar, LIDAR, and GPS, allowing them to sense and understand their surroundings instantly. These cars analyze enormous volumes of data using AI algorithms and machine learning models, making snap judgments regarding their speed, direction, and interactions with their surroundings. Self-Driving Cars seek to improve efficiency and safety on the roads by continuously assessing and responding to shifting traffic patterns, road conditions, and unforeseen obstacles. They may also help reduce accidents that result from human error (Cui et al., 2019; Hodge et al., 2021; Tong et al., 2019).

Drones and Robotics

Drones and robotics are two important subcategories in the field of autonomous systems that have been applied in a wide range of industries. Unmanned aerial vehicles (UAVs), or drones, are small, unmanned aircraft that are outfitted with sophisticated sensors and artificial intelligence capabilities. These aerial vehicles are used for a variety of tasks, such as delivery and logistics, aerial photography, and surveillance. Drones have shown to be extremely useful in a variety of industries, including entertainment, infrastructure inspection, agriculture monitoring, disaster response, and entertainment (Hodge et al., 2021; Mehta et al., 2021). This is because they can fly independently and maneuver through a variety of environments. They are essential tools in many industries because of their small size, agility, and effective access to dangerous or remote areas.

Simultaneously, robotics has advanced significantly, with sophisticated machines capable of performing intricate tasks with precision and efficiency. For example, industrial robotics has become an integral part of manufacturing processes, streamlining production lines and performing tasks such as welding, assembly, and quality control. These self-driving robots are guided by AI algorithms and sensor feedback, increasing productivity and ensuring consistent quality. Robots designed as companions or assistants in healthcare, hospitality, and household chores have demonstrated adaptability in service-oriented domains, assisting humans in tasks ranging from assisting surgeons in operating rooms to assisting the elderly with daily activities.

Drones and robotics have a lot in common: they are both autonomous beings powered by AI algorithms that use sensor technology to understand their surroundings and carry out preprogrammed tasks (Aouf et al., 2019). But they serve different environments and uses; robotics functions in terrestrial environments, while drones are primarily used in aerial spaces. Research is still being done on both categories to improve their capabilities, give them more autonomy, and integrate them into new domains. Despite the wide range of potential uses for drones and robotics, there are still many unanswered questions in these fields, including those pertaining to safety, legal requirements, morality, and public opinion. All things considered, their integration promises greater productivity, safety, and innovation in a variety of operational environments, signifying a paradigm shift in a number of industries.

Industrial Automation

One of the most important uses of autonomous systems is industrial automation, which integrates cutting-edge technology to transform manufacturing processes and increase productivity and efficiency in industrial environments. Fundamentally, industrial automation is the use of robotics, AI-driven systems, and autonomous machinery to carry out tasks that have historically been performed by humans. These self-contained systems use sensors, actuators, and AI algorithms to carry out precise, repetitive tasks at different production stages.

Industrial robots are a key component of industrial automation because they are designed to complete certain tasks quickly and precisely. Numerous tasks are carried out by these robots, such as welding, painting, assembling, handling materials, and quality assurance. They increase production output, maintain consistent quality, and decrease operational errors whether they work independently or in tandem with human labor.

Moreover, autonomous mobile robots (AMRs) and guided vehicles (AGVs) traverse factory floors while moving supplies and products between various workstations. These cars optimize logistics and reduce manual handling by navigating safely in dynamic environments through the use of sensors and AI-based navigation systems (Aguirre & Rodriguez, 2017).

Predictive maintenance has advanced thanks to the integration of AI and machine learning in industrial automation. In order to anticipate equipment breakdowns or maintenance requirements, AI-driven algorithms evaluate sensor data. This allows for proactive interventions, lowers unscheduled downtime, and increases Overall Equipment Effectiveness (OEE).

There are several advantages to industrial automation. These include higher productivity, lower operating expenses, better-quality products, safer workplaces thanks to automated dangerous tasks, and quick response to shifting consumer demands.On the other hand, there are still some drawbacks which includes the upfront costs associated with implementation, the requirement for specialized knowledge to oversee and maintain automated systems, worries about employment displacement, and the necessity of strong cybersecurity defenses to shield linked systems from possible attacks.

ROLE OF AI IN AUTONOMOUS SYSTEMS

Artificial Intelligence plays a fundamental role in Autonomous Systems, acting as the cognitive engine that allows these systems to operate autonomously and decide for themselves. Artificial Intelligence serves as the central nervous system of autonomy, enabling the processing of sensor data and directing subsequent actions accordingly.

Data Processing and Analysis: The enormous volumes of data gathered by sensors integrated into autonomous systems are processed in large part by AI algorithms. These algorithms sort through intricate datasets, finding pertinent information, interpreting the surrounding context, and extracting patterns. AI algorithms effectively process data, regardless of the type—visual data from cameras, spatial data from GPS systems, or other sensory inputs.

AI-Powered Decision-Making: AI powers autonomous systems' decision-making. These systems can learn from data and experiences through deep learning, reinforcement learning, machine learning models, and other AI techniques, gradually strengthening their decision-making capabilities. Whether it is navigating a road for an autonomous vehicle, identifying objects in the path of a drone, or figuring out the best course of action for an industrial robot on a manufacturing line, they analyze the interpreted data to make informed decisions or take actions.

Adaptability and Learning: Adaptability is a key component of artificial intelligence in autonomous systems. These systems are always picking up new skills and adjusting to their ever-changing surroundings. They can improve their performance and responsiveness by honing their decision-making through continuous learning processes based on historical data, real-time data, and feedback loops.Increasing Efficiency and Autonomy: High levels of autonomy are made possible for Autonomous Systems to function by AI algorithms. AI greatly improves these systems' efficacy and efficiency by automating decision-making procedures and lowering the need for continual human intervention. Their independence allows them to operate in a range of situations and surroundings, which makes them flexible and able to adjust to different circumstances.

Increasing Efficiency and Autonomy: High levels of autonomy are made possible for Autonomous Systems to function by AI algorithms. AI greatly improves these systems' efficacy and efficiency by automating decision-making procedures and lowering the need for continual human intervention. Their independence allows them to operate in a range of situations and surroundings, which makes them flexible and able to adjust to different circumstances.

Safety and Risk Mitigation: AI helps to improve autonomous systems' safety. These systems can improve overall safety in applications like drones, industrial automation, and autonomous vehicles by analyzing potential risks, predicting outcomes, and making decisions to mitigate risks.

AI Algorithms and Techniques

AI algorithms play a pivotal role in automation by enabling machines and systems to perform tasks, make decisions, and adapt to changing circumstances without constant human intervention. In automation, artificial intelligence algorithms are what power the intelligence and decision-making that are built into systems. Together, these algorithms provide a broad toolkit that makes automation possible in a number of sectors (Da Silva Assis et al., 2016). The foundation of artificial intelligence, Machine Learning (ML) algorithms, makes automation easier by enabling systems to learn from data patterns and make predictions or decisions without explicit programming. Supervised learning algorithms in ML facilitate automation by training models on labeled data to predict or classify outcomes. This is an essential component in automating processes like manufacturing predictive maintenance or image recognition for quality control. Furthermore, unsupervised learning algorithms find hidden structures or patterns in data, which helps automate the process of grouping related data points. An advanced subset of machine learning called deep learning (DL) uses neural networks to process large amounts of data and perform tasks like computer vision, natural language processing, and autonomous decision-making in industrial settings. In order to automate sequential decision-making processes, Reinforcement Learning (RL) algorithms enable systems to learn optimal behaviors through interactions with an environment. By handling repetitive tasks, streamlining workflows, and enabling systems to learn and adapt to changing conditions, these AI algorithms collectively empower automation and revolutionize efficiency, accuracy, and adaptability in automated workflows across industries.

Deep Learning and Neural Networks

Neural networks and deep learning (DL) are sophisticated subsets of machine learning that are essential to the development and operation of autonomous systems. These technologies have allowed Autonomous Systems to perform sophisticated tasks and make judgments based on complex patterns and data, revolutionizing their capabilities.Using artificial neural networks with several layers, Deep Learning is a branch of machine learning. From enormous volumes of data, these deep neural networks are able to identify patterns, extract features, and make predictions or classifications. Because this technique can handle unstructured data like text, audio, and images, it has had a significant impact on autonomous systems.

Neural Networks use linked nodes, or neurons, arranged in layers to simulate the composition and operations of the human brain. These networks can process complex information in a hierarchical manner because of their input, hidden, and output layers. Numerous neural network architectures, including generative adversarial networks (GANs), recurrent neural networks (RNNs), and convolutional neural networks (CNNs), have proven useful in a range of autonomous systems applications. CNNs are excellent at analyzing images and videos, which enables Autonomous Systems to carry out activities like segmentation, object detection, and image classification. They're widely used in applications like lane markings, traffic signs, and pedestrian detection in autonomous vehicles. Recurrent neural networks, or RNNs, are utilized in situations where temporal dependencies are present because they are efficient at processing sequential data.

Natural Language Processing and Perception

Natural Language Processing (NLP) and perception stand as pivotal components in the realm of Autonomous Systems, playing integral roles in facilitating communication with humans and understanding the surrounding environment. NLP techniques empower these systems to comprehend, process, and generate human language, encompassing tasks such as speech recognition, language understanding, text generation, and dialog systems. By interpreting spoken or written language, NLP enables autonomous vehicles to respond to voice commands, chatbots to engage in conversations, and systems to analyze textual data for insights or decision-making. Simultaneously, perception in Autonomous Systems involves the fusion of data from various sensors like cameras, LIDAR, and radar, enabling the systems to comprehend their environment. Computer vision algorithms decode visual information, aiding in object recognition, scene understanding, and environment interpretation, vital for autonomous vehicles to navigate safely or for robots to interact seamlessly in dynamic settings. The integration of NLP and perception empowers Autonomous Systems to communicate effectively with humans and gain insights from their surroundings, fostering advancements in applications like autonomous vehicles, robotics, and smart environments, paving the way for more intelligent and intuitive interactions between machines and humans.

APPLICATIONS OF AUTONOMOUS SYSTEMS AND AI

Autonomous Systems integrated with Artificial Intelligence applications have permeated diverse industries, reshaping operations and introducing unprecedented efficiency, precision, and innovation. Within transportation, the emergence of self-driving cars and autonomous vehicles powered by AI algorithms has redefined mobility, promising safer and more efficient travel. Healthcare has seen a revolution with AI aiding in diagnostics, personalized treatment plans, and even drug discovery, improving patient care and outcomes.

AI in Automobiles

Industries have embraced AI-driven automation, transforming manufacturing processes in Industry 4.0 by optimizing production lines, predictive maintenance, and smart logistics. Agriculture benefits from precision farming techniques, employing drones and IoT sensors to monitor crops and resources, maximizing yields sustainably. Financial services utilize AI for fraud detection, risk assessment, and algorithmic trading, enhancing operational efficiency and security. Retail experiences AI-driven advancements with personalized customer service, recommendation systems, and inventory management. Smart city initiatives harness AI for traffic management, energy optimization, and public safety enhancement. Even in scientific exploration and space missions, AI supports navigation, data analysis, and robotics, enabling groundbreaking discoveries. These applications underscore the breadth of AI-driven Autonomous Systems, catalyzing advancements across industries and significantly impacting how we live, work, and explore the world around us.

Self-Driving Cars

Self-Driving Cars, which aim to transform transportation by enabling vehicles to navigate and operate autonomously without constant human intervention, represent the convergence of cutting-edge technology, artificial intelligence, and automotive engineering. These vehicles are fundamentally equipped with an advanced array of sensors, cameras, radar, LIDAR, and GPS systems to enable real-time perception and interpretation of the surrounding environment. The data gathered from these sensors is processed by AI algorithms, specifically machine learning and deep learning models, which allow the car to make decisions instantly about navigation, route planning, object detection, and reacting to changing traffic conditions (Brown, 2016; Da Silva Assis et al., 2016; Kumar et al., 2019; Preethiya et al., 2018; Preethiya et al., 2019a; Preethiya et al., 2020).

Advanced Driver-Assistance Systems (ADAS)

Modern cars are equipped with advanced driver-assistance systems (ADAS), which are an intermediate step toward complete autonomy. ADAS features include automatic emergency braking, adaptive cruise control, and lane-keeping assistance.

Perception systems that interpret sensor data, mapping technologies that offer a digital depiction of the surroundings, decision-making algorithms that evaluate data and choose the best course of action for driving, and actuation mechanisms that convert decisions into movements of the vehicle are important parts that enable these vehicles to operate autonomously.

When it comes to the development and application of self-driving cars, safety is still the top priority. For these cars to be safe and dependable, rigorous testing, simulations, and ongoing AI algorithm development are essential. Significant obstacles that must be overcome include liability issues, regulatory frameworks, public acceptance, and ethical considerations.

The implications of self-driving cars on society could be extensive. They promise to improve road safety by lowering human error-related accidents, streamline traffic, make transportation more accessible to people with limited mobility, and possibly even change the look of cities by rearranging transportation infrastructure (Amrutkar et al., 2024; Koubaa et al., 2021).

AI in Healthcare and Medicine

The application of artificial intelligence to medicine and healthcare promises ground-breaking discoveries and breakthroughs that will change treatment regimens, patient care, diagnosis, and healthcare delivery systems. Artificial intelligence, which is distinguished by its capacity to evaluate enormous volumes of data, spot patterns, and forecast outcomes, has shown itself to be a potent ally in enhancing medical procedures and resolving enduring issues in the healthcare sector.

Fundamentally, artificial intelligence in healthcare spans a wide range of uses, utilizing deep neural networks, computer vision, natural language processing, and machine learning algorithms to extract knowledge from complicated medical data. Healthcare practitioners can now extract actionable insights from a variety of sources, including genomic sequencing, medical imaging, electronic health records (EHRs), and real-time patient monitoring.

Furthermore, remarkably accurate diagnosis of conditions such as neurological disorders, cancer, and cardiovascular diseases has been demonstrated by AI-powered medical imaging systems, helping radiologists to interpret imaging results more precisely. Through the extraction of valuable information from unstructured medical texts, natural language processing aids in clinical decision-making and promotes more effective documentation (Bajwa et al., 2021).

The application of AI in healthcare has great potential, but there are drawbacks as well, including the need for transparent and understandable AI models, regulatory compliance, data privacy, and ethical issues with AI-driven decision-making in patient care.

Robotics in Surgery

Robotics in surgery is a game-changing development in medical technology that combines advanced technology and surgical techniques to improve accuracy, dexterity, and patient outcomes. In contrast to conventional surgical techniques, robotic surgery uses robotic systems under the control of surgeons to carry out minimally invasive procedures with increased accuracy and control (Chopra et al., 2022).

Surgical robots, which are operated by surgeons via a console and furnished with cutting-edge technology like robotic arms and specialized instruments, are the foundation of robotic surgery. These systems include micro-scale instruments that can perform complex movements and maneuvers that are difficult for human hands to perform alone, as well as high-definition cameras that provide three-dimensional views of the surgical site.

AI in Diagnosis and Treatment

With the use of cutting-edge algorithms and data analytics, artificial intelligence in diagnosis and treatment has become a ground-breaking area in healthcare, revolutionizing patient care, diagnosis, and treatment approaches. Artificial Intelligence technologies employ machine learning, deep learning, and natural language processing to analyze large datasets that include genetic data, clinical research findings, medical images, and electronic health records. Through the analysis of imaging scans, laboratory results, and patient histories, these sophisticated algorithms provide healthcare providers with enhanced diagnostic capabilities, assisting in the early detection of diseases with remarkable accuracy. Additionally, by incorporating patient-specific data, AI-driven systems facilitate the development of individualized treatment plans, enabling the use of customized drugs and therapies that maximize efficacy while minimizing side effects.

AI in Industrial Automation

Artificial Intelligence (AI) is a key player in the transformation of industrial automation through the advancement of conventional manufacturing processes (Kautish et al., 2024). AI algorithms bring about the era of smart factories by optimizing efficiency, improving precision, and enabling predictive maintenance in industrial settings. Large-scale sensor and machine datasets are analyzed by AI-powered systems, which enable real-time decision-making to maximize efficiency, minimize errors, and cut down on downtime. By identifying patterns in production lines, machine learning algorithms optimize workflows, anticipate equipment failures, and proactively schedule maintenance, thereby averting expensive disruptions. AI-enabled collaborative robots, or cobots, operate alongside people to accomplish complex tasks safely and precisely. AI-powered computer vision systems supervise quality control, checking goods to make sure they meet strict requirements. While AI increases productivity, there are still issues that need to be addressed, including workforce adaptation to AI-integrated environments, cybersecurity threats, and ensuring ethical AI deployment. Notwithstanding these obstacles, artificial intelligence's (AI) incorporation into industrial automation is redefining manufacturing and holding out the prospect of greater productivity, affordability, and innovation in the industrial sector.

Other Applications

AI algorithms are essential for fraud detection, risk assessment, and trading decision automation in the banking and finance industries. Healthcare uses AI for medical image analysis, treatment optimization, and illness diagnosis, while retail uses AI for personalized recommendations, inventory control, and chatbots for customer service. AI is having an impact on education through chatbots for student support, adaptive tutoring systems, and personalized learning experiences. In addition, AI improves cybersecurity protocols, transforms the way utilities distribute energy, and maximizes agricultural yield through precision farming methods. AI is being used by the entertainment sector for content creation, personalization, and recommendation. AI also has an impact on human resources, logistics, transportation, and other areas. It makes things easier for autonomous cars, automated hiring, predictive maintenance, and route optimization. The fact that AI is being widely used in these fields demonstrates how it can spur innovation, improve operations, and bring about revolutionary changes that have a significant impact on a wide range of businesses and daily life (Mohan et al., 2023; Preethiya et al., 2019b; Venkat et al., 2023).

Real-Time Case Studies

Waymo, a subsidiary of Alphabet Inc., is developing autonomous driving technology, with plans for commercial deployment. However, a 2018 collision in Arizona raised concerns about safety and readiness. Amazon's warehouse robots have improved efficiency but have also caused accidents, including collisions with human workers and inventory damage. The integration of AI and robotics in warehouse automation raises concerns about workplace safety, human-robot collaboration, and job displacement. IBM Watson, an AI-powered platform used in healthcare, has faced criticism for its accuracy, scalability, and integration with existing systems. Despite its potential to improve clinical decision-making and patient outcomes, data quality, interoperability, and trust remain significant barriers to adoption. Amazon Go Stores use AI and sensor fusion technology to provide cashier-less shopping experiences, but face challenges such as technical reliability, accuracy in tracking, job displacement, privacy concerns, and surveillance. Autonomous drones are being developed for agricultural applications, but face technical issues like flight stability, obstacle avoidance, and data processing. Regulatory hurdles and socio-economic impacts on traditional farming practices and rural communities are also a concern. Robotic Process Automation (RPA) in finance automates repetitive tasks like data entry, reconciliation, and report generation, mimicking human interactions with computer systems. Challenges include ensuring accuracy in complex financial processes, integrating RPA with existing systems, and addressing security risks. Additionally, there are concerns about job displacement and the need for upskilling finance professionals to work effectively with RPA systems.

ETHICAL AND SOCIAL IMPLICATIONS

The rapid integration of Artificial Intelligence into various facets of daily life brings forth a spectrum of ethical and social implications that warrant careful consideration. One primary concern lies in the potential bias embedded within AI algorithms, leading to unfair treatment or discrimination. Biased datasets might perpetuate societal prejudices, reflecting in AI decisions across sectors such as hiring, lending, and criminal justice. The opacity of some AI models poses challenges in understanding their decision-making processes, raising questions about accountability and transparency. Privacy breaches and data security vulnerabilities arise from the extensive collection and utilization of personal data by AI systems, necessitating stringent safeguards to protect sensitive information. Additionally, the potential for job displacement due to automation and shifts in workforce dynamics triggers discussions about retraining programs and equitable access to opportunities in an AI-driven economy (Daniel, 2023; Vesnic-Alujevic et al., 2020).

Autonomous agriculture and precision farming revolutionise food production, improving efficiency, resource use, and crop yields. However, autonomous systems in agriculture create challenges and ethical issues. Because autonomous machinery collects vast data on crop conditions, soil health, and agricultural processes, data ownership and privacy are major considerations. Farmers must handle data ownership, usage, and exploitation. AI-driven decision-making also raises algorithm transparency and accountability concerns. Farmers seek independent, reliable, and in their best interests autonomous machinery algorithms. Autonomous agricultural technology may replace conventional farming practices and influence rural communities, which might have socioeconomic effects. Technical innovation, ethical values, regulatory supervision, and stakeholder interaction are needed to deploy autonomous farm systems properly. Farmers, researchers, legislators, and industry stakeholders must collaborate to use autonomous farming for sustainable food production.

The maritime industry is developing autonomous ships with AI-driven navigation systems to improve productivity and reduce human error. Autonomous vessels in busy waterways and their ability to adapt to unexpected environmental conditions pose safety risks. Diverse strategies are needed to address these concerns. First, create robust sensor systems that can recognise and manoeuvre around barriers to prevent collisions. Coordination and collision avoidance need good communication with other boats. Safe and lawful operation requires international marine standards. Establishing public trust in autonomous maritime technology requires fail-safe techniques to intervene in crises and avert tragedies. To tackle these complexities, reliable communication systems and extensive testing are essential. By proactively addressing these issues and prioritising safety, the maritime sector can safely deploy autonomous ships and develop trust in this revolutionary technology.

Delivery drones might change urban logistics as Amazon and Google lead attempts to deliver things in congested areas. Drone integration into cities is difficult. Due to the rising number of drones, airspace management is a major concern, raising questions about safety legislation and accidents with other aircraft or people. Safe and ethical delivery drone deployment requires overcoming technology difficulties like collision avoidance systems. Navigating air traffic control and privacy laws is vital. To establish urban drone operations guidelines, local authorities, communities, and aviation groups must collaborate. Collaboration and coordination are needed to achieve delivery drones' transformative potential while prioritising safety and public welfare.

Automotive innovations like Tesla's Autopilot and Full Self-Driving (FSD) capabilities assist drivers with lane-keeping and autonomous parking. Despite promised safety and convenience gains, these features have been challenging to implement. Multiple crashes have occurred due to drivers overusing Tesla's self-driving technologies. Tesla's autonomous driving strategy is challenged by the accidents. The degree of human supervision needed for these duties and the system's capabilities and constraints have been questioned. Concerns remain concerning driver complacency while using semi-autonomous features, which may diminish concentration and hinder reaction. Tesla and other companies struggle to promote advanced capabilities while ensuring customer comprehension and proper use. Continuous refinement of autonomous systems, honest disclosure of their powers, and aggressive misuse prevention are needed to address these issues. Tesla and other innovators may enhance autonomous driving by solving safety and user awareness issues.

CONCLUSION

Autonomous systems and artificial intelligence have the potential to completely transform industries in the future, but they also face a number of obstacles that must be addressed for responsible and successful integration. The advancement of deep learning and natural language processing will be the main focus of AI's future development, opening up new possibilities for analysis and communication. However, ethical issues with AI decision-making transparency, bias reduction, and accountability present serious difficulties that call for careful regulation and moral frameworks. Though moral and legal challenges still exist, autonomous systems, particularly in the transportation sector, will prioritize infrastructure development and safety improvement. The impending difficulty is adjusting to AI's impact on the workforce, which calls for significant work redefinition and reskilling. Strong security measures to protect AI systems and user data will be required in tandem with cybersecurity threats and the requirement for data privacy regulations. Furthermore, proactive policies and international cooperation will be necessary to address societal effects like bias perpetuation and socioeconomic disparities as well as ensure equitable AI accessibility. The future of AI and Autonomous Systems depends on proactive ethical thinking, legal frameworks, and cooperative efforts to fully realize their potential while reducing the inherent difficulties for a tech future that is both responsible and inclusive.

REFERENCES

Aguirre, S., & Rodriguez, A. (2017). *Automation of a Business Process Using Robotic Process Automation (RPA): A Case Study*. Springer. . doi:10.1007/978-3-319-66963-2_7

Amrutkar, C., Satav, A., Sonawwanay, P. D., & Pawar, A. H. (2024). Overview of Autonomous Vehicle and Its Challenges. Techno-Societal 2022. *ICATSA*, *2022*, 243–251. doi:10.1007/978-3-031-34648-4_25

Aouf, A., Boussaid, L., & Sakly, A. (2019). Same fuzzy logic controller for two-wheeled mobile robot navigation in strange environments. *Journal of Robotics*, *2019*, 2465219. doi:10.1155/2019/2465219

Bajwa, J., Munir, U., Nori, A., & Williams, B. (2021, July). Artificial intelligence in healthcare: Transforming the practice of medicine. *Future Healthcare Journal*, *8*(2), e188–e194. doi:10.7861/fhj.2021-0095 PMID:34286183

Brown, B. (2016). The social life of autonomous cars. *MIT Technology Review, 50*(2).

Chaurasia, A., Parashar, B., & Kautish, S. (2024). Artificial Intelligence and Automation for Industry 4.0. In S. Kautish, P. Chatterjee, D. Pamucar, N. Pradeep, & D. Singh (Eds.), *Computational Intelligence for Modern Business Systems. Disruptive Technologies and Digital Transformations for Society 5.0*. doi:10.1007/978-981-99-5354-7_18

Chopra, H., Baig, A. A., Cavalu, S., Singh, I., & Emran, T. B. (2022, August 17). Robotics in surgery: Current trends. *Annals of Medicine and Surgery (London)*, *81*, 104375. doi:10.1016/j.amsu.2022.104375 PMID:36051814

Cui, Q., Wang, Y., Chen, K.-C., Ni, W., Lin, I.-C., Tao, X., & Zhang, P. (2019). Big data analytics and network calculus enabling intelligent management of autonomous vehicles in a smart city. *IEEE Internet of Things Journal*, *6*(2), 2021–2034. doi:10.1109/JIOT.2018.2872442

Da Silva Assis, L., da Silva Soares, A., Coelho, C. J., & Van Baalen, J. (2016). An evolutionary algorithm for autonomous robot navigation. *Procedia Computer Science*, *80*, 2261–2265. doi:10.1016/j.procs.2016.05.404

Daniel, T. (2023). Autonomous AI Systems in Conflict: Emergent Behavior and Its Impact on Predictability and Reliability. *Journal of Military Ethics*, *22*(1), 2–17. doi:10.1080/15027570.2023.2213985

Gupta, S., Upadhyay, D., & Dubey, A. K. (2019). Self-Driving Car Using Artificial Intelligence. In M. Kumar, R. Pandey, & V. Kumar (Eds.), *Advances in Interdisciplinary Engineering. Lecture Notes in Mechanical Engineering*. Springer. doi:10.1007/978-981-13-6577-5_49

Hodge, V. J., Hawkins, R., & Alexander, R. (2021). Deep reinforcement learning for drone navigation using sensor data. *Neural Computing & Applications*, *33*(6), 2015–2033. doi:10.1007/s00521-020-05097-x

Khamis, A., Patel, D., & Elgazzar, K. (2021). Deep Learning for Unmanned Autonomous Vehicles: A Comprehensive Review. In A. Koubaa & A. T. Azar (Eds.), *Deep Learning for Unmanned Systems. Studies in Computational Intelligence* (Vol. 984). Springer. doi:10.1007/978-3-030-77939-9_1

Mehta, P. L., Kalra, R., & Prasad, R. (2021). A Backdrop Case Study of AI-Drones in Indian Demographic Characteristics Emphasizing the Role of AI in Global Cities Digitalization. *Wireless Personal Communications*, *118*(1), 301–321. doi:10.1007/s11277-020-08014-6 PMID:33424130

Mohan, P., Sabarwal, T., & Preethiya, T. (2023). Indian Sign Language Character Recognition System. *2023 4th International Conference on Electronics and Sustainable Communication Systems (ICESC)*. IEEE. 10.1109/ICESC57686.2023.10193309

Preethiya, T., Muthukumar, A., & Durairaj, S. (2018) Providing Secured Data Aggregation in Mobile Wireless Sensor Network. *Proceedings of 4th IEEE International Symposium on Robotics and Manufacturing Automation*. IEEE. 10.1109/ROMA46407.2018.8986735

Preethiya, T., Muthukumar, A., & Durairaj, S. (2019a). Double Cluster Head Heterogeneous Clustering for Optimization in Hybrid Wireless Sensor Network. *Wireless Personal Communications. Wireless Personal Communications*, *110*(4), 1751–1768. doi:10.1007/s11277-019-06810-3

Preethiya, T., Muthukumar, A., & Durairaj, S. (2019b). Mobility Handling in Cluster based Mobile Wireless Sensor Network. *Proceedings of 2019 IEEE International Conference on Clean Energy and Energy Efficient Electronics Circuit for Sustainable Development (INCCES).* IEEE. 10.1109/INCCES47820.2019.9167692

Preethiya, T., Muthukumar, A., & Durairaj, S. (2020). An energy efficient clustering and multipath routing for mobile wireless sensor network using game theory. *International Journal of Communication Systems*, *33*(7), 1–18.

Tong, W., Hussain, A., Bo, W. X., & Maharjan, S. (2019). Artificial intelligence for vehicle-to-everything: A survey. *IEEE Access : Practical Innovations, Open Solutions*, *7*, 10823–10843. doi:10.1109/ACCESS.2019.2891073

Venkat, Y., Chand, K. P., & Preethiya, T. (2023). An intrusion detection system for the Internet of Things based on machine learning. *2023 International Conference on Recent Advances in Electrical, Electronics, Ubiquitous Communication, and Computational Intelligence (RAEEUCCI).* IEEE. 10.1109/RAEEUCCI57140.2023.10134432

Vesnic-Alujevic, L., Nascimento, S., & Pólvora, A. (2020). Societal and ethical impacts of artificial intelligence: Critical notes on European policy frameworks. *Telecommunications Policy*, *44*(6), 2020. doi:10.1016/j.telpol.2020.101961

Chapter 2
Explainable AI-Based Semantic Object Detection for Autonomous Vehicles

Elamathiyan A.
Karpagam Academy of Higher Education, India

G. Dhivya
https://orcid.org/0000-0001-7058-7917
Karpagam Academy of Higher Education, India

ABSTRACT

A goal is to facilitate the recognition and segmentation of the route driven by autonomous vehicles through the use of machine learning (ML) models. The pixel-wise road detection task, the semantic segmentation architectures underwent training and comparison. By using XAI, the authors are able to interpret and read the predictions generated by these abstract models. They generated arguments for the recommended segmentation model for autonomous vehicle road detection using a range of XAI approaches. Supervised learning is enabled by KNN, Decision Tree, and Random Forest, which are the current algorithms used for comparison. On the other hand, the newly built K means clustering function best when paired for image processing since they are good with images. To display the results of computations for the evaluation parameters of each algorithm, including accuracy, sensitivity, recall and precision, tables, and the necessary features from the evaluation matrices are utilised. The k-means clustering system for explainable AI-based semantic object detection in automated cars achieves 94.58% accuracy on both train and test sets.

INTRODUCTION

Autonomous car technology has advanced significantly in the last few years, changing the way people travel and paving the way for more efficiency, convenience, and safety in the future. Securing the reliable and transparent detection of objects in the vehicle's surrounds is a major problem in the development of

DOI: 10.4018/979-8-3693-1962-8.ch002

autonomous vehicles, and integrating state-of-the-art technologies is a crucial part of this transformative journey. According to Tyagi, A. K., &Aswathy, S. U. (2021), in this perspective, the intersection of semantic object recognition and explainable artificial intelligence appears as a vital frontier, where advances are paving the way for autonomous systems that are safer and easier to understand.

In order to sense and manoeuvre through their surroundings, autonomous cars depend on complex sensor arrays, such as radar, LiDAR, and cameras. This perceptual framework is centred on semantic object detection, which is the process of identifying and categorising items according to their semantic meaning.But in safety-critical applications like autonomous driving, questions have been raised regarding the interpretability of some deep learning models' decisions due to their opaque nature. Presenting Explainable AI, a paradigm that improves prediction accuracy while demystifying decision-making by offering explanations for why a given object was recognised or a specific action was performed.

Antoniadi, A. M., et al (2021), for autonomous vehicles to be widely used, trust must be developed among users, authorities, and the general public. Acceptance may be hampered by the opacity of conventional deep learning models. Explainable AI promotes trust in the technology by providing transparency, allowing consumers to understand how the car interprets its environment and makes judgements.

The above figure 1 described the general architecture of object detection. The ability of an artificial intelligence system to give clear and intelligible justifications for its choices and behaviours is known as explainable artificial intelligence, or XAI. Semantic object detection is the process of locating and categorising things in a scene or image according to their semantic significance. The use of XAI with semantic object detection in autonomous cars is important in order to maintain the transparency and reliability of the system. Autonomous vehicles have developed so quickly because of artificial intelligence (AI), which has the potential to significantly change the transportation landscape.

As these vehicles venture into ever-more complex and dynamic environments, the synergy between Semantic Object Detection and Explainable AI (XAI) represents a critical frontier that promises not only

Figure 1. Architecture of object detection

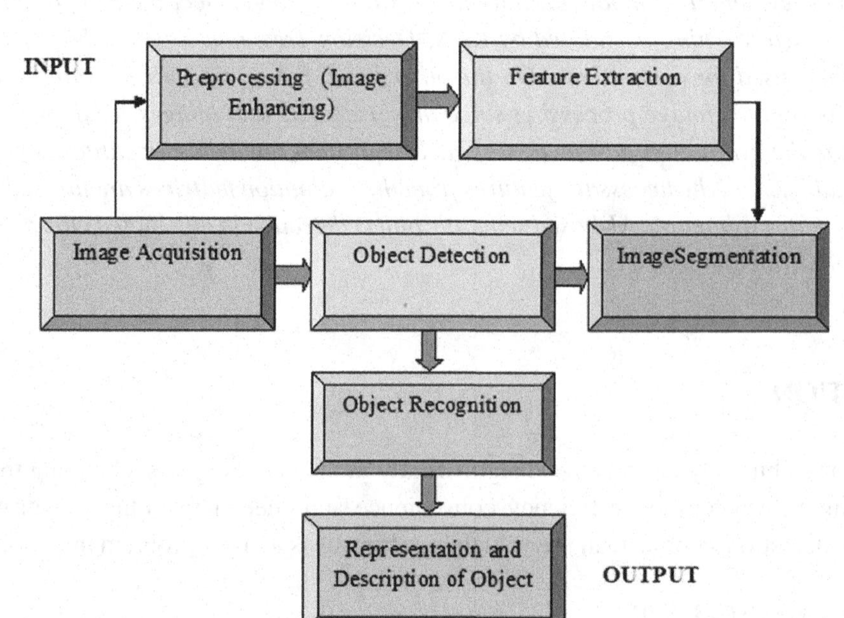

technological advancement but also a paradigm shift in how we view, trust, and incorporate autonomous systems into our daily lives.Using input from input image acquisition, pre-processing, also known as picture enhancement, is carried out first. For object identification, the output of this stage is delivered. The feature extraction is notified simultaneously with the preprocessed image output. In order to segment the image that needs to be identified, the segmentation stage gets the output from these two steps. These procedures are the fundamental and main ones used to identify parts of images. the object identified in the subsequent object detection step. Forwarded to the representation and description step is the input from object recognition. Upon completion, the outcome is presented in an easy-to-understand format according to the specifications.

Semantic Object Detection, an essential part of autonomous vehicle perception, broadens the car's understanding of its surroundings beyond simple identification. This sophisticated ability, according to Zhao, J., et al. (2021), comprises assigning semantic importance to objects in the visual field in addition to their recognition. Traffic signs, people, cars, and other contextually significant entities enable the automobile to make decisions based on its perception of the scene's semantic content. These entities are more than just pixels on a screen.

Object Detection in Neural Network

Object detection networks give the bounding box with the object's coordinates and the kind of object that can be found in the picture. Similar to image classification networks, object detection networks utilize convolution layers to identify visual features. These basic procedures should be followed in order to develop a neural network for object detection: Start by compiling a range of topic-representative pictures. Next, label and provide a name to every picture. After that, a neural network is constructed by Google Colab using these pictures, which are referred to as a "training set".

Explainable AI-based Semantic Object Detection works with humans and autonomous systems in addition to models and algorithms. Designing user-friendly and instructive human-machine interfaces becomes essential. The interface should provide succinct explanations in addition to communicating the AI's decisions in order to facilitate successful collaboration between the automobile and its human operators.

With the convergence of Semantic Object Detection and Explainable AI, the development trajectory of autonomous cars is being fundamentally altered. Its implications extend beyond the fine technical specifics of algorithms and models, into the fields of safety, trust, and societal integration. We must investigate the transparent possibilities of a future where autonomous vehicles live in harmony with humans, given the complex landscape of self-driving cars. This confluence becomes a societal responsibility that goes beyond technology. This goes beyond simply a convergence of technologies since it also represents a convergence of understanding, transparency, and the possibility of a safer and more comprehensible autonomous travel experience, Zablocki, E., et al (2022).

Autonomous vehicles (AVs) have advanced significantly in the last several years, thanks in large part to advances in artificial intelligence (AI).Wang, J., et al (2022), Autonomous driving relies heavily on object detection, which is the process of identifying and classifying objects in the vehicle's surroundings. Autonomous cars are becoming more and more dependent on reliable and safe object detection as they maneuver through complex and dynamic environments. This work aims to investigate the issues, possible remedies, and significance of explainable AI-based semantic object identification for driverless vehicles.

Autonomous vehicles, often known as self-driving cars, sense their surroundings using a range of sensors, including cameras, radar, and lidar. The cornerstone of the perception system, object recognition, enables the vehicle to recognize and respond to a wide range of items, including bikes, pedestrians, cars, and obstructions. Recent breakthroughs in artificial intelligence (AI), particularly in deep learning, have revolutionized traditional object recognition approaches, with a focus on computer vision algorithms. Recurrent neural networks (RNNs) and convolutional neural networks (CNNs), two deep learning models, have demonstrated superior performance over other models in object detection tests. However, because these models are opaque and difficult to understand how they make decisions, it is difficult to trust and employ them in safety-critical applications such as autonomous driving. As a result, the explainable AI paradigm has emerged, focusing on building models whose conclusions are understandable and interpretable by people. Semantic object detection is not only capable of identifying objects; it can also give those descriptive labels such as "car," "pedestrian," or "traffic light." This semantic knowledge is essential for the car to make sensible decisions and navigate safely in a range of environments. Arrieta, A. B., et al (2020), Explain ability combined with semantic object recognition ensures that decisions made by AI models are transparent and provide justifications for how an object was recognized and classified.

To solve the issues with semantic object detection in autonomous automobiles, engineers and researchers are looking into a variety of approaches. One way is to use complex deep learning networks designed specifically for object detection tasks. Models like the You Only Look Once (YOLO) and Single Shot MultiBox Detector (SSD) have gained popularity because to their accuracy and real-time performance. Another crucial strategy to employ is multi-sensor fusion. Autonomous vehicles have many sensors fitted, each of which records a different aspect of their environment. Data from cameras, lidar, radar, and other sensors are combined to improve overall perception capability, leading to more reliable and strong object detection, as assessed byAziz, L., et al (2020).

Explain ability is achieved by integrating interpretable machine learning models with post-hoc explanation techniques. Interpretable models, such as rule-based systems or decision trees, require transparent decision-making processes at their core. Post-hoc explanation techniques, on the other hand, aim to explain the decisions made by complex models after they have been trained. Simple, locally relevant explanations for specific model predictions are generated by methods like LIME (Local Interpretable Model-agnostic Explanations).For semantic object detection in autonomous automobiles, explain ability integration offers a number of benefits. To begin with, it allays concerns regarding the mystery surrounding deep learning algorithms by enhancing the legitimacy of AI models. Autonomous vehicle acceptance is likely to increase as manufacturers, regulators, and end users are able to understand and trust the AI's decision-making process. Secondly, systems are easier to troubleshoot and improve when they have explanation capability. By understanding how the model reacts to different scenarios, developers can identify and fix potential biases or weaknesses in the system. This repeated feedback loop is crucial for the creation of trustworthy object detection systems as well as continuous improvement.Explainable AI also helps with regulatory compliance. As the usage of autonomous vehicles spreads, regulatory frameworks are being developed to ensure safety and accountability, Omeiza, D., et al (2021), legal criteria are met by explainable AI, which offers a transparent framework that is auditable and validateable.

There are five parts to this paperwork. The introductionportion work that was covered in Section I. The shortcomings and restrictions of the current systems are examined in Section II. In Section III, we explained our recently proposed system design work for explainable AI-based semantic object detection for autonomous vehicles. The explanation and thorough output result for our recently built system are provided in Section IV. The end of this study is covered in Section V.

LITERATURE REVIEW

This literature review explores the rapidly evolving field of autonomous vehicle technologies, with a focus on integrating XAI concepts to improve transparency and interpretability in semantic object identification algorithms. Understanding the methods, challenges, and advancements in explainable AI is crucial as the demand for prudent and astute decision-making in self-driving cars grows. This survey aims to illustrate the crucial relationship between state-of-the-art technology and the crucial need for transparency in autonomous systems by analyzing the major contributions, strategies, and emerging patterns in the field of explainable AI-based semantic object identification.

Mankodiya et al., (2022), describes a method for semantic object detection in autonomous cars using explainable AI (XAI). The authors propose a method called OD-XAI to enhance the interpretability of object detection algorithms in autonomous vehicles. The goal of this work is to demonstrate that autonomous vehicles (AVs) can identify and segment their path using deep learning (DL) models. Three semantic segmentation architectures were trained and evaluated with the goal of pixel-wise road detection. The maximum IoU scores on the test and train sets were 0.9621 and 0.9459, respectively. These deep learning algorithms are referred to as "black box models" due to the difficulty in understanding their exceedingly intricate design.With XAI, they can interpret and comprehend the predictions of these abstract models. Using a range of XAI methods, they produced justifications for the proposed segmentation model for road detection in autonomous vehicles.

Ponn, et al (2020),uses cameras to tackle problems with automated vehicle object detection. Finding and comprehending the difficult situations that come up during this process is the primary objective of the study. In this study, these additional variables that impact camera-based object detection accuracy are thoroughly investigated for the first time. To make matters worse, it is challenging to appropriately characterize the detection performance and explain specific detection findings because the existing algorithms are based on artificial intelligence. In order to evaluate and explain the detection performance of various object detection techniques, a modeling technique based on the examined effect variables is provided, and the recently developed SHapley Additive exPlanations (SHAP) methodology is utilized.The findings demonstrate that, independent of the detection technique, many factors affecting detection performance always have the same impact. Two examples of these parameters are the object's positioning inside the image and its relative rotation toward the camera. The findings demonstrate that, irrespective of the detection technique employed, a number of significant parameters, such as an object's location on the picture or its relative rotation towards the camera, significantly impact the detection performance. Specifically, the vulnerabilities found in the analyzed object detectors could be leveraged to create important and demanding situations for automated car testing and type certification processes.

Thakker et al (2020) detailed in the paper, which offers a novel approach utilizing Semantic Web technologies and is demonstrated with a smart cities flood monitoring application inside the context of an EU-funded project. Overviewing "explainable deep learning" as a subset of the "explainable AI" challenge, the paper provides an outline of the concept. Monitoring of drainage and gullies in strategic areas that are vulnerable to flooding issues should be a component of any flood monitoring system. Cameras are used to display the impacted areas in real-time after a DL-based classifier is often developed to detect things such as leaves, plastic bottles, and other objects. The existence and coverage of these objects in the input data are then used to train the classifier to identify blockages. They inventively provide an Explainable AI solution in this article by combining DL and Semantic

Web technologies to build a hybrid classifier. In this hybrid classifier, the DL component determines item existence and coverage level, while the semantic rules, developed in close collaboration with experts, do categorization. Our hybrid classifier makes use of expert knowledge specifically linked to flooding, providing flexibility in classifying the image based on elements and their coverage links. The experimental findings, shown with a real-world use case, showed that the hybrid approach to image classification performed 11% better on average (F-Measure) than the DL-only classifier. Another noteworthy advantage is that it applies experts' knowledge in developing guidelines for decision-making to account for complex circumstances.

Atakishiyev et al. (2021), makes intelligent decision-making entirely intelligible to humans, yet current autonomous vehicles lack a component that accomplishes this. As a result, society views technology less favorably. Because of this, AI systems for driverless cars need to be able to defend its conclusions in real time and make safe decisions while still adhering to international rules and regulations. This research provides important insight into the process of developing explainable artificial intelligence (XAI) solutions for self-driving cars. In specifically, they offer the following contributions. The most recent XAI research for autonomous driving is first given in-depth overview form.They continue by offering a XAI paradigm that considers the social and legal conditions necessary for autonomous driving systems to be rational. Lastly, they provide a range of XAI methodologies as prospective research topics for later work that can improve operational safety and transparency to assist regulators and other stakeholders in embracing autonomous driving technology.

Moradi et al. (2023) emphasizes model-independent explainable AI techniques for object recognition in image data. By using a unique masking methodology for AI-based object identification systems, we propose and implement a black-box explanation approach called Black-box Object identification Explanation by Masking (BODEM). To generate several versions of an input image, we advise using both local and remote masking techniques. Distant masks measure the impact of disturbing pixels outside of an object on the detection model's conclusions, whereas local masks perturb pixels inside a target item to see how the object detector responds to these changes. A saliency map is subsequently produced, indicating the relative importance of each pixel based on the comparison of the detection output pre- and post-masking.To show how important each pixel in the original image is in relation to the objects that were located, a heatmap is then constructed. Experiments on different models and datasets of object detection have demonstrated that BODEM is a useful tool for interpreting object detector behavior and identifying shortcomings. This enables BODEM to be used in settings such as black-box software testing to evaluate and explain AI-based object detection methods. Data augmentation tests also conducted by BODEM suggest that the local masks generated by the system might be utilized to further train the object detectors and enhance the robustness and precision of their detection services.

Hussain et al. (2021), eXplainable Artificial Intelligence (XAI) is defined as a collection of approaches and procedures for turning so-called "black-box" AI algorithms into "white-box" algorithms, which produce results that are transparent and explicable and provide information about the parameters, factors, and steps the algorithm used to arrive at those results. The authors of this work take a `engineering' approach to illustrate the principles of XAI, adding to the body of previous research on the subject. They go over the stakeholders in XAI while providing an engineering viewpoint on the mathematical structure of the system. After that, we use an autonomous car as a case study to discuss the numerous uses of XAI for its various components, including perception, object recognition, control, action decision-making, and so on.

Olszewska (2022) the paper “Snakes in Trees: An Explainable Artificial Intelligence Approach for Automatic Object Detection and Recognition” is presented, outlining a novel explainable system that makes use of snakes that reside inside trees to sense and identify objects automatically. The suggested approach uses recursive snake computation, or parametric active contours. This results in multi-layered snakes, with the first layer representing the main object of interest and the subsequent levels defining the various foreground sub-parts. These snakes divide the world into areas based on visual clues that they interpret into semantic notions. Decision trees are generated based on these attributes, which leads to efficient semantic categorization of the pieces and automatic scene annotation. In relation to smart cities,

Dong, J.,et al(2023)aims to develop explainable Deep Learning (DL) models that will improve the dependability of autonomous driving systems by looking into the literature. The work expresses the decision-making process of the AV system as an image-based language creation (or image captioning) problem, as opposed to the conventional classification objective. To provide human-understandable explanations, the suggested method creates textual descriptions of the driving events before making any driving decisions. This results in the suggestion of a revolutionary multi-modal deep learning architecture that can mimic the correlation between an image (driving state) and a language (descriptions) at the same time. The Transformer-based structure of its complete system allows it to effectively simulate the learning processes of human drivers and perform worldwide attention.The results show that, in addition to producing appropriate driving judgments for autonomous vehicles (AVs), the suggested model can also produce coherent and valid words to explain specific driving scenarios. Additionally, it is noted that the suggested model performs noticeably better in terms of creating explanations and motivating actions than many baseline models. Given that the suggested model explains the behavior of an antivirus program, it may help boost end-user confidence.

Fujiyoshi, H., et al (2019),discusses deep learning for image recognition in the context of autonomous cars. It provides insights on the development and challenges of using deep learning models to recognize and interpret visual data—which is crucial for self-driving cars—and specializes in complex computational techniques. The purpose of the paper “Deep learning-based image recognition for autonomous driving” is to provide light on the advancements and challenges in the field of autonomous driving by utilizing deep learning techniques for picture recognition.

Muhammad, K.,et al(2022)the paper’s conclusions are as follows as a consequence: Scene recognition for vision-sensor-based autonomous vehicles is still a field that requires significant work. That’s why the majority of the scene understanding research that is now being done involves computationally sophisticated deep learning models. This assessment defines, examines, and assesses the current state of the area. We also look at state-of-the-art performance and temporal complexity studies of state-of-the-art modeling alternatives, in addition to covering the fundamental scene understanding pipeline. Not to add, distinguished achievements as well as noteworthy shortcomings of current research projects are emphasized. Furthermore, the paper offers an extensive summary of the datasets that are currently available along with an analysis of the difficulties that researchers continue to encounter despite their late achievements.Finally, to welcome researchers and practitioners to this exciting area, our work suggests directions for future study.

Bourdon, P., et al. (2021),Together with the details of the broader medical imaging background, Explainable Artificial Intelligence (XAI) is used to evaluate these types of imaging-related challenges. They also go over the rise in popularity of deep learning in this field and examine machine learning techniques created for breast cancer diagnosis. In actuality, despite the encouraging outcomes of the past few years, cutting edge research indicates that deep learning systems confront numerous significant hurdles. Talk about the latest developments and solutions to these problems as well.

Shen et al. (2022) examines how semantic AI may affect security in the setting of autonomous cars. In the rapidly developing field of semantic artificial intelligence (AI) security studies, they conduct the first thorough knowledge organization. 53 of these publications are collected, examined, and arranged by them according to important study facets for the security field. Based on quantitative comparisons between security works from closely comparable areas and current AD AI securities works vertically, they have compiled a list of the six most significant scientific gaps that we have found. These could offer perceptions and possible paths for the community, research purpose, methodology, and design in addition to design.They take the lead in bridging the largest methodology-level gap in science by creating PASS, an open-source, standard, and extensible system-driven evaluation platform for the semantic AD AI security research community. They moreover use representative semantic AD AI assaults on our created platform prototype to illustrate the capabilities and benefits of such a platform.

SYSTEM DESIGN

This chapter assessed the K-Means clustering algorithm for object segmentation against the other system segmentation and classification approaches currently in use in order to establish the most efficient method for explainable AI-based object identification of driverless autos. Figure 2 below the architecture displays the general block diagram for analysis. The stage that comes after the initial image pre-processing is called image enhancement. The process of segmenting photos is then initiated with the finished output. This stage's output is sent to the categorization phase. After this is done, the total performances are compared with the existing methods.

Explainable Artificial Intelligence

It is important to comprehend the decision-making process behind a particular item detector detection in order to provide context. Yet, due to their numerous trainable weights and hierarchical nonlinear struc-

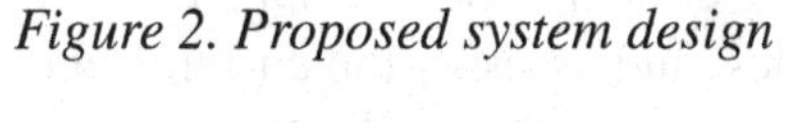

Figure 2. Proposed system design

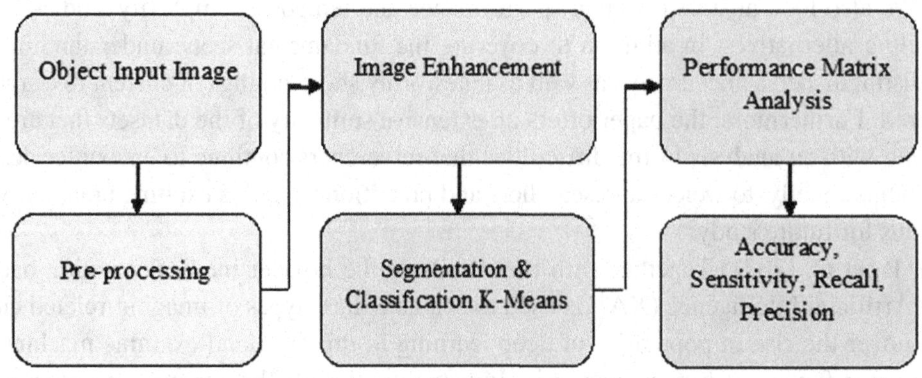

ture, deep neural networks—the foundation of all currently available object detectors—are inherently opaque. Therefore, developing methods that improve decision transparency and provide understandable explanations for individual actions is the aim of the recently established field of research known as eX-planatory Artificial Intelligence (XAI). In recent years, numerous methods have been proposed for this goal. The so-called model-agnostic interpretation method is explained in this section. Next, a K-Means is displayed, allowing for a quicker and easier approximation of the explanation.

K-Means Clustering Algorithm

K-means clustering is a simple, unsupervised technique for segmentation item discovery. Using this K-means approach, the total number of clusters and cluster centers is computed using the following image. It is necessary to choose cluster locations that maintain a safe distance from the cluster core in order to achieve effective clustering. Once the connecting of pixels in several clusters is finished, the pixels are prematurely aggregated. Next, based on distance measurements, new cluster centers are found and data points are assigned to them.

The pseudocode for K-Means clustering algorithm is shown in the above figure 3. New pixels are added or removed from each cluster throughout each iteration phase, which also affects the cluster's centroid. This iteration 75 procedure is repeated until there is no variance in the cluster centers. The final image shows the clustered output of the K-means algorithm. The K-means algorithm is used iteratively in two steps to minimize the distance between a data point and its centroid. This technique determines whether the image has any naturally occurring clusters by compiling a set of data characteristics from the vector space. It divides the pixels into a preset number of clusters using relatively straightforward processes.

The iterative two-step K-means technique can be used to reduce the distance between data points and the centroid of k-clusters.In the first step, the c centroid is calculated, and in the second phase, each data point is assigned to the cluster whose centroid is closest to that specific data point. The Euclidean distance is one of the methods most frequently used to determine the distance to the nearest centroid. Following grouping, the cluster points are assigned to the Euclidean distance that minimizes the distance between each center and each data point based on the centroid, and each cluster's new centroid is recalculated. The clusters of a partition are defined by its centroid and member items. The point at which all of the

Figure 3. Pseudocode for K-means algorithm

Algorithm 4: K-means Clustering

Step 1: Input image

Step 2: Get the values of k and choose the cluster centers

Step 3: Calculate the distance between pixels and cluster centers

Step 4: Assign each pixel to the nearest cluster center

Step 5: Compute the mean i.e. Cluster for each cluster

Step 6: Check whether there is any variation in cluster centers

Case1: If yes, then set the new mean as cluster centers and follow step 3

Case2: If no, then compute the statistics and separable information items in a cluster are separated from one another by their total distance is known as the cluster's centroid. The centroid-based technique is an iterative procedure that minimizes the overall distances between each item and its cluster centroid across all clusters. Assuming a collection of centers, V = {v1,v2,…..,vc}, and a set of data points, X = {x1,x2,x3,…..,xn}.

1. Choose "c" cluster centers at random.
2. Determine how far apart each data point is from the cluster centers.
3. Assign the data point to the cluster center that has the shortest distance between it and the other cluster centers.
4. 4. Utilizing, recalculate the new cluster center:

$$vi = (1ci) \sum xci\, j=1\ j \tag{1}$$

Where the number of data points in the ith cluster is denoted by "ci".

5. Compute the distances between every data point and the recently discovered cluster centers once more.
6. Stop if no data point is reassigned, and then go back to step 3 if not.

If the original cluster center is chosen wrongly and the method converges to local minima, the resulting cluster will be erroneous. Typically, the first cluster centers are chosen at random from the incoming photographs. Throughout the clustering process, it is imperative to take into account the local link between the data points. Using the histogram approach, the mean intensity value for each tissue class is initialized.

Diagram 4 illustrates the pre- and post-K-means clustering algorithm operation. Image intensities are re-distributed inside the cluster upon the convergence of the cluster means recalculation. This means

Figure 4. K-means clustering

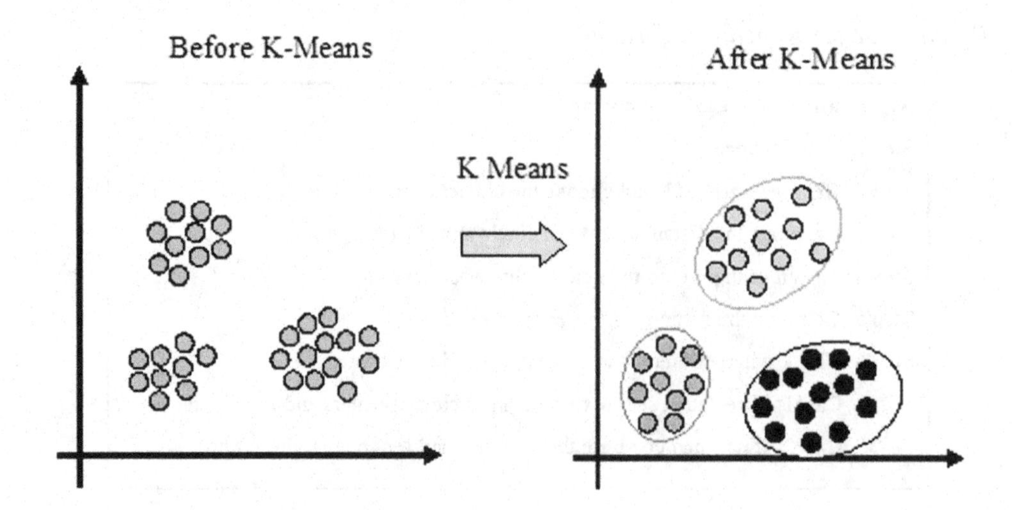

Figure 5. Flowchart of K-means clustering algorithm

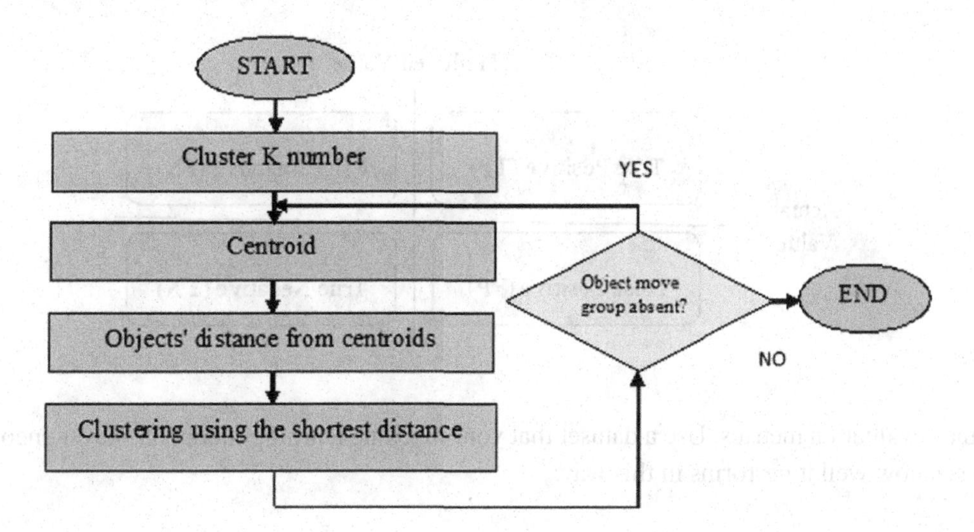

that the output mask contains tissue class labels ranging from one to K. The K-means algorithm has a disadvantage despite being a straightforward clustering technique.

The ideal centroid is attained by resolving a difficult optimization issue, and its initial position dictates the quality of the ultimate solution. Although randomization is the most popular initialization, other techniques can be applied depending on the situation to choose the initial centroid.The K-means clustering model algorithm is displayed in Figure 5. The following part covered the overall comparison result between the suggested system and the current system.

RESULT AND DISCUSSION

In addition to presenting contemporary object detectors, this section gives an overview of several assessment metrics. Images from the automated driving domain must be gathered for the continuing research. The performance of the object detectors will be analyzed, modeled, and explained using this data set. The production process demands a large time and resource commitment because the data set needs to be as large as feasible and every item in the pictures needs to be tagged. Thus, the most relevant data set is chosen after looking through those that are available to the general publicKang, Y.; et al (2019). Additionally, a large number of the relevant data sets for automated driving that have been released into the public domain for research are also somewhat recent. When applying K-Means techniques for object detection, Random Forest, Decision Tree, and KNN algorithms are used. The recently proposed K-Means study makes use of MATLAB 2013A to calculate a classifier mean's efficacy.

Performance Metrics

Figure 6, the confusion matrix, provides an example and synopsis of a categorization technique's efficacy. The performance analysis of the K Means machine learning classification models can be evaluated using

Figure 6. Confusion matrix

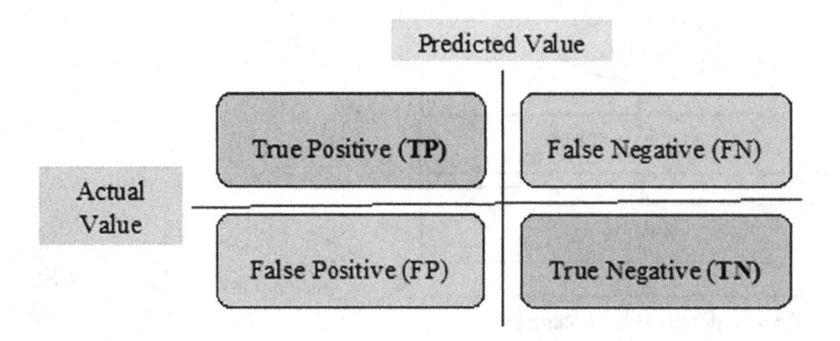

performance evaluation metrics. Use a dataset that your machine learning model has never encountered before to see how well it performs in this way.

Accuracy

A figure representing an error rate that deviates from 100% is displayed by the accuracy formula. Prior to assessing accuracy, the error rate must be evaluated. Error rate as a percentage can also be obtained by dividing the observed value by the actual value.

Accuracy = (TP+TN)/(TP+TN+FN+FP)

Refer to figure 7 above and table 1 for the accuracy result using the recommended and current technologies. By comparing the K Means with other existing systems, they are 94.58% accurate.

Sensitivity

By dividing the total number of positives by the number of precise positive forecasts, one can calculate sensitivity.

Sensitivity = TP/(TP+FN)

Table 1. Accuracy

Algorithm	Accuracy (%)
KNN	83.36
DT	84.46
RF	86.35
K Means	94.58

Figure 7. Accuracy graph

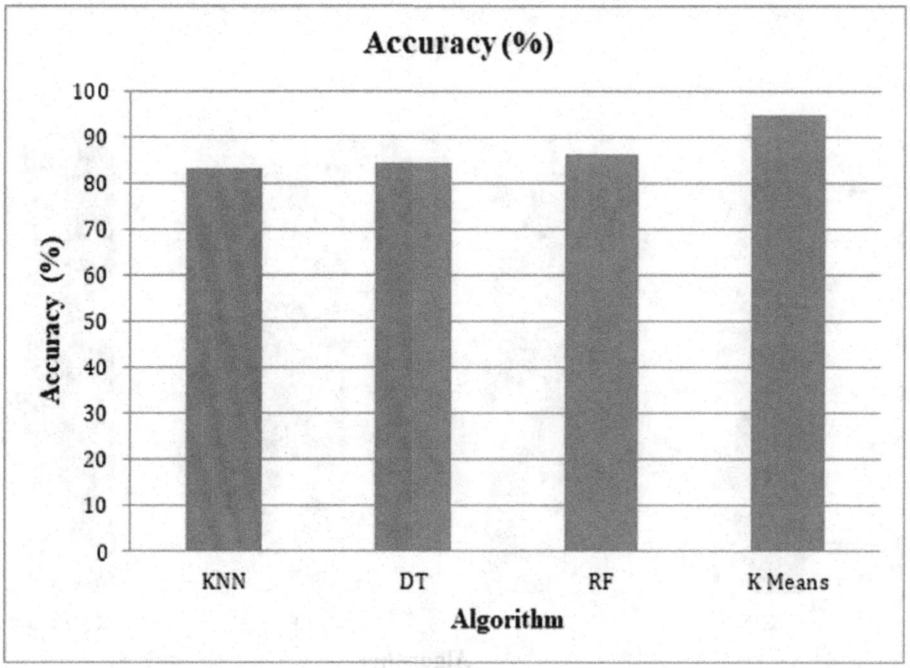

Table 2. Sensitivity

Algorithm	Sensitivity (%)
KNN	81.17
DT	83.52
RF	85.76
K Means	92.14

Table 2 yields the sensitivity results with the current and recommended systems, which are shown in the previously described figure 8. It shows that the K Means produces a higher sensitivity result of 92.14%, which is better than the comparative existing approaches.

Recall

Recall provides information about the model's ability to identify true positives. This instance shows how few patients with the condition were actually found because there are no genuine positives. There are two false negatives accessible.

Recall = (TP) / (TP+FN)

The results of the recall utilizing the recommended and current systems are shown in Figure 9, which was previously addressed and is derived from Table 3. According to the plotted graph, our

Figure 8. Sensitivity graph

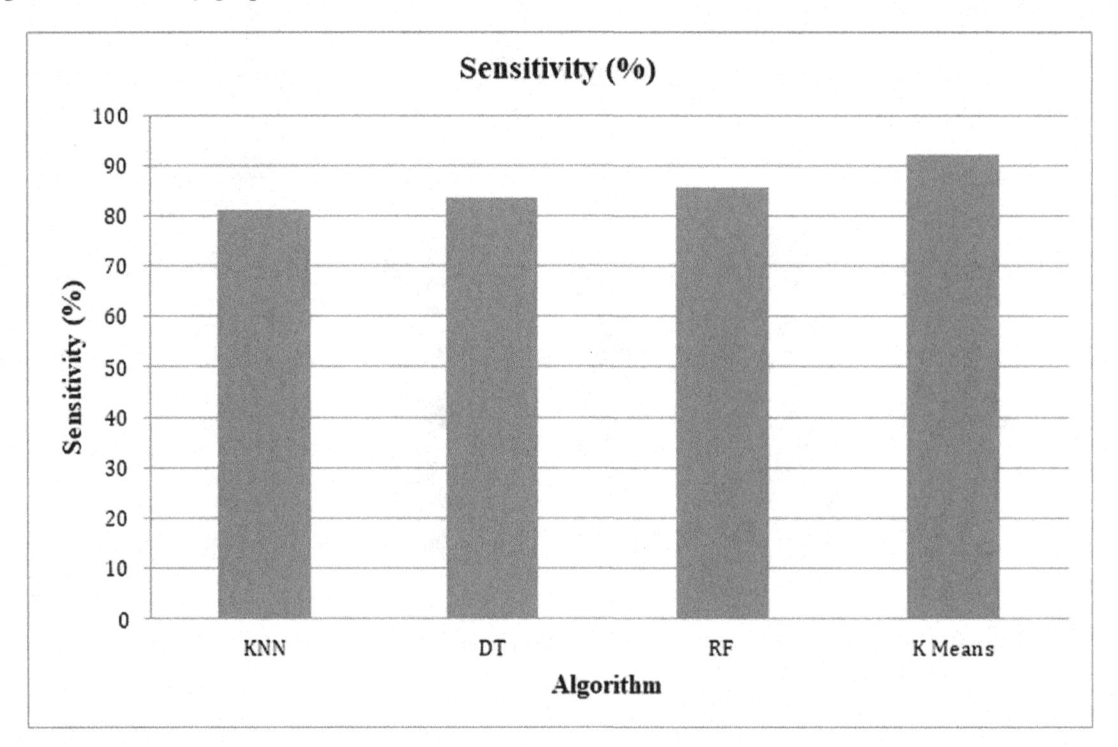

Table 3. Recall

Algorithm	Recall (%)
KNN	80.45
DT	82.36
RF	86.24
K Means	90.36

newly constructed K Means system offers a recall of 90.36% higher than the previously described current systems.

Precision

The accuracy level indicates how well many measures agree with one another. You can evaluate precision by computing the average deviation, which sums together the measurement errors.

Precision =TP/((TP+FP))

A screenshot taken from Table 10 is shown in Figure 10, which shows how precise the recommended and available technologies are. Plotting the results shows that our newly designed K Means system outperforms the previously described existing approaches with a precision of 93.69%.

Figure 9. Recall graph

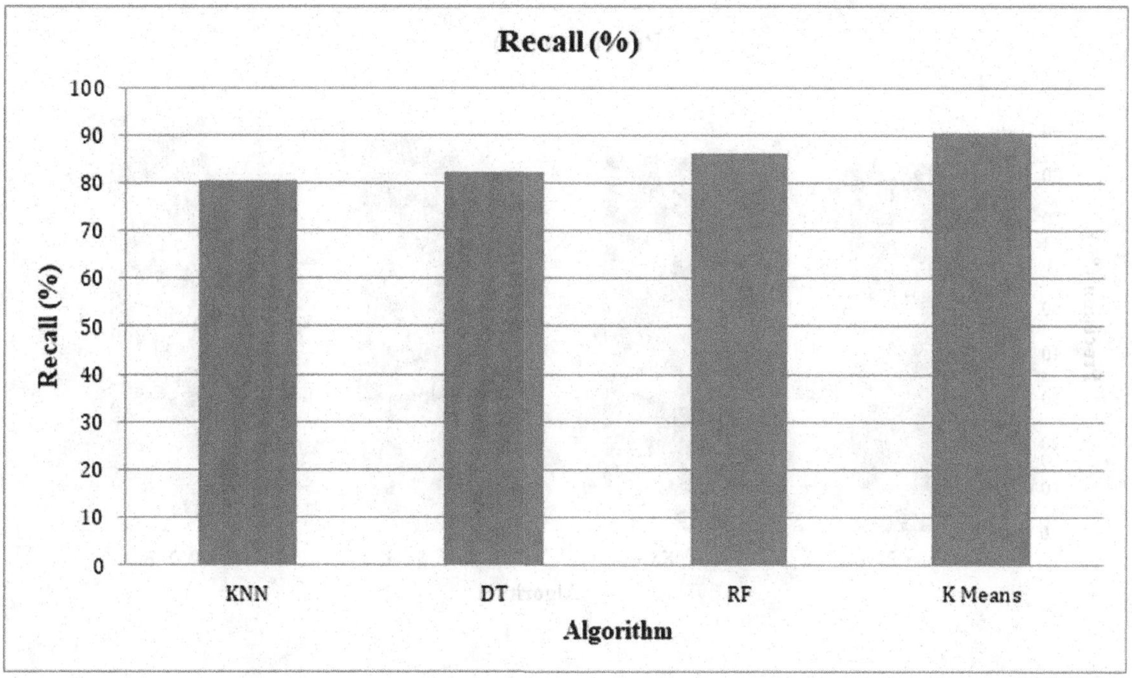

Table 4. Precision

Algorithm	Precision (%)
KNN	84.52
DT	87.18
RF	89.72
K Means	93.69

CONCLUSION

This work provides a comprehensive, interpretable performance analysis of object detectors used in automated driving using AI. Then, a unique method is shown that enables the modeling and even explanation of the behavior of machine learning-based object detectors using simple models (K-Means). The very high accuracy shown by the trained K-Means for forecasting the detection performance supports the assumption that an object's detection result may normally be connected to its presented information. Therefore, a simple K-Means algorithm might replicate the detection ability of an object detector with high accuracy by combining data about the object's characteristics and the surrounding environment from a large collection of autonomous vehicles.By modeling and explaining the object detectors using AI data, it will be feasible to discover challenging scenarios for automated vehicle testing in the future. An further use for these models is in virtual simulation. The established K-Means methodology may be used to any type of object detector and achieves 94.58% accuracy, as demonstrated by the nearly similar

Figure 10. Precision graph

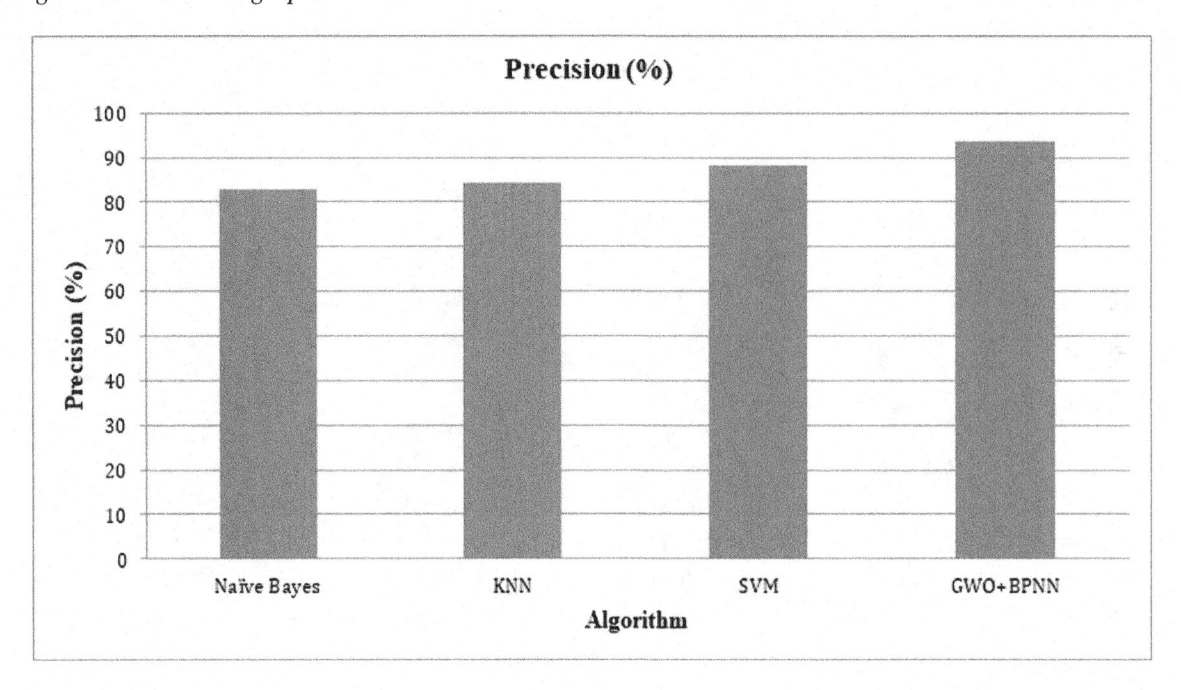

results for all object detectors studied. Eventually, other algorithms will be added to the XAIsystem to provide textual explanations to users.

REFERENCES

Antoniadi, A. M., Du, Y., Guendouz, Y., Wei, L., Mazo, C., Becker, B. A., & Mooney, C. (2021). Current challenges and future opportunities for XAI in machine learning-based clinical decision support systems: A systematic review. *Applied Sciences (Basel, Switzerland), 11*(11), 5088. doi:10.3390/app11115088

Arrieta, A. B., Díaz-Rodríguez, N., Del Ser, J., Bennetot, A., Tabik, S., Barbado, A., & Herrera, F. (2020). Explainable Artificial Intelligence (XAI): Concepts, taxonomies, opportunities and challenges toward responsible AI. *Information Fusion, 58*, 82–115. doi:10.1016/j.inffus.2019.12.012

Atakishiyev, S., Salameh, M., Yao, H., & Goebel, R. (2021). Explainable artificial intelligence for autonomous driving: A comprehensive overview and field guide for future research directions. *arXiv preprint arXiv:2112.11561.*

Aziz, L., Salam, M. S. B. H., Sheikh, U. U., & Ayub, S. (2020). Exploring deep learning-based architecture, strategies, applications and current trends in generic object detection: A comprehensive review. *IEEE Access : Practical Innovations, Open Solutions, 8*, 170461–170495. doi:10.1109/ACCESS.2020.3021508

Bourdon, P., Ahmed, O. B., Urruty, T., Djemal, K., & Fernandez-Maloigne, C. (2021). Explainable ai for medical imaging: Knowledge matters. *Multi-faceted Deep Learning: Models and Data.*

Dong, J., Chen, S., Miralinaghi, M., Chen, T., Li, P., & Labi, S. (2023). Why did the AI make that decision? Towards an explainable artificial intelligence (XAI) for autonomous driving systems. *Transportation Research Part C, Emerging Technologies*, *156*, 104358. doi:10.1016/j.trc.2023.104358

Fujiyoshi, H., Hirakawa, T., & Yamashita, T. (2019). Deep learning-based image recognition for autonomous driving. *IATSS Research*, *43*(4), 244–252. doi:10.1016/j.iatssr.2019.11.008

Hussain, F., Hussain, R., & Hossain, E. (2021). Explainable artificial intelligence (XAI): An engineering perspective. *arXiv preprint arXiv:2101.03613*.

Kang, Y., Yin, H., & Berger, C. (2019). Test Your Self-Driving Algorithm: An Overview of Publicly Available Driving Datasets and Virtual Testing Environments. [Google Scholar] [CrossRef]. *IEEE Transactions on Intelligent Vehicles*, *4*(2), 171–185. doi:10.1109/TIV.2018.2886678

Mankodiya, H., Jadav, D., Gupta, R., Tanwar, S., Hong, W. C., & Sharma, R. (2022). Od-xai: Explainable ai-based semantic object detection for autonomous vehicles. *Applied Sciences (Basel, Switzerland)*, *12*(11), 5310. doi:10.3390/app12115310

Moradi, M., Yan, K., Colwell, D., Samwald, M., & Asgari, R. (2023). Model-agnostic explainable artificial intelligence for object detection in image data. *arXiv preprint arXiv:2303.17249*. doi:10.2139/ssrn.4429462

Muhammad, K., Hussain, T., Ullah, H., Del Ser, J., Rezaei, M., Kumar, N., & de Albuquerque, V. H. C. (2022). Vision-based semantic segmentation in scene understanding for autonomous driving: Recent achievements, challenges, and outlooks. *IEEE Transactions on Intelligent Transportation Systems*, *23*(12), 22694–22715. doi:10.1109/TITS.2022.3207665

Olszewska, J. I. (2022, February). Snakes in Trees: An Explainable Artificial Intelligence Approach for Automatic Object Detection and Recognition. In ICAART (3) (pp. 996-1002).

Omeiza, D., Webb, H., Jirotka, M., & Kunze, L. (2021). Explanations in autonomous driving: A survey. *IEEE Transactions on Intelligent Transportation Systems*, *23*(8), 10142–10162. doi:10.1109/TITS.2021.3122865

Ponn, T., Kröger, T., & Diermeyer, F. (2020). Identification and explanation of challenging conditions for camera-based object detection of automated vehicles. *Sensors (Basel)*, *20*(13), 3699. doi:10.3390/s20133699 PMID:32630350

Shen, J., Wang, N., Wan, Z., Luo, Y., Sato, T., Hu, Z., & Chen, Q. A. (2022). Sok: On the semantic ai security in autonomous driving. *arXiv preprint arXiv:2203.05314*.

Thakker, D., Mishra, B. K., Abdullatif, A., Mazumdar, S., & Simpson, S. (2020). Explainable artificial intelligence for developing smart cities solutions. *Smart Cities*, *3*(4), 1353–1382. doi:10.3390/smartcities3040065

Tyagi, A. K., & Aswathy, S. U. (2021). Autonomous Intelligent Vehicles (AIV): Research statements, open issues, challenges and road for future. *International Journal of Intelligent Networks*, *2*, 83–102. doi:10.1016/j.ijin.2021.07.002

Wang, J., Li, Y., Zhou, Z., Wang, C., Hou, Y., Zhang, L., & Chen, S. (2022). When, where and how does it fail? A spatial-temporal visual analytics approach for interpretable object detection in autonomous driving. *IEEE Transactions on Visualization and Computer Graphics*. PMID:36040948

Zablocki, É., Ben-Younes, H., Pérez, P., & Cord, M. (2022). Explainability of deep vision-based autonomous driving systems: Review and challenges. *International Journal of Computer Vision*, *130*(10), 2425–2452. doi:10.1007/s11263-022-01657-x

Zhao, J., Zhao, W., Deng, B., Wang, Z., Zhang, F., Zheng, W., & Burke, A. F. (2023). Autonomous driving system: A comprehensive survey. *Expert Systems with Applications*, 122836.

Chapter 3
AI-Based In-Cabin Monitoring System for Autonomous Vehicles

Ajith K.
Karpagam Academy of Higher Education, India

R. Sharmila
Karpagam Academy of Higher Education, India

ABSTRACT

Fully autonomous vehicles (FAVs) require internal monitoring in order to function without a human driver. The far-from-sufficient FAV, sufficient in-cabin monitoring is a prerequisite to ensure both people and vehicles. On public roadways, there are a lot of accidents that happen, most of them are the result of reckless driving. Modern driver monitoring systems evaluate driver behavior and, if necessary, highlight risky driving behaviours using special sensor technologies. The result accurately predicted bounding boxes and the real data show a considerable amount of overlap. Unlike most past efforts, the authors use a random forest to learn a template-based model. This way, forecast the object probability of a window in a sliding window technique and regress its aspect ratio using a single mode at the same time. Examined mobility services at increasing degrees autonomy, including the exercise caution and the best ways.

INTRODUCTION

Systems for intelligent surveillance and monitoring are commonly employed to guarantee security and safety. Closed-circuit television (CCTV) and video surveillance cameras are common forms of public monitoring. Intelligent transportation systems also use video monitoring for tasks like road traffic and in-cab monitoring. Public visual information is also necessary for driving automation to do various duties. However, these AVs may be operated by the driver a human if necessary. The cars meant solely for passenger use; they lack drivers. As a result, nobody is responsible for such AVs. Moreover, the occupants of shared and public transportation (such ridesharing, car-sharing, and car-full services in

DOI: 10.4018/979-8-3693-1962-8.ch003

autonomous vehicles) are unfamiliar with each other. Therefore, it's imperative to ensure each individual sat in such AVs is secure and healthy. The vehicle should also be protected from external dangers and/ or any malicious behavior on the part of its occupants. As a result, FAVs essentially need a complex, real-time in-cabin monitoring task.

In-Cabin Monitoring

Bosch et al. have proposed visible light camera's eye-tracking. The works that the authors have reviewed have shown signs of anger, fear, melancholy, curiosity, humiliation, urgency, boredom, annoyance, and other emotions. In level 4 autonomous vehicles, driver monitoring plays a major role in advanced driving (ADAS). In the authors, Yifang, et al (2020) have examined problems to developments linked automobiles several efforts and the AI techniques employed to guarantee the safety of AVs within the cabin. Popular AI techniques for three distinct tasks driving assistance, takeover preparedness, and driver status monitoring have been enlisted. Emotion, tiredness, distraction, and attention detection are all included in driver status monitoring. Intention analysis and warnings about driving risks are features of driver assistance, while evaluation of takeover preparation is a feature of takeover readiness. They have included several works using AI for these tasks in their survey. There has been an evaluation of the driving monitoring and assistance systems. They investigated driving while taking the driver, the vehicle, and the road environment into consideration. This review focuses on the following factors: driver distraction, driver weariness, and aggressive driving style.

Safety and Security in Level Four and Beyond AVs

Protecting someone or something from harm, risk, or danger is what it means to be safe. Protection of people and property against outside dangers and illegal activity is referred to as security. Security and safety measures must be included in an antivirus program. Ensuring the security of an individual or object from inadvertent mishaps is known as safety. In a similar vein, security is defence against intentional harm by Mallaboyev, N. M., et al (2022). When it comes to the main issues about vehicle needs from being mishandled, but the occupants also need safety.

Occupants and Objects

Many categories seen inside the AV's cabin are taken into account when creating the database. Passengers, kids, pets, smoking materials, mobile phones, car seats, drinks, luggage, personal possessions, and hazardous or destructive objects are among them by Rathi, B. Senthil et al (2021). Volunteers were observed doing one activity at a time and sporadically executing numerous activities while their photographs were being taken.

Cellular and Electronic Devices

It's possible that the car's interior has electronic and cell phone devices. These could be tablets, laptops, cell phones, or any combination of these. Nonetheless, require identification in order trace misplaced or lost device since they can identify certain images that include gadgets and cell phones.

Smoking Items

Since smoking is harmful to your health, it should not be allowed in public areas. Public transportation should be applied to vehicles that are shared or utilized for transportation. As a result, smoking is not allowed within this kind of vehicle. Therefore, smoking item detection needs to be a part of the IMS for the security and safety of drivers and passengers. We have separated out smoking products into four categories: cases, lighters, e-cigarettes, and cigarettes.

Food and Beverages

To maintain the cleanliness and safety of the in-cabin, detection is crucial when it comes to shared vehicle food and beverages. A passenger's eating or drinking during the voyage could annoy other passengers. For instance, spilling on others can occur from negligent eating or drinking. Additionally, the lingering food and drink debris could pose a risk to the subsequent occupants of the car. Additionally, this kind of litter can interfere with car controls, seriously impairing vehicle safety. In order to alert passengers in the event that using food or beverages within a car causes any issues.

Harmful Objects

Weapons, dangerous equipment, and harmful items must be identified in order prevent stop any acts of violence or vandalism to protect the higher-end autonomous driving vehicles. Tools including screwdrivers, scissors, baseball bats, knives, and so on have been confiscated as potentially hazardous materials. Here are a handful of these pictures.

Face Anonymization

For public surveillance systems, privacy is a major concern. As an open monitoring system, IMS experiences the same issue. Many nations have placed stringent limitations on the use of facial recognition technology in public spaces in order to protect personal data. By protecting, facial anonymization effectively resolves this problem. An effective approach for the face anonymization of the inhabitants has been proposed in our earlier studies. For face swapping and re-enactment, GAN is employed. We advise utilizing the virtual human faces produced by GANs as the source image. By doing this, improved anonymization is guaranteed. Preserving the face features of the target image is important.

Face Detection

The majority of object detection research in static pictures has been on face and person detection. These domains were chosen because face and person detection are critical steps in most systems involving human-computer interaction. Basic forms and constraints are used by the majority of early face detection programs. Kumar, Ashu et al (2019) developed a technique that can recognize face components by using local restrictions on an image pyramid. It explains an analogous rule-based approach. Rule-based approaches are beneficial in terms of processing cost and work well for faces due to their regular internal structure. By storing a human face as a set of binary relationships between the mean intensities of eleven distinct regions, the concept of the "template ratio" was extended. The template ratio method worked on the assumption that these correlations would hold true in the face of significant changes in illumination direction and intensity.

AI Camera Positioning

There are two primary issues with FAVs and two primary issues with the front-facing cameras that are placed for external influences. Besides, it's hard to watch the second row from the front of the car because of obstructions in your line of sight. Two front-facing cameras are required to fully monitor the in-cabin entertainment system in FAVs, as depicted in these drawings Mishra, Ashutosh et al (2022). Outside, it will also double the difficulty of the same challenge. There will be similar issues with every other AI camera. That being said, the recommended position place powerful IMS activates something strange it becomes necessary to conduct thorough observation or surveillance in addition to obtaining evidence.

Comparison of Popular Object Detection Models

We have compared several AI algorithms to determine which strategy is optimal item identification methods are enumerated and surveyed. One example is "you only look once (YOLO or yolo)", which classification for one of them. An example of a two-stage detector is R-CNN. Even though there are many variations of both types, the algorithms' effectiveness depends on how they are implemented. Regarding inference delay, one-stage techniques are considered, and in terms of accuracy, two-stage detectors are thought to be suitable explain by Faniadis, Efstathios, and Angelos Amanatiadis (2020). These methods' retrained weights are currently accessible for a wide range of object categories. However, using the pre trained weights of these algorithms directly was producing false results. This failure can be related to the viewing angle of the AI camera.

In-Cabin Measurement

This technique gauges the concentration of pollutants within the cabin in particular environmental conditions. Depending on ISO 16000-6:2011 requirements, either a Summa canister, Tedlar bag, or Tenax-TA tube is used for this. SVOCs are examined using GC/MS utilizing a polyurethane foam (PUF) sample used by Lan, Hangzhen (2020). Volatile carbonyls, such as formaldehyde, are sampled using DNPH cartridges and evaluated using high-performance liquid chromatography (HPLC), following ISO 16000-3:2011. When it comes to the identification and measurement of VOCs SVOCs, on-site monitoring method may be suitable in addition to the off-site analysis. For instance, portable tools such typically used the field to detect the PM concentrations in-cabin. Because they are simple to use in the field, in-cabin measurements are frequently utilized.

Machine Learning

A key characteristic that unites a wide variety of techniques, heuristics, and algorithms is machine learning: they somehow endow a system with the capacity to carry out a certain task by means of a collection of cases. By giving the system a collection of example patterns from both classes from a set of training data, we hope to automatically teach the system the characteristics that distinguish one class from the other within the context of our pattern classification problem. The positive and negative instances have labels of +1 and -1, respectively. The degree of performance that the trained system attains on a collection of instances that were absent from the training set, or test set, is the objective and a key indicator of success. By using the test set to assess performance, we are able to ascertain the system's ability to generalize to data that it has never seen before.

A model of the domain, or an input/output mapping, is what is meant to be extracted by the system from a set of training examples. This is the definition of the task of learning from examples. This kind of tactic appeals to me especially for several reasons. Most importantly, we can learn a problem's attributes via instances, which spares us from having to explicitly build a solution. A customized solution may suffer when a user imposes what he considers to be the essential components or characteristics of a decision problem by Greener, Joe G., et al (2022). A learning-based approach abstracts away the important components and relationships in a decision problem automatically as a trained model. However, learning-based techniques may have the issue of poor generalization due to overfitting, in which the model has "too well" mastered the decision problem and is unable to generalize to new data.

The five sections of this essay are listed below: Session 2 of the current system shows its shortcomings in object detection using kaggle datasets and other machine learning and artificial learning techniques. The third session was a demonstration of the suggested sentiment analysis method. It blends Random Forest (HCSO RF) with Hybrid Chicken Swarm Optimization concepts. The outcomes of the anticipated system are shown in Session 4. An examination of the recommended approach for object detection in vehicle cabins rounds off Session 5.

RELATED WORKS

Artificial and machine learning approaches are covered in this section. A summary of pertinent literature is provided prior to using this evaluation to pinpoint knowledge gaps and make clear particular research goals.

Several businesses have lately revealed that their in-cabin surveillance systems use deep learning models like YOLO versions to track driver behavior in an effort to improve passenger safety and security. The car owners have also been surveyed regarding the AI methods employed for in-cabin surveillance. They claim that services applications car is made feasible by artificial intelligence (AI). The release of an in-car monitoring dataset enables interior vehicle monitoring byKatrolia, Jigyasa Singh, et al (2021). A hardware implementation for monitoring a car's occupants and driver has been completed. Poon, Yen-Sok, et al (2022) developing a driver monitoring system with deep learning and in-cabin surveillance; they have also gathered a dataset of telemetry and video data for this use.

Person Identification

Among the multipurpose objectives of facial recognition technology are classification, discrimination, and recognition. Urbanization and smart cities demand facial recognition for a multitude of purposes. Consequently, a range of person-identifying facial recognition systems have been given by earlier researchers. Three types of face recognition methods exist: local, holistic, and hybrid. While holistic approaches use the full face, including the background, local methods only use a subset of the features on the face for facial identification. Hybrid approaches integrate both local and comprehensive techniques, as their name suggests. With the advent of artificial intelligence (AI) that makes use of CNNs and DL, facial recognition algorithm performance has grown. Dixit, Priyanka, and Sanjay Silakari (2021) made public DeepFace, a deep neural network-powered facial recognition system. Furthermore, multiple further enlarged forms of DeepFace have been verified by independent investigations. Thorough examination of face recognition techniques, comparative analysis, and prospective uses in the future were conducted. Common facial recognition techniques were divided into three categories: hybrid, holistic, and local techniques. Different approaches were compared for accuracy, complexity, and robustness. The advantages and disadvantages of each tactic were also examined. Wang et al. successfully reviewed deep learning-based facial recognition techniques in their work. They investigated in detail a number of popular DL-based techniques, including autoencoder-, CNN-, and GAN-based methods have produced a summary of the primary technology roadblocks and emerging real-world issues with deep facial recognition.

In-cabin monitoring was used in an experiment to examine how drivers interacted with automated driving. This study aimed to apply a similar methodology with FAVs. The detection of violence in vehicles is reviewed by an Authors Saxena, Anvita et al (2020) have investigated a range of research papers and artificial intelligence techniques for emotional recognition in cars in order to identify violent situations in carpooling services. An analysis of the in-vehicle monitoring system (IVMS) revealed the need for in-cabin passenger anomaly detection in order to curb unsafe driving behaviors. Their system included a camera and a computer, according to a patent filing, and it was deemed appropriate for a modern car. Among its responsibilities were identifying the seat, the individuals occupying it, their orientation, and the orientation of the seat itself. To avoid foreign objects like bottles, cans, electronic devices, keys, books, etc. from interfering with car controls or breaking them down, an in-cabin safety and cleaning issue has been considered. Similarly, numerous studies and works of literature have been published in the past about the monitoring of a car's interior. Safety and averting hazardous conditions in the event of an accident become critical at high degrees of autonomous driving. Due to the limitations and restrictions found inside the vehicle's interior, Priyadarshi, Rahul (2020) found it can be challenging to decide which sensor placement is best for them. Artificial data was produced in order to assess person position in cabin environment with specific people' postures to create photorealistic scenes. On the other hand, their generated dataset contains scenes when people are looking forward. The safety of the occupants in the case of an accident has been their main concern. They produced a synthetic dataset with many scenes in order to comply with European safety rules. In-cabin monitoring systems have lately been developed by numerous companies and sectors.

It consists of many AI-assisted intelligent monitoring cameras that draw attention to an important point about where these cameras should be placed inside the AV's cabin. It contrasts the in-cabin monitoring experiences that result from different camera placements. The surveillance camera records both inside and outside of the cabin when it is installed in front-facing (FF) mode by Ramos-Sorroche, Emilio, et al (2023). Consequently, the person outside the car is mistakenly identified by the object detector algorithm as the car's occupant. The installation of rooftop (RT) cameras, however, makes it simple to avoid such circumstances. The only thing captured by the camera in the middle of the rooftop is the AV's inside.

Passenger Discomfort and Health Issues

The cabin of an airplane is much more airtight than a building and can accommodate considerably more people per square foot. Another problem is the discomfort patients experience from decreased cabin pressure. These changes may have an effect if the body contains trapped air. Ascending altitude causes air to escape from the sinuses and middle ear, which is why many travellers describe feeling like their ears are popping. This is generally not thought to be a problem. Pressure equilibrium must be restored to the middle ear and sinuses during the aircraft's altitude decrease before to landing Milledge, James (2020). In the event that this doesn't happen, discomfort could feel like your ears or sinuses are closed. It not only makes your ears pop, but it also severely harms the sinuses in your nasal canal. In order to investigate these aspects of human comfort, state-of-the-art computational fluid dynamics technologies are required. A non-invasive method for assessing how lower pressure affects the sinuses cabin and ear lobes and for helping find a way to reduce discomfort is the use of CFD.

Multi-Scale Feature Representations

Representing and processing multi-scale data efficiently is one of the primary challenges in object detection. Previous detectors frequently use the pyramidal feature hierarchy that is taken from backbone networks to make predictions directly. Feature Pyramid Network (FPN) is one of the first papers that suggests a top-down method for combining features from different scales. More recently, feature network topology is automatically designed by NAS-FPN using neural architecture search. Performance-wise, NAS-FPN is more effective, but the search process requires hundreds not coherent erratic. In this research, we aim to improve the principled and intuitive aspects. In a recent study, outstanding model efficiency for image categorization is demonstrated by simultaneously increasing network width, depth, and resolution.

The deformable part-based model is the most well-known example and one of the object detection paradigms that has been studied the most. This method integrates several discriminatively learnt components by using a star model called pictorial structure. The components model is the top layer, while the star model is the bottom layer. This paradigm can be conceptualized as having two layers. However, DNNs' general layers, the work by Trewartha, Amalie, et al (2022) takes advantage of domain knowledge the sections are structured with a cinematic aesthetic and are based on personally created Histogram of Gradients (HOG) descriptors.

Deep Architectures

Compositional models are deep models for object detection and parsing that express the object as a layered composition of image primitives. Part-based models serve as an inspiration for these models. The And/or graph is a well-known example that uses a tree to represent an item, and nodes to represent separate pieces, and or nodes to show alternative modes of the same part. Similar to DNNs, the And/or graph consists of several layers, with the lowest layers representing tiny generic picture primitives and the uppermost levels representing object components. These compositional models are easier to interpret than DNNs used by Tarnawski, Jakub M., et al. (2020). On the other hand, they require deduction, while the DNN models that are studied in this work are merely feed-forward and do not need latent variable deduction. Segment-based compositional models for detection are also examples, whereas primitives focus on shape and use Gabor filters or larger HOG filters. These solutions are usually hindered by the complexity of training and use specially designed learning mechanisms. Additionally, they combine bottom-up and top-down methods for drawing conclusions.

Neural Networks

Neural networks (NNs) are compositional models with less interpretable and broader nodes than the models previously stated. It has been several years since NNs of which convolutional NNs are the most well-known example were originally used to vision-related problems. Until recently, few people were aware of how effective these models known as DNNs were for large-scale picture categorization applications used by Tian, Yongqiang, et al. (2021). However, their usefulness for detection is limited. A more complex type of detection called scene parsing has been attempted with multi-layer convolutional neural networks.

Medical image segmentation has been investigated using DNNs. However, the NNs are used in all approaches as super pixel-level, pixel-by-pixel local, or semi-local classifiers. Our localization technique is regression, but it requires the full image as input. As a result, using NNs in this way is more efficient. This is maybe the closest answer to ours; it makes use of a much smaller network, different characteristics, and a different loss function to distinguish between numerous instances of the same class.

SYSTEM DESIGN

This section explains the recommended research architecture and technique. The hybrid machine learning strategy uses data from the vehicle interior camera covering a wide range of activities, including different doing so, to train the system using Google's Kaggle dataset (Human, bag, pet, etc.). In order to integrate and use the suggested methods practically and improve the results, a set of activities that allow drivers to predict a specific person's car with increased sensitivity, specificity, accuracy, and precision must be established. Because they have different budgets and preferences, scientists also employ machine learning techniques that are more effective and less costly. An outline of the suggested methodology is provided in Figure 1.

Training

Segmentation instances are used to generate data on the observed shape variance. This data is encoded by the shape model such that the shape-variant Hough transform can use it. The established border appearance model provides a measure of the probability given is part of sought-after defining the boundaries with precision in a training set, the user establishes the segmentation task. The primary variations in object shape and local border appearance that may arise during segmentation must be captured in the training set's images. In addition to borders, landmarks are used show the relationship boundaries various items.

Pre-Processing (Mean Filter)

The median filter in image processing is most frequently used to remove noise, such as salt and pepper. Median filters have several applications in image processing. On the other hand, median filters struggle to remove heavy-tailed noise, obliterate fine details like lines and crisp edges, as well as perform badly when noise that is dependent on the input is present. The investigation into nonlinear median-based filters has produced excellent findings and opened up some exciting new research directions. Because versatility, ability to preserve edges, and resistance to spurious noise, the traditional median filter continues to be

Figure 1. Flow chart for the proposed techniques

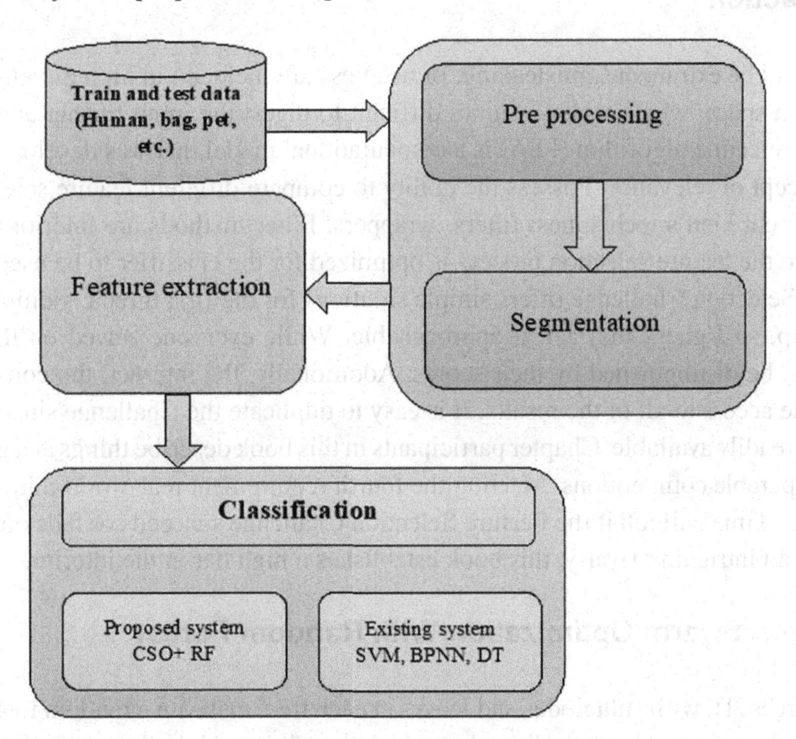

the standard in applications including image processing. However, little elements in the image, like tiny lines and borders, are frequently lost when using the median filter. More median-type filters, including stack and weighted filters, have become accessible in recent years.

Segmentation (Edge Detection)

When transitioning from image processing to picture analysis, image segmentation is a crucial step. On the one hand, it forms the basis of the target expression and has a major impact on the feature measurement. On the other hand, by means of segmentation, feature extraction, and parameter measurement, it serves as the intended expression, reducing the original image to a more abstracted and simplified form. Consequently, high-level visual analysis and comprehension are enabled. Real-world production uses image segmentation extensively, and it has essentially taken over all related image processing fields.

The accuracy of edge detection is critical to many applications of vision systems. The reason for this is because edges aid in defining the form and placement of the objects being studied. Edge-based methods are widely used for image segmentation because they provide the data needed for image analysis and pattern recognition processes carried out during high level image processing. Edge identification is a challenging problem, especially when dealing with low contrast, noisy, and fuzzy images. Consequently, scholars discussed it in great detail over time. Although several edge detection methods have already been proposed, most of them belong to one of two categories: moment-based methods or image-based methods.

Feature Extraction

Features that could be extraneous, misleading, or needless are included in high dimensional data. This widens the search space, which makes it more difficult to digest the input further and impedes learning. The feature selection algorithm (FSA) is a computational model that was developed in response to a particular concept of relevance. Possess the ability to compare different feature selection techniques empirically different kind's techniques: filters, wrappers. Filter methods are inferior to wrapping approaches because the feature selection process is optimized for the classifier to be used.

The Feature Selection Challenge offers simple solutions for the first three conditions. Seventy-five teams showed up, so I guess they felt it approachable. While everyone stayed on the scale, the top performers could be distinguished by their scores. Additionally, the internet, the companion CD, and this book provide access to all of the results. It is easy to duplicate the Challenge since all the data and Matlab code are readily available. Chapter participants in this book describe things at a greater level than in previous comparable competitions. Meeting the fourth requirement real-world relevance may be the most challenging. Time will tell if the Feature Selection Challenge succeeds or fails on this one. As the public record of an intriguing rivalry, this book establishes a high bar in the interim.

Hybrid Chicken Swarm Optimization With Random Forest

A collection of trees, Tt, with split nodes and leaves in each tree, make up a random forest. Every image patch that arrives is evaluated by the split nodes, which then forward it to the left or right child based on how the patch appears. During training, the picture patches' statistics are stored in each leaf L.

Chicken Swarm Optimization

The suggested approach uses the chicken swarm optimization (CSO) technique in conjunction with a limited number of carefully chosen characteristics to identify feature combinations that optimize classification accuracy. With respect to training and validation sets, the CSO's fitness function seeks to optimize maintaining fewest amount chosen in below Figure 2.

This allows each chicken to be described by its position. Each of the three has a unique position update formula: a hen, a chick, and a rooster. Based on the swarm's fitness function values, identify the position of the chickens with this equation:

$$X_{i,j}(t+1)=X_i-r_{Num-hNum,j}^{(t)} \quad (1)$$

Procedure for Removal and Distribution The following are the possible contents of the concrete, where the eggs are dispersed around the search area Pop do if i=(rNum+hNum +1). If then.

$$X_{i,j}(t+1)=lb+(ub-lb)Xrand \quad (2)$$

In this context, pop refers to population size. The search range's lower and upper bounds are represented by the symbols lb and ub, respectively. Here is a description of the primary steps:

Figure 2. Algorithm for chicken swarm optimization

```
Initialize RN, HN, CN, M N, G;
Randomly initialize each chicken in the swarm
Xi(i = 1, 2, ..., N ).;
Initialize the max numbers of iteration Tmax;
while T < Tmax do for each iteration
        if T % G equals 0 then
end
        foreach chicken Xi in the swarm do
                if Xi is a roster then
                        Update Xi's location using equation 1;
                end
                if Xi is a hen then
                        Update Xi's location using equation 3;
                end
                if Xi is a chick then
                        Update Xi's location using equation 6;
                end
                Evaluate the new solution using equation 7;
        end
end
```

1. Setting the population's initial value. The first individual determination between the CSO and PSO algorithms and parameter parameters is the main topic of discussion.
2. An evaluation of fitness. Splitting up into smaller groupings. The two halves of the original population that are divided using the same scale are known as subgroups 1 and 2.
3. To determine the global optimum value, we simulate a swarm of chickens foraging by allocating responsibilities to subgroup 1 using the CSO technique. Figure 3 below describes the CSO optimization flow diagram.

The PSO method adjusts the particle's position and velocity in subgroup 2 in order to determine subgroups 1 and 2 in order to share information while updating the optimal swarm value. The revised Subgroup ideal values are computed and used to determine the allotted accomplished and desired obtained.

Figure 3. CSO flow diagram

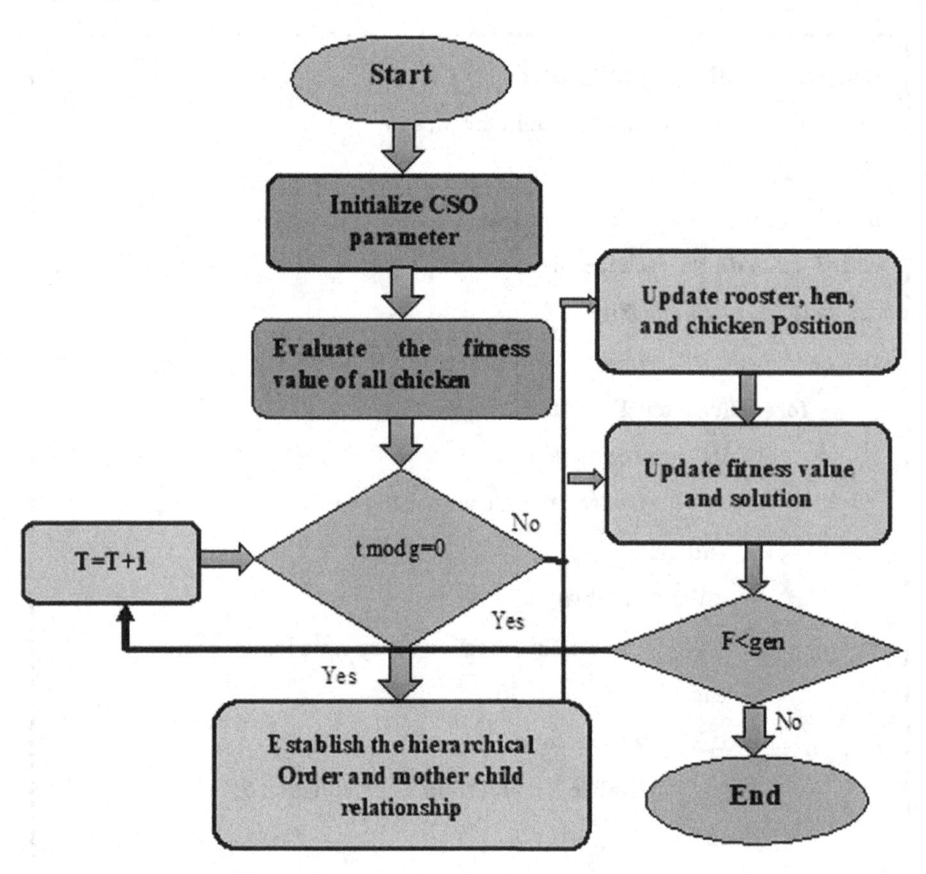

Random Forest-RF

It is the likelihood for each class c in a classification issue, denoted by the symbol p(c|L). The distribution that one wants to estimate in a regression job is over the continuous parameter x ∈ RH. To achieve both objectives, random forests for object identification must be trained to identify and classify patches that belong to an object as well as use those patches to regress the location and scale of the object.

Depending on whether the sample should move left or right, a binary-valued function decides which internal node in the tree should receive it. The set of decision trees is represented by T dt given any sample of data, the elements in Y will reach the tree's leaf. For this (admittedly simple) tree, let the notation be LF (Q) ∈ T td. If this is not present, a decision tree consists of a node that has a left and a right subtree. A data sample x ∈ X is split function φ: X → {0, 1} determines whether it is routed to the left decision sub-tree tl∈ T dt (if φ(x) = 0) or the right one tr∈ T dt (if φ(x) = 1). ND (φ, tl, tr)∈ T td is one technique to depict such a tree. Finally, to generate a prediction about a data sample, a decision forest is an ensemble F ⊆ T td of decision trees that averages the individual forecasts obtained from each tree. Figure 4 shows the pseudo code for the random forest algorithm.

Uncertain Regression Forests often represent a non-linear mapping M: RM → RK, where an example x is mapped to a goal prediction y. A portion of the training set is used to train each binary decision tree in the ensemble, {Tt} T t=1 (where T is the number of trees), in order to get this mapping. To lower the

Figure 4. Pseudo code for the random forest algorithm

```
Algorithm 1: Pseudo code for the random forest algorithm
To generate c classifiers:
for i=1 to c do
    Randomly sample the training data D with replacement to produce Di
    Create a root node, N, containing Di
    Call BuildTree(Ni)
end for
BuildTree(N):
if N contains instances of only one class then
    return
else
    Randomly select x% of the possible splitting features in N
    Select the feature F with the highest information gain to split on
    Create f child nodes of N,N1,...,Nf where F has f possible values (F1,...,Ff)
for i=1 to do
    Set the contents of Ni to Di, where Di is all instances in N that match Fi
    Call BuildTree(Ni)
    end for
end if
```

uncertainty of the target variables in the resulting subsets, the given training data is recursively divided into two divisions using a single decision tree Tt.

More specifically, the data is divided into two distinct subsets, L and R, by the splitting functions φ(x), from which a random sample is taken by each node in a tree. We measure the information gain first, and then evaluate all splitting functions. By using the splitting function φ∗(x) that produces largest information gain L and R subsets. All samples that fall within this leaf are used to estimate a density model p(y), which is then used leaf node target if any these conditions are met. Taking target and obtaining is the simplest method for estimating the probability distribution p(y). But there are even more complex variations, such as nonparametric densities or using a Gaussian kernel density estimate.

RESULT AND DISCUSSION

In this presentation, the findings of the suggested system, Hybrid Chicken Swarm Optimization with Random Forest (HCSORF), are explained. The recently built classification algorithm—which uses datasets from the Kaggle website is validated and introduced using object detection in the cabin of a single vehicle. In the training raw dataset, there are many characteristics for both letters and integers. MATLAB 2013a is used in study recommended system with HCSORF in object detection in the cabin. A 13th generation Intel i3 CPU with 16GB RAM and 512GB ROM powers the machine. The SVM, BPNN, DT, and recently constructed CSO with RF are compared in terms of sensitivity and accuracy metrics.

Confusion Matrix

To summarise the performance ML, a confusion matrix employed this method employed categorization matrix shows how much TP, TN, FP, and FN the model generated using the test set of data. Picture 6 above shows the structure of the confusion matrix for performance evaluation.

Accuracy

Accuracy is calculated only a small percentage of patients may have cancer in an unbalanced dataset such as the cancer dataset, accuracy is misleading and cannot be used to assess a model's performance.

Accuracy = (TP+TN)/(TP+TN+FN+FP)

Examining Figure 6, Table 1 shows that the correctness of the suggested system, determined by HCSORF, is 88.16, 1.82 times greater than the accuracy of the existing system, determined by 86.34.

Sensitivity

Sensitivity (SN) is calculated as the ratio of accurately predicted positives to all positives. The ratio of correctly predicted actual yeses to all actual yeses is the definition of sensitivity.

Sensitivity = TP/(TP+FN)

Figure 5. Confusion matrix

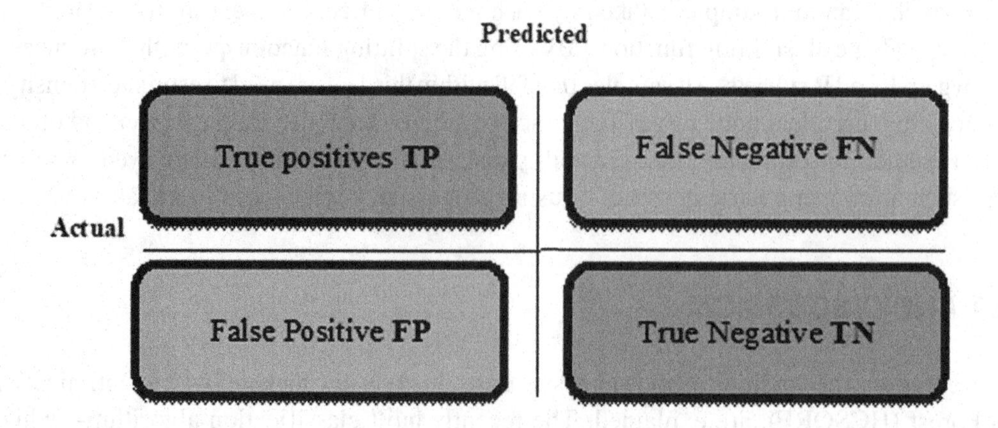

Table 1. Result values for accuracy with proposed and existing system

Algorithm	Accuracy
SVM	82.31
BPNN	84.65
DT	86.34
HCSORF	88.16

Figure 6. Accuracy graph of the proposed and current systems as a result

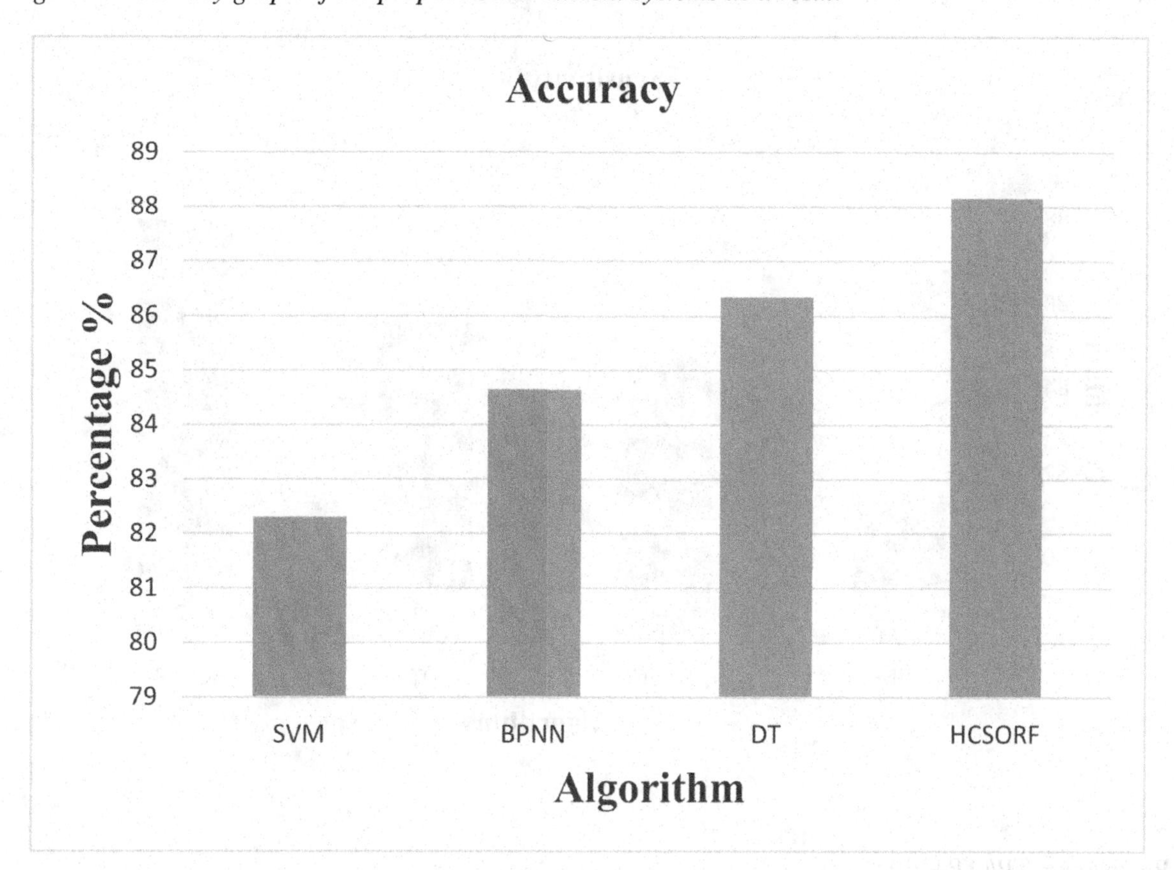

Based on Table 2's findings and with reference to Figure 7, the suggested system's HCSORF-determined sensitivity is 87.98, 2.43 times higher than the current system's 85.55 sensitivity.

Precision

A class's accuracy was defined instances that classifier identified as belonging to that specific class to categorized in that class purpose spam filtering and web search results, precision a measure of relevance or exactness matters. Decrease the amount of legitimate emails that are labeled as spam using the spam filtering algorithm.

Table 2. Result values for sensitivity with proposed and existing system

Algorithm	Sensitivity
SVM	81.36
BPNN	84.65
DT	85.55
HCSORF	87.98

Figure 7. Result sensitivity graph for proposed and existing system

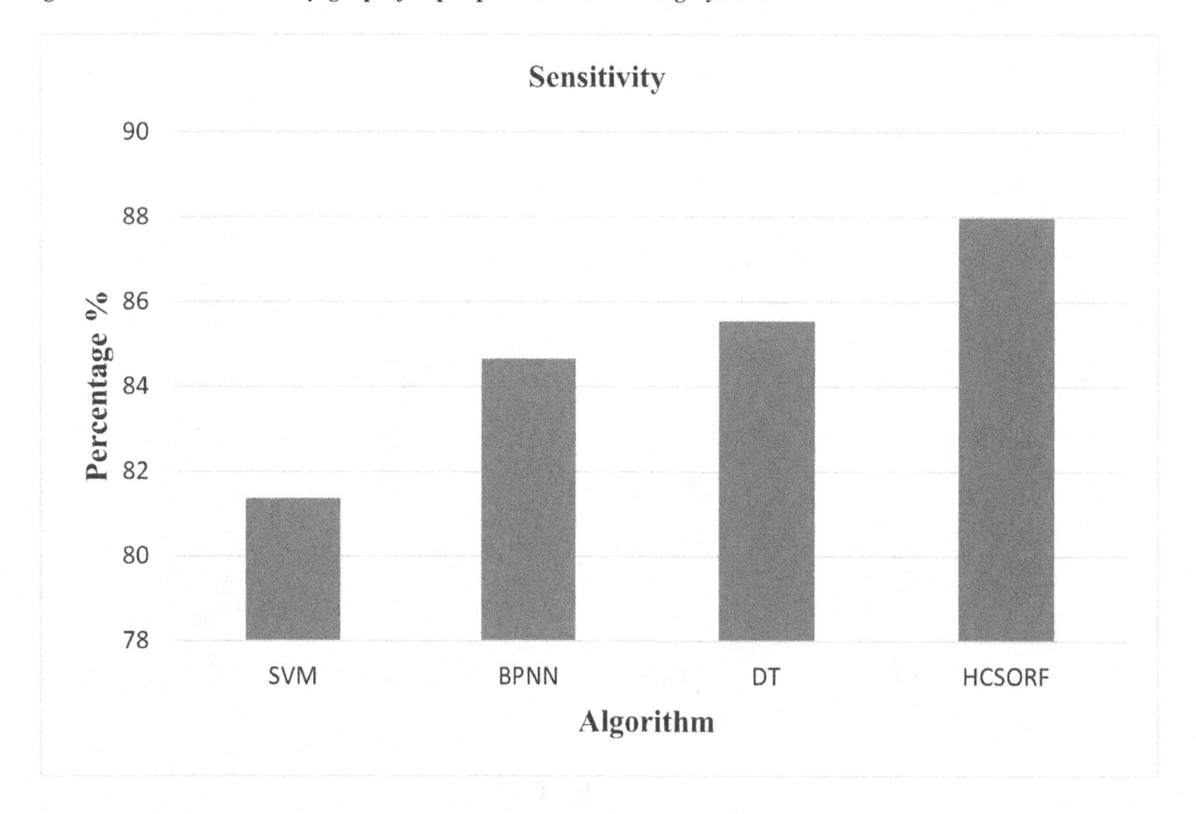

Precision = TP/(TP+FP)

The results in Table 3 show that the recommended system's precision, based on HCSORF assessment, is 90.65. This is 4.46 times more exact than the current system, according to the 86.19 evaluation of Figure 8.

F1 Score

Recall and precision are its harmonic means. Recall and precision scores for a model are aggregated. Over the course of the dataset, the accuracy statistic documents how frequently a model has produced accurate predictions.

Table 3. Result values for precision with proposed and existing system

Algorithm	Precision
SVM	84.16
BPNN	85.43
DT	86.19
HCSORF	90.65

Figure 8. Result precision graph for proposed and existing system

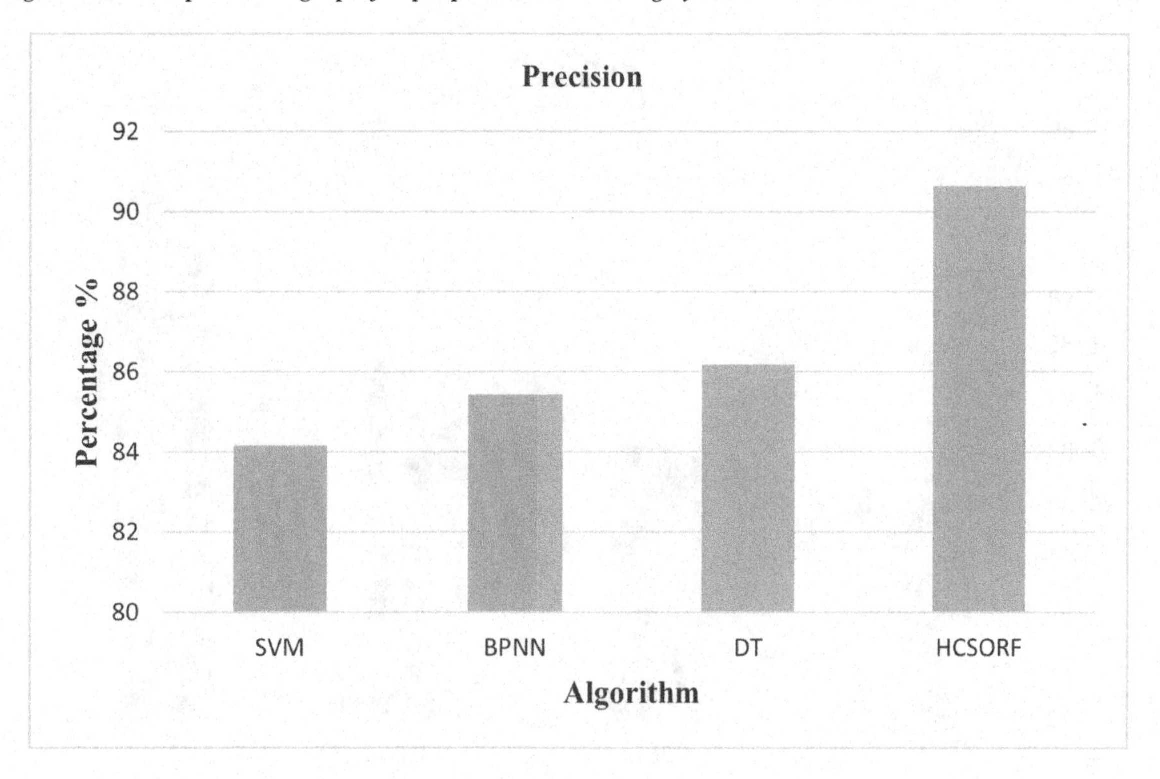

F1-score = 2 * (percision*recall)/(precision+recall)

In Figure 10, the F1 score of the proposed system is 88.23, which is 3.27 times higher than the F1 score of the current system, which is 84.96, according to HCSORF. Table 5 displays these outcomes.

Specificity

The specificity indicator is the ratio of all real nodes to precisely anticipated actual nodes. To calculate SP, divide the entire amount of negatives divided by the total amount of successfully anticipated. It is also known as TNR.

Specificity = TN/(TN+FP)

Table 4. Result values for F1-score with proposed and existing system

SVM	F1-score
SVM	82.06
BPNN	83.15
DT	84.96
HCSORF	88.23

Figure 9. Result F1-score graph for proposed and existing system

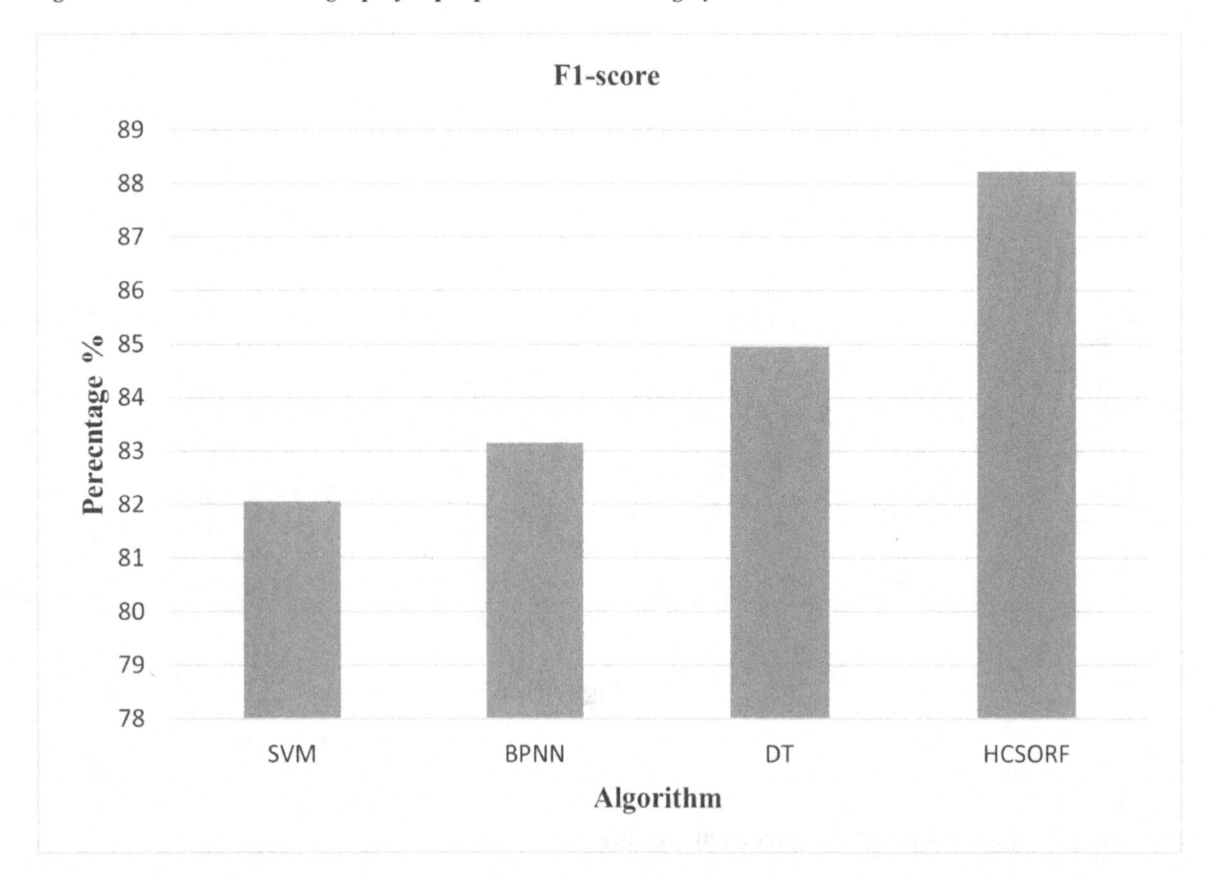

Based on Table 5's results, the suggested system's Specificity (91.65) is 4.4 times higher than the current system's Specificity (87.25) in reference to Figure 10. This is based on HCSORF data.

The graph below and the Tables 1, 2, 3, 4, and 5 above show that the performance of the suggested system is significantly better than the present system's. The results for Specificity are shown in the previous Figures (6, 7, 8, 9, 10). This means that the suggested method works better when the HWODT technique is applied to Drug Analysis for the Healthcare method. According to this, the recommended tactics performed better in the object detection system in vehicle cabin.

Table 5. Result values for specificity with proposed and existing system

Algorithm	Specificity
SVM	83.57
BPNN	85.87
DT	87.25
HCSORF	91.65

Figure 10. Result specificity graph for proposed and existing system

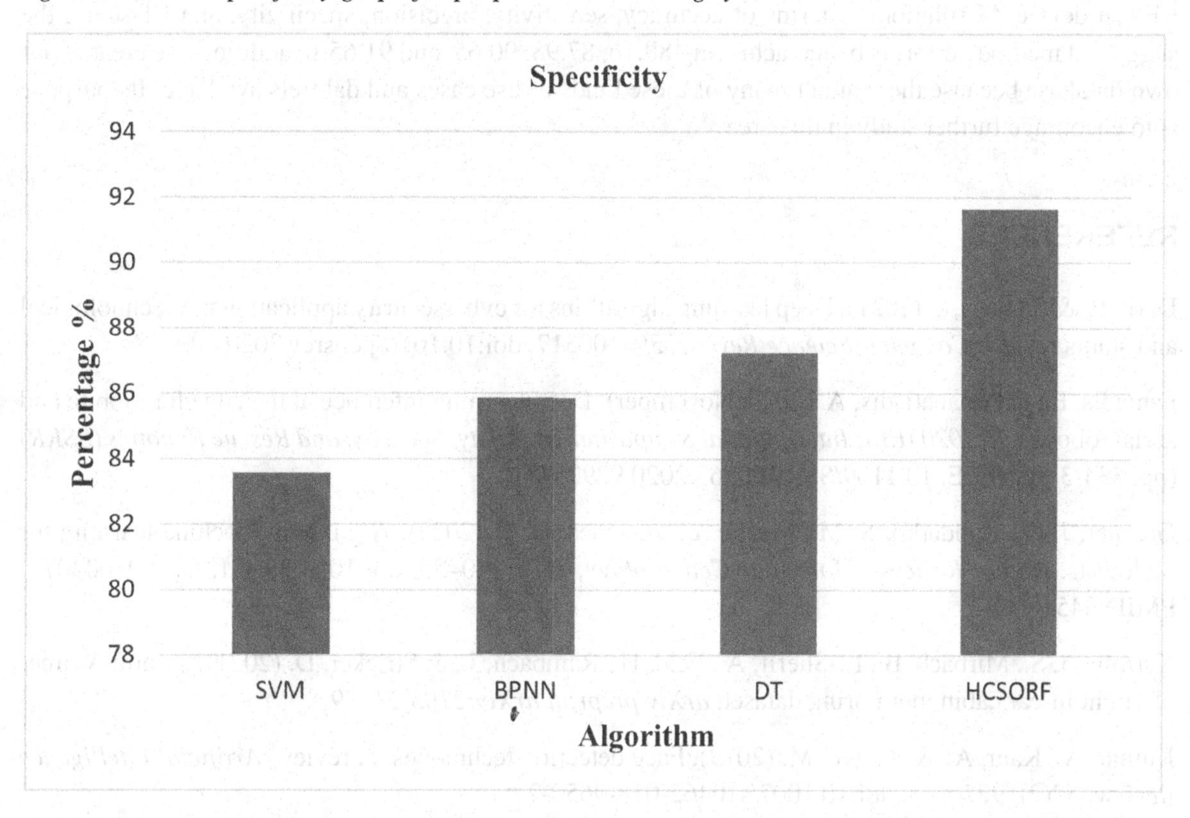

CONCLUSION

In this research, propose ML strategy add learning component to the current system so that it can be tailored to the specific needs of each driver. Long-term goals include learning from the gathered data and adjusting the previously strict thresholds to greater flexibility. This will be accomplished by gaining new insights and establishing correlations between the observed behavior and the underlying data using machine learning techniques. To improve the performance of each driver in accurately identifying potentially risky situations, these findings ought to be assessed and subsequently included into the threshold-based system. Extensive experiments were conducted using the data to establish how an implementation may be done in order to match this need. Businesses and academic institutions want a reliable IMS and a large enough database. A wide range of potential outcomes, including preferred variants for AV in-cabin monitoring, are included in our database. An RT camera is used in the first level to carry out routine monitoring tasks, while further monitoring is required in the second level to hunt down the offender and victim and gather any remaining pieces of evidence. Regarding private AVs, privacy is not a big concern; in public AVs, on the other hand, privacy is crucial. On-device AI, on the other hand, is the foundation of the proposed IMS and has information security advantages built in. In addition, communication of only insignificant details like the number of passengers, occupancy, etc., is required during the travel. It is possible to identify privacy threats that arise in unusual circumstances. We have suggested using facial anonymization to protect people's privacy. Users' privacy has been enhanced by our suggested Hybrid Chicken Swarm Optimization with Random Forest (HCSO RF) on-device AI solution. In terms of accuracy, sensitivity, precision, specificity, and F1-score, the suggested method performs better, achieving 88.16, 87.98, 90.65, and 91.65.In addition, we created our own database because there aren't many of these kinds of use cases and datasets available. Its purpose is to encourage further study in this area.

REFERENCES

Dixit, P., & Silakari, S. (2021). Deep learning algorithms for cybersecurity applications: A technological and status review. *Computer Science Review*, *39*, 100317. doi:10.1016/j.cosrev.2020.100317

Faniadis, E., & Amanatiadis, A. (2020, November). Deep learning inference at the edge for mobile and aerial robotics. In *2020 IEEE International Symposium on Safety, Security, and Rescue Robotics (SSRR)* (pp. 334-340). IEEE. 10.1109/SSRR50563.2020.9292575

Greener, J. G., Kandathil, S. M., Moffat, L., & Jones, D. T. (2022). A guide to machine learning for biologists. *Nature Reviews. Molecular Cell Biology*, *23*(1), 40–55. doi:10.1038/s41580-021-00407-0 PMID:34518686

Katrolia, J. S., Mirbach, B., El-Sherif, A., Feld, H., Rambach, J., & Stricker, D. (2021). Ticam: A time-of-flight in-car cabin monitoring dataset. *arXiv preprint arXiv:2103.11719.*

Kumar, A., Kaur, A., & Kumar, M. (2019). Face detection techniques: A review. *Artificial Intelligence Review*, *52*(2), 927–948. doi:10.1007/s10462-018-9650-2

Lan, H., Hartonen, K., & Riekkola, M. L. (2020). Miniaturised air sampling techniques for analysis of volatile organic compounds in air. *Trends in Analytical Chemistry*, *126*, 115873. doi:10.1016/j.trac.2020.115873

. Ma, Y., Wang, Z., Yang, H., & Yang, L. (2020). Artificial intelligence applications in the development of autonomous vehicles: A survey. *IEEE/CAA Journal of AutomaticaSinica,* 7(2), 315-329.

Mallaboyev, N. M., Sharifjanovna, Q. M., Muxammadjon, Q., & Shukurullo, C. (2022, May). INFORMATION SECURITY ISSUES. In Conference Zone (pp. 241-245). IEEE.

. Milledge, J. (2020). Hypobaria: high altitude, aviation physiology, and medicine. *Cotes' lung function*, 615-637.

Mishra, A., Cha, J., & Kim, S. (2022). Privacy-preserved in-cabin monitoring system for autonomous vehicles. *Computational Intelligence and Neuroscience*, *2022*, 2022. doi:10.1155/2022/5389359 PMID:35498178

Poon, Y. S., Lin, C. C., Liu, Y. H., & Fan, C. P. (2022, January). YOLO-based deep learning design for in-cabin monitoring system with fisheye-lens camera. In *2022 IEEE International Conference on Consumer Electronics (ICCE)* (pp. 1-4). IEEE. 10.1109/ICCE53296.2022.9730235

Priyadarshi, R., Gupta, B., & Anurag, A. (2020). Deployment techniques in wireless sensor networks: A survey, classification, challenges, and future research issues. *The Journal of Supercomputing*, *76*(9), 7333–7373. doi:10.1007/s11227-020-03166-5

Ramos-Sorroche, E., Rubio-Aparicio, J., Santa, J., Guardiola, C., & Egea-Lopez, E. (2023). In-cabin and outdoor environmental monitoring in vehicular scenarios with distributed computing. *Internet of Things : Engineering Cyber Physical Human Systems*, 101009.

Rathi, B. S., Kumar, P. S., & Vo, D. V. N. (2021). Critical review on hazardous pollutants in water environment: Occurrence, monitoring, fate, removal technologies and risk assessment. *The Science of the Total Environment*, *797*, 149134. doi:10.1016/j.scitotenv.2021.149134 PMID:34346357

Saxena, A., Khanna, A., & Gupta, D. (2020). Emotion recognition and detection methods: A comprehensive survey. *Journal of Artificial Intelligence and Systems*, *2*(1), 53–79. doi:10.33969/AIS.2020.21005

Tarnawski, J. M., Phanishayee, A., Devanur, N., Mahajan, D., & Nina Paravecino, F. (2020). Efficient algorithms for device placement of dnn graph operators. *Advances in Neural Information Processing Systems*, *33*, 15451–15463.

Tian, Y., Ma, S., Wen, M., Liu, Y., Cheung, S. C., & Zhang, X. (2021). To what extent do DNN-based image classification models make unreliable inferences? *Empirical Software Engineering*, *26*(5), 84. doi:10.1007/s10664-021-09985-1

Trewartha, A., Walker, N., Huo, H., Lee, S., Cruse, K., Dagdelen, J., Dunn, A., Persson, K. A., Ceder, G., & Jain, A. (2022). Quantifying the advantage of domain-specific pre-training on named entity recognition tasks in materials science. *Patterns (New York, N.Y.)*, *3*(4), 100488. doi:10.1016/j.patter.2022.100488 PMID:35465225

Chapter 4
Real-Time Fault Detection and Condition Monitoring for Industrial Autonomous Vehicles

Thirusudhan M.
Karpagam Academy of Higher Education, India

Sasikumar S.
Karpagam Academy of Higher Education, India

ABSTRACT

Real-time status monitoring and early defect diagnosis are becoming more and more necessary for modern industrial systems. Developing intelligent remote diagnostic technologies and integrating on-board and off-board diagnosis are two more unresolved research projects in the automotive industry. The automated transfer vehicle (ATV) equipment condition monitoring example in this chapter is part of a smart industrial use case. The suggested method successfully permits ATV fault scenarios to be seen in real time by expanding to a fleet of devices in an actual production plant. The application of a statistical threshold is the initial stage in creating a high-performance detection model for defect detection on stacked long short-term memory networks. For better computation speed, a second model trimming approach based on principal component analysis is suggested. The improved fault detection technique is ultimately applied by the airborne embedded computer platform with field-programmable gate arrays.

INTRODUCTION

The manufacturing industry domains have seen significant changes since the advent of the Industry 4.0 concept. The sophistication of conventional manufacturing plans and procedures has increased as a result of industrialization and information technologies including robotics, big data, the Internet of Things (IoT), and intelligent measuring techniques. Automating and integrating these technologies into industrial processes makes production setups more sustainable and manageable. These advancements increased productivity and adaptability but decreased the amount of human interaction on production

DOI: 10.4018/979-8-3693-1962-8.ch004

sets. However, automation and artificial intelligence (AI) shouldn't take the place of workers on the manufacturing floor; instead, man-hours from these technologies should be applied to more difficult jobs with thoughtful solutions.Intelligent technologies such as autonomous robots are usually assigned repetitive jobs in smart industrial settings, with people being assigned more complex and creative activities by Verma, Amar Kumar, et al (**2021**). In today's sophisticated industrial environments, a lot of autonomous machinery performs routine tasks like patrolling or the cautious and secure transfer of supplies and parts without the need for human assistance. They can also work in areas where labor laws and safety standards are in effect. That manner, they can ensure the safety of workers even when they are carrying out dangerous jobs.

Since autonomous equipment has become a vital part of the fleets of equipment used in industrial facilities, high uptime and continuous operation are needed. This means that your car can now run on its own. Over twenty years have gone by since reality replaced science fiction. This is made possible by recent developments in the fields of radars and microprocessors10. There has also been a notable advancement in the creation of portable and lightweight technologies. This has led to the development of ultra-light technology, which enables quick decision-making in emergency scenarios.

Concepts of Autonomous Driving

Over the past decade, a lot of progress has been made in the field of autonomous vehicles. Autonomous vehicles, or simply AVs for short, are self-driving cars. They may follow preprogrammed instructions, perform automated chores, and observe their surroundings. Systems for autonomous cars serve a variety of functions. an exhaustive list of all the jobs and functions an autonomous system is capable of doing. Three main components make up the strategy.

- Sensing and perception,
- Planning,
- Control.

Every system needs sensors to collect and process the ambient data needed for it to work, which makes them an indispensable part of the system. In order to give the system a safe and sensible course, the sensed data is employed in planning. Actuator: The actuator is guided by the control component of the system along the intended, practical path. Every single one of these components needs to be arranged while building an AV system. With or without assistance from a human driver, an autonomous vehicle (AV) system can park a car on its own thanks to a feature by Van Wyk, Franco, et al (**2019**). Two major factors have led to the increased interest in autonomous parking in recent times. First of all, it provides a high level of safety for novice drivers who could find parkingurban planners and managers are particularly concerned about the second issue, which is the scarcity of parking spaces, because of the traffic. For this reason, creating an automated parking system for smart cars has become imperative. Table 1 shows the Universe of faults and list of monitoring variables.

Artificial Intelligence

Computers and robots with artificial intelligence can do tasks that are normally done by humans. These include the ability to recognize speech, translate between languages, analyze visual stimuli, and make

Table 1. Universe of faults and list of monitoring variables

Fault List		Monitored Variables	
Name	**Notation**	**Name (Units)**	**Notation**
Air Flow Sensor Fault (F1)	AFS	Air Pressure (KPa)	Pm
Leakage in AIS (F2)	AIS_leak	Amount of Fuel Injected (mg)	FC
Blockage of Air Filter (F3)	AF_blockage	Air to Fuel Ratio	A/F
Throttle Angle Sensor Fault (F4)	TAS	Engine Speed (rpm)	NE
Less Fuel Injection (F5)	LFI	Vehicle Speed (km/h)	SPD
Added Engine Friction (F6)	AEF	Throttle Angle (deg)	TA
Air/Fuel Sensor Fault (F7)	AFuel_S	Air Flow Meter (Volts)	VG2
Engine Speed Sensor Fault (F8)	ESS		

decisions. Presently, artificial intelligence can be classified into three categories: artificial general intelligence (AGI), artificial super intelligence (ASI3), and restricted AI. Strong AI, or AGI, is machine intelligence that is able to do all tasks at a level almost equivalent to human intellect, as opposed to ASI, which is machine intelligence that is always more intelligent than human cognition. Artificial intelligence is often generated in the current world in order to accomplish the task needed.Having said that, it's plausible that certain companies are secretly researching AI. In many aspects, artificial intelligence (AI) has changed our lives significantly throughout the years. With enough training, it can now do tasks that humans were previously unable to accomplish by Cai, Yingfeng, et al (**2021**).

Intelligent machines in an industrial setting, such as robots, are physical systems with varying degrees of independence that collaborate in a range of dynamic physical situations. The intrinsic capability of these autonomous robots to operate in a variety of unforeseen conditions makes them susceptible to a variety of mechanical and operational issues. Maintaining operations and minimizing unscheduled downtime for these equipment necessitates fast fault detection, diagnosis, and monitoring. Because of advancements in deep learning (DL), hardware system processing power, and advanced measurement technologies, data-driven defect detection approaches have immense potential in production contexts. Utilizing data-driven fault detection methods that extract features from recorded sensor inputs, equipment operational irregularities are identified. A diagnosis is established in relation to a reference.Smart sensors can be used to carry out intricate measurements that yield precise information on the operating environment of the devices. This is why proper analysis of such data produces useful information that allows human operators to track the current status of the devices in real time and monitor component levels said by Yao, Lei, et al. (**2020**).

Machine Learning

Mathematical and analytical models and algorithms are developed and automated using machine learning (ML) to improve system performance for a given task. Uncovering hidden insights in data requires the application of machine learning (ML), which uses methods from operations research, statistics, and heuristic procedures but does not provide explicit guidance on what to look for or how to proceed. The three primary subfields of machine learning are unsupervised learning, supervised learning, and reinforcement learning by Chen, Zhenpeng (**2019**). These are explained in the section below.

Unsupervised Learning

Understanding data that is generated solely from input data is known as unsupervised learning. Data points are arranged according to a set of criteria in a process called clustering, which is a method for unsupervised learning. All data sets can be grouped into groups using the clustering approach by Liu, Ze, et al (**2021**).

Supervised Learning

Supervised learning (SL) is used to build an output and input prediction model. Among SL approaches, regression analysis is the subgroup that searches for a relationship between variables. By utilizing the correlation between variables gathered from the dataset, machine learning uses this to predict the outcome of an event. A classifier that accurately determines which category a new observation falls into and forecasts the target class for each data category by Yao, Lei, et al (**2021**).

Reinforcement Learning

Artificial intelligence (AI) will progress faster into machine learning in the real world thanks to a new technique called reinforcement learning (RL), which focuses on decision-making. We present a quick comparison of reinforcement learning, unsupervised learning, and supervised learning.

Effective diagnostics and monitoring are possible when all of the equipment is networked with all of the other industrial operational systems and equipment. Or, to put it another way, every component of the industrial environment must be viewed as an Internet of Things (IoT) device and connected to other nearby IOT devices. Because of this, these gadgets are linked together and use virtual networks like the cloud to continuously and quickly exchange data. In many scenarios, however, where real-time data transfer and analysis are essential, bottlenecks may result from the volume of data, the computational load on the centralized cloud, and the bandwidth utilized by several devices.It is consequently impractical to transfer all of the raw data to the centralized cloud server for processing needs. One potential remedy for the aforementioned obstacles is edge artificial intelligence, or edge computing combined with artificial intelligence said by Deng, Wanghua, and Ruoxue Wu (**2019**). Using smart sensors attached to Internet of Things devices, Edge AI does preprocessing or inference at the edge nodes, near to the source, before data is transferred to the central cloud. This approach provides lower latency, quicker responses, data security, and scalability even if it transfers less data. Therefore, Edge AI offers beneficial solutions for industrial Internet of things devices, including automated transfer vehicles (ATVs), which depend on fast data rates and low transmission latency to monitor their condition in real time.

An open-source FIWARE-based data distributor middleware platform, an industrial ATV, an edge AI accelerator, a data storage unit, a visualization tool, and other components have been combined into a freshly constructed testbed. The suggested method is widely applicable and can be modified for any production tool that has Internet of Things connection protocols and sensors installed by Gupta, Abhishek, et al. (**2021**). We have customized our ATV use case to fit the robot fleet condition monitoring platform solution. Notable conclusions drawn from this research include:

- A general-purpose Edge AI framework that provides defect diagnostics and real-time status tracking has been unveiled.
- Assessing the design and identifying operational issues is aided by the use of industrial ATV use cases.
- Sensor fusion and the data-driven DL algorithm are used for inference by edge AI accelerator hardware.
- Real-time inference results from edge AI devices are sent to the data storage via a data pipeline powered by the open-source middleware platform FIWARE.

The five sections of this essay are listed below: Session 2 of the current system shows its shortcomings in data prevention in IoT and other techniques. The third session was a demonstration of the suggested System for data security in vehicle. The outcomes of the anticipated system are shown in Session 4. An examination of the recommended approach for vehicle data in IoT rounds off Session 5.

LITERATURE REVIEW

The Internet of Things, smart measurement technologies, and the Internet can all be used to connect physical environments or equipment into the digital world. Because of their different hardware and other components, these settings and devices frequently have problems interacting and communicating with one another. Standards and software have been created to enable this kind of transformation. The author Ouyang, Zhenchao, et al (**2019**) platform FIWARE, for example, has components that allow for effective communication between Internet of Things devices.

The researcher Bakdi, Azzeddine, et al (**2021**) FIWARE platform is a great tool for creating smart environments, according to a number of studies in the literature. For instance, consider developing cloud- and Internet-of-things (IoT)-based health system applications and services that interface with FIWARE platform components via a range ofUtilizing a platform that integrates data from the FIWARE platform, Internet of Things technologies, and port measurement equipment, an architecture for monitoring and decision-making for seaport settings was highlighted. The FIWARE platform was utilized for irrigation operations to test the design of precision agriculture software in an authentic agricultural setting. The efficacy of the FIWARE platform for smart city applications is demonstrated by building a new testbed and modeling huge IoT installations. Based on preliminary studies conducted in multiple areas, a variety of Internet of Things devices can safely and efficiently connect with one another using the FIWARE platform and its components. FIWARE was utilized to give this project a flexible and all-inclusive base.

A author Yaqoob, Ibrar, et al (**2019**) central server processes the environmental data collected by Internet of Things (IoT) sensors in the environment that most data-centric smart solutions offer. However, this approach can have certain drawbacks with regard to data security, network resource usage, and latency. Numerous studies that circumvent these limits are reported in the literature, making use of edge AI technologies. Edgent is an edge AI system designed to operate at low latency by utilizing deep neural networks for edge differencing, based on a digital twin architecture for anomaly detection and industrial system condition monitoring. The edge processing methods based on single-shot multi box detectors for real-time video smart parking surveillance applications were evaluated in terms of detection accuracy, adaptability, and system dependability.An innovative artificial intelligence (AI) system utilizing lightweight convolutional neural networks (CNNs) and multi-sensor data fusion techniques for real-time event recognition is being tested on the LiBr absorption chiller, a typical air conditioner component. Two distinct deep learning models and a reference air quality dataset were used to evaluate the proposed structure. Extensive research has been conducted on edge AI frameworks and approaches; however, most of the studies are either domain-specific or lack real-world testing environments.

According to Sun, Lei, et al(2020) Universe of Watched Variables and Engine Failures Eight different engine failures are being looked into, including one black actuator fault, three blue plant faults, and four red sensor faults. Three steady-state operational conditions are used in order to replicate the model: The range of pedal angle (PA) is 15 to 20 degrees. The tracked data contained seven variables that were monitored by the engine. Using a 3-D plot of engine speed, throttle angle, and air pressure signals, pre- and post-fault data is gathered for 10 seconds, with the fault on-set time set at 2.5 seconds. This allows for the separation of the nominal data's operational domains. To replicate all of the 10 severity levels (6 to 15%) for each operational scenario, a 1% step size variation is utilized.

The author Bae, Ilat al(**2019**) outlines the procedure for identifying and diagnosing a problem, regardless of the operating region in which it occurs. Industrial equipment may work better and last longer thanks to developments in sensor and measuring technologies. The data gathered from a range of sensors is then examined using signal processing techniques to enable further uses, such as equipment failure detection and condition monitoring during operation. Since machine learning is widely used in many academic domains, data-driven modeling approaches are used in industrial environments for defect classification and detection. Deep learning (DL) techniques are mostly beneficial for these applications due to their ability to produce high-capacity models through direct learning from raw data.Various papers describe the use of data from sensors and deep learning algorithms to identify anomalies, diagnose issues, and track the health of Internet of Things (IoT) devices. Another study uses time/frequency domain sound data with DL to diagnose gear breakdowns. For instance, they suggest using the wavelet transform of vibration signals from acceleration sensors as the input for the DL model to determine the failure states of a planetary gearbox. Integrating several signals can aid in a more accurate and dependable diagnostic and health monitoring of industrial machinery since different sensors can detect the environment with different attributes.

Robot Operating

The researcher Simon, Martin, et al (**2019**) majority of mobile robots, such as ATVs and other Internet of Things gadgets, are composed of numerous physical parts, such as wheels, engines, cameras, and other sensors. Building flexible and efficient software interfaces for utilizing and operating robots is more challenging due to the variety of these components. Furthermore, these limitations might make communication between IoT components more difficult. Robot hardware is made up of many sensors and actuator elements. To make controlling this hardware easier, ROS middleware provides a standardized development environment with common sensor drivers and interface techniques. The ROS ecosystem makes it possible for several robots with similar parts to use the same code blocks with very little modification.Robots can be controlled and operated by a suite of software applications known as the Robot Operating System (ROS). Nodes, services, messages, and topics are the four main ideas that ROS provides. Nodes are software components that communicate and perform computations together. A node can publish messages over a topic and receive published messages from other nodes that are subscribed to the same subject. Similar to standard web services, the service paradigm is composed of requests from client nodes and responses from server nodes. Moreover, all nodes on a ROS platform must be linked to the master node, which controls and coordinates all communication.

Edge Computing

A network resource or machine that sits between cloud servers and a system's data sources is called an edge. In general, edge devices are located close to data sources due to their compact, decentralized, and power-efficient nature. Together, edge computing and artificial intelligence create edge artificial intelligence. A good example of this combination is the local application of machine learning or deep learning algorithms to data collected from physical devices via edge node computer capabilities said by Chiu, Yu-Chen, et al. (**2020**). Edge AI achieves this by bringing processing power and artificial intelligence closer to data sources. There is a difference in the amount of raw data that is moved to centralized cloud servers from Internet of Things device hardwareAll inference models running on edge nodes generate it as their only output. Decentralized processing of received data at edge nodes reduces the amount of computing power and network bandwidth needed for centralized servers. Hence, implementing edge AI techniques reduces latency and transmission costs while enhancing data security, privacy, scalability, and dependability.

FIWARE Platform

A middleware that facilitates the development of smart manufacturing solutions is called a Generic Enabler (GE) on the FIWARE platform. Generic Entires (GEs) are software components designed to work with several applications and/or architectures. Arranging suitable GEs based on a system's needs is the platform's concept. A context broker is at the heart of any FIWARE system. Processing, analyzing, and displaying context information, as well as interacting with robotics and the Internet of Things, are just a few of the tasks that specific GEs are made to perform explained by Liu, Liangkai, et al (**2020**). API and context data management are also under their purview. The foundational components of the FIWARE platform are a CB and a database that stores context data. In Internet of Things scenarios, CB is responsible for the context entities, which consist ofto enable FIWARE CBs to use and publish NGSI v2 entities generated by ROS messages, an additional GE known as FIROS establishes a communication channel between the robotic domain and cloud.

The Six Levels of Automation

With the increasing prevalence of autonomous vehicles (AVs) across the globe, it is necessary to assess our progress in realizing the vision of AVs by assessing their performance on their dynamic driving task (DDT) through the use of an extensive hierarchical table. SAE's comprehensive taxonomy (Level 0–Level 5) covers the six automation steps. Below is a list of them:

Level 0 (No Automation)

Almost all cars on the road belong to this class. Whoever supplies the DDT has complete influence over them. Mechanisms to provide the driver with temporary support in an emergency might exist. Since human involvement is necessary to operate the emergency systems, the vehicle cannot be termed automated by Langarica, Saúl (**2019**).

Level 1 (Driver Assistance)

This one has the least automation. In some situations, the car's electronics might provide the driver with some temporary assistance. Because it allows the human driver to focus on other driving-related tasks like steering and braking, adaptive cruise control, which allows an automobile to maintain a safe distance from the vehicle in front of it, may be regarded as Level 1 by Li, Da, et al (**2020**).

Level 2 (Partial Driving Automation)

Advanced driving assistance systems (ADAS) are similar to Level 2, which allows steering and speed control of an automobile remotely but still requires a human driver to be present. The term "self-driving" is misleading, even if a person can still operate an automobile at any time. Tesla Autopilot enables certification up to level two by Guo, Ningyuan, et al. (**2020**).

Level 3 (Conditional Driving Automation)

Compared to Level 2 vehicles, the car can now operate with greater autonomy. They are equipped with several sensors that allow them to sense their environment and make choices for themselves. However, a human driver needs to be prepared to take over in the unlikely event that the technology malfunctions. A few other features that aren't included in Level 2 cars are automated emergency braking (AEB), driver monitoring (DM), and traffic jam assistance (TJA) by Manoharan, Dr Samuel. (**2019**).

Level 4 (High Driving Automation)

In the case of a problem or system failure, Level 4 cars are equipped to take control. There is a distinction between cars classified as Level 4 and Level 3. Even though they can function independently, they are now limited to a low-speed, geofenced metropolitan area. This will result from any future changes made to the infrastructure and laws. Human drivers usually do not need help from others, however they always have the option to intervene and take over by Carranza-García, Manuel, et al (**2021**).

Level 5 (Full Driving Automation)

With this, the dream of driverless vehicles has come true. Dynamic driving is not necessary since Level 5 cars totally do away with the need for human interaction. There won't be any kind of accelerator or brake pedal or steering wheel. These will be capable of doing all driving maneuvers that a typical human driver can. A level 5 vehicle is exempt from geo-fencing since it is autonomous and able to move anywhere. Even though Level 5 vehicles have already undergone a great deal of worldwide study and testing, it will take some time before they are made available to the general public by Rohan, Ali et al (**2019**).

Infeasible Sensing

However, the author Yin, Shen(**2019**) due to their basic characteristics, embedded sensors usually have very limited perception capabilities and are unable to sense their surroundings. The primary drawback of using visual-based object detection as the primary sensing technique is that the inference's success is usually dependent on the caliber of the collected photos. The image quality can be affected by a number of factors, including brightness, weather, distance of the camera from the subject, and image resolution. Neither the road lighting nor the bicycle lights quickly lit the affected section of the path. Due to erroneous assumptions on the pedestrian, the pre-trained deep learning model initially misclassified them as an unknown entity, a car, and finally a bicycle.This experience has demonstrated that decisions made solely on the basis of one kind of sensor can often lead to disastrous outcomes. One viable potential approach to stop this is to combine different sensing modalities for perception, such as those from LiDAR sensors, sensors placed at intelligent traffic signals, and sensors from other cars. Such technique is referred to as "sensor fusion".

Trade-Off Between Reliability and Latency

Usually, the researcher Jabbar, Rateb, et al. (**2020**) explain an inference's performance includes both correctness and delay. Because the accuracy of the model mostly depends on the caliber of the input data and the model's competency, shallow neural network models and "bad quality" data can both reduce the inference accuracy of a pre-trained deep learning model. A further communication delay is caused by moving the workload from the mobile device to a more powerful edge server using a more advanced neural network model. On the other hand, the offloading could sometimes become inefficient due to the channel dynamics. Some apps may not be able to handle the additional latency, even if it didn't exist.The trade-offs between latency and inference accuracy must therefore be carefully considered while developing edge intelligence. Achieving an appropriate trade-off between increased latency and dependability requires minimizing the wireless transmission delay between the devices and the edge server.

Limited Resources

In contrast to a large number of strong GPUs and CPUs integrated at the cloud, edge servers sometimes cannot handle a high rate of device offloading requests due to restrictions in computation, cache, and power resources, as well as limitations in connection bandwidth. In particular, cooperative resource allocation and optimal offloading decisions on the limited resources at the edge server are critical for edge intelligence said by Hong, Yuanduo, et al (**2021**).

Data Security and Privacy

In many edge intelligence application domains, data security and privacy considerations are crucial since it is likely that the mobile device data needed for processing and inference contains private and sensitive information that the user may not have wanted to be collected. Using autonomous driving as an example, consider the sheer number of images the AVs take, many of which contain sensitive personal data used by Muhammad, Khan, et al (**2020**). A direct transfer of this type of data to the edge server for processing could jeopardize user privacy. This problem usually occurs during the training and inference phases of the model. This problem can be fixed by uploading the intermediate features to the edge server after the device has processed the fundamental data (or training).

SYSTEM DESIGN

The efficacy of the proposed architecture is assessed in conditions similar to an actual industrial workstation. However, the proposed technique has not yet been applied in real production. Integration can be accomplished using the current architecture without necessitating major modifications to the system's constituent parts.

Proposed System Architecture

This work proposes an architecture for a real-time defect detection and condition monitoring system that is edge AI-based and general-purpose. There are five primary layers in the architecture:

- The IoT layer,
- The edge layer,
- The FIWARE layer,
- The data storage layer
- The visualization layer.

ROS-installed devices comprise the Internet of Things layer. For data collection, some devices have installed sensors. After the ROS topic publication system receives information from the master node on certain topics, relevant messages are disseminated. Part of the edge layer are edge AI devices that have deep learning algorithms for inference. All these little decentralized devices are placed close to one other in the IoT layer. The chosen IoT device is connected to each edge device via an Ethernet wire. The IoT device releases topics, and the edge device uses ROS to subscribe to them.

The results, acquired through the use of real-time sensor data collected from subjects by the DL-based defect detection algorithm, are disseminated over another topic on the master node. FIROS, a database, and GEs acting as a CB make up the FIWARE layer. FIROS transforms the message comprised in the topics broadcast by the master node into NGSI v2 entities before it is sent from IoT devices to the CB. After these entities and attributes are published from the CB, they are kept up to date in the database. The primary components of the data storage layer are those related to data persistence, specifically a database server and processing engine capable of transforming data delivered by CB into the appropriate format for database storage.

Device Components

The system should scale for a fleet of ATV equipment and a myriad of other IoT devices for status monitoring, even if the trials are conducted with a single ATV. It should be easy for any hardware running ROS to communicate with the system. Additionally, there isn't a built auto-model updating mechanism in the system, and the DL model utilized at the edge for inference is assumed to be static. As such, it is assumed that an operator may manually update the edge model as necessary, and the edge model's long-term resilience to equipment state drifts is not explored.

Fault Diagnosis of Battery Cells

Battery cells can malfunction due to four main reasons: overcharging, overdischarging, overcurrent, and overtemperature. The battery cells could sustain irreversible harm from them. They are included in the problem diagnosis since these fault scenarios are connected to the battery's properties. Batteries must go through complex chemical and physical processes in order to charge and discharge. That being said, the cells may explode and catch fire if they are overcharged, overdischarged, or overcurrent. Battery charging and discharging voltages must be continuously and in real-time monitored to avoid these issues. Using the over-charged/over-discharged/over-current protection, the system controller will halt charging or discharging the battery cells if the current or the upper or lower cut-off voltage approaches the maximum value.

In the same vein, overheating is an additional error that must be prevented. The prototype 12S system makes use of six thermistors, which are fastened to the spaces between each pair of batteries, to track the temperature within each battery cell. Two-level reporting and protection mechanism and temperature-window self-recovery strategy are proposed. The controller will not automatically shut off or charge in order to warn users if the temperature rises above 50 °C. However, the controller will initiate the second-level protection as soon as the cell temperature over 60 °C, thereby halting the charging and discharging procedure. If the temperature of each cell falls below 40 oC, the controller will initiate the self-recovery procedure immediately. The faults listed above can be found using the flowchart.

Robot Operating System

Mobile robots, including ATVs and other Internet of Things devices, often include a large number of components in their hardware, including wheels, engines, cameras, and other sensors. The multiplicity of these elements makes it challenging to design software interfaces for robot operation and control that are both efficient and adaptable. Furthermore, these kinds of limitations might affect how well-integrated IoT entities are. To overcome these difficulties, ROS middleware offers a standardized development platform with shared sensor drivers and interface solutions, which simplifies the use of robot hardware with a variety of sensors and actuator elements. With very small changes, numerous robots with comparable components can use the same code blocks thanks to the ROS ecosystem.Robot control and operation are made easier by a group of software applications known as ROS. ROS comprises four primary concepts: nodes, messages, topics, and services. Nodes are software elements that interact with one another and do calculations.

Any node can receive published messages by having the capacity to subscribe to and post messages over a topic. The service model's client node requests and server node answers are comparable to those of conventional web services. On a ROS platform, all nodes also need to be connected to the master node, which manages and plans out all communication.

Edge Artificial Intelligence

"Edge" describes any network or computational device positioned in between cloud servers and system data sources. Edge devices are often tiny, distributed, and low-power devices that are situated in close proximity to data sources. The fusion of artificial intelligence and edge computing is known as edge AI. The utilization of edge node computing resources for local machine learning or deep learning algorithm processing of hardware device data explains this combination. In other words, edge AI brings processing power and artificial intelligence closer to the data. The outcome of inference models operating on edge nodes is provided to centralized cloud servers instead of all the raw data collected from the hardware of IoT devices. The amount of data that is obtained will increase, so the processing power and network of centralized servers.

FIWARE Platform

The middleware FIWARE platform and its generic enabling components facilitate the building of smart industrial systems. Software components referred to as GEs are designed for various applications and/or architectures. The idea behind the platform is to compile relevant GEs according to system requirements.

A context broker is the central component of any FIWARE system (CB). Application-specific GEs manage context data/API, process, evaluate, and display context information, and communicate with robots and the Internet of Things, among other tasks. The FIWARE platform's most basic version consists of the CB and database that store context information. In IoT settings, context entities are managed by CB and are where the data generated by IoT devices is kept. Moreover, a distinct GE known as FIROS establishes the link between the robotic domain and cloud in order to transform ROS messages into NGSI v2 entities that FIWARE CBs may use and publish.

On-line Module

There is constant system monitoring for the online FDD module. The generalized likelihood ratio test, or GLRT, is a method used to process the sensor data in order to identify problems in real time. The reference feature set from training across a moving window of data is recovered once the issue has been identified. The following blocks, the severity estimations and the classification blocks, use training parameters to categorize defects and estimate severity levels, respectively. The name of the failing component, the problem's on-set time, its severity, and the required repair action are all provided to the user via the fault diagnostic decision block. To facilitate their use in the online and offline modules.

RESULT AND DISCUSSION

This section presents the experiment results related to the use of the ATV in different fixed-to-the-ground barriers that could result in low- or high-level anomalies. First, a detailed explanation of the testing setup and real-time inference findings is provided. The data storage interface is then shown. The dashboards and built-in monitoring system are finally shown. A comprehensive analysis of the collected data is provided in the conclusion.

Testing Environment and Real-Time Inference

The previously mentioned pre-trained DL model is used by the edge AI unit to infer information about the ATV's state while it performs a predetermined load-carrying task over a route. In the simulated testing environment, the Python client code designed for the ATV edge AI unit is used to execute the DL model and draw conclusions. It also subscribes to the ROS topics, which supply the sound and vibration data transmitted by the ROS master node from both motors. After being tagged, the ATV moves across the first three metal objects in this environment, which are known to produce high-level anomalies when the offline model is being trained.Next, by drawing a rectangle on the ground and following it until it reaches the starting point's coordinates, the ATV path planning task directs the vehicle to continue along the path. While traveling to the terminal, the ATV also crosses three copper cables that are known to cause low-level irregularities for ATV operating circumstances shows in figure 1.

Most notably, the high- and low-level abnormalities in this study can be tailored by the end users to fit any behavior that occurs within particular production environment settings. The ATV uses sound and vibration sensors near the DC motors to collect data while it completes the preprogrammed duty. The Edge AI unit then uses these data to filter and process them in order to make decisions about the ATV's status in real time.

Interface of the Data Storage

FIWARE is the middleware platform that is used to implement data transfers. Elastic search data persistence stack maintenance and backups then take place. Within the data storage, each subject released by the ATV and to which the FIWARE platform subscribes has its own index pattern. In our testing environment, the Kibana tool is used to find and refine the data captured in Elastic search. It displays the results obtained when the ATV performs a specific load-carrying operation four times. When the ATV is found to be in normal operating condition, the chart's first row displays the time-stamps corresponding to those hits on the x-axis and the quantity of hits on the y-axis.Examples of

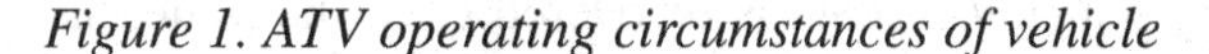
Figure 1. ATV operating circumstances of vehicle

low-level anomalies found in the ATV are shown in the center graph. Time stamps that correspond to the high-level abnormalities detected by the ATV are plotted on the x-axis of the bottom graph. For the most part, the trials are conducted in open, unobstructed locations using the ATV. The bulk of data input results in the dashboard show healthy hits, indicating that the fused data acquired by sensors accurately depicts the state of the field in figure 2. After the data is examined, a high-level abnormal state is discovered when the metal bumpers that are fastened to the ground come into touch with the ATV's primary wheels.

Moreover, since this physical state was included in the model as a low-level anomaly, it appears that the ATV crosses copper cables more frequently, increasing low-level anomaly hits. Along with anomalous cases, the time axis displays the DL model's high model capacity. The experimental verification ensures that the timestamp data displayed in the dashboard is accurate and represents actual, physical occurrences of the anomaly events, even with slight delays caused by network latency. High and low anomalous strike times are influenced by turning velocity, ATV pace, and physical separation between cables.

Data Visualization and Alert Tool for Condition Monitoring

Dashboards for inferred conditions and real-time condition monitoring are generated based on the findings and the data utilized for the inference. Grafana is used to visualize elastic search data and to provide real-time email alerts to end-user stakeholders in the case that data alterations are made accidentally. Grafana's versatility makes it an easy tool to set up for a range of data formats. It can also be used by end-user operators to develop dashboards that show information from monitored environments and outcomes from Edge AI inference. This page displays the most recent, real-time statistics on ATV operations. The data is refreshed every five minutes. Two-dimensional graphs provide the raw vibration and sound data from both motors, while a gauge-style chart shows the ATV's current

Figure 2. Fused data acquired by sensors accurately

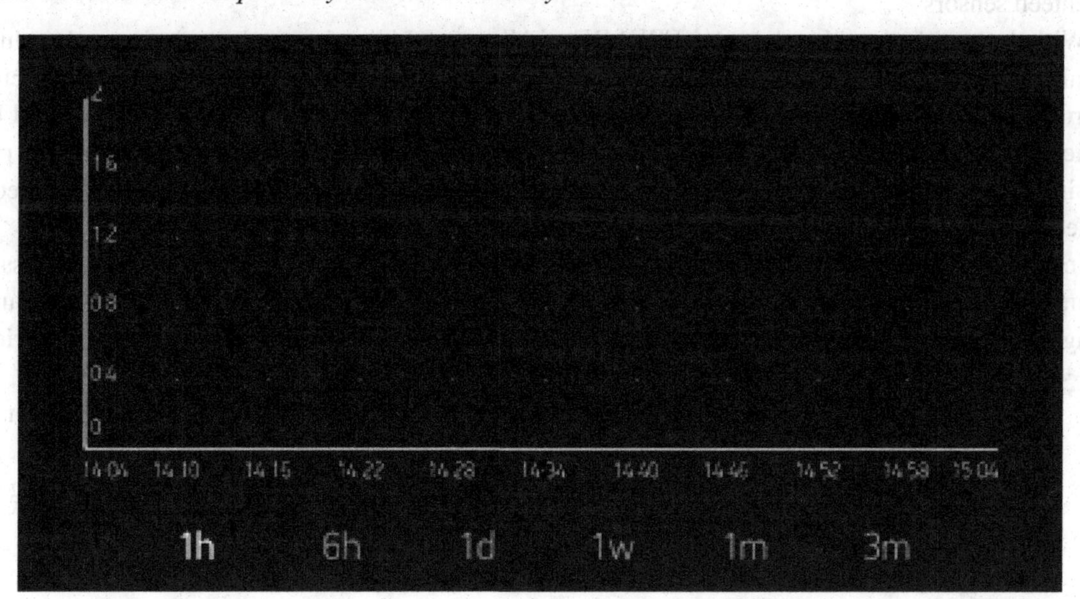

Figure 3. Dashboard is updated

state.Every 5 mi, 30,000 data points are added to each graph, and 100 Hz is the frequency at which the data are collected.

Every time the dashboard is updated, the 500 most recent data points are added to the right side of each graph, and the 500 oldest data points are taken out of the left shows in figure 3. The 512 data points are represented by the findings of the DL model, thus the changes in anomaly hits exactly match the additional 512 data points. Consequently, each motor's sound and vibration data are displayed in the graphs below, which also serve to illustrate the differences between the data from anomalous and healthy samples. These graphs are useful tools for tracking the system's hardware condition thirteen of seventeen sensors

With the use of edge AI and custom FIWARE middleware, this project seeks to design and evaluate a real-time fault diagnosis and condition monitoring system. Since the suggested system architecture is broad in nature, it can be readily scaled and configured for various use cases, enabling several IoT devices to be status monitored at the same time. Furthermore, the potential power benefits of processing data using edge resources have been demonstrated using experimental data employing edge AI technology.

In terms of latency, privacy, communication costs, and storage costs, using the edge processing technique yields significant performance advantages. The benefits offered by edge AI are measured using a variety of criteria. The computation of these metrics takes into account two distinct scenarios:

- All raw vibration and audio data obtained from the sensors is transmitted to a central server for processing;
- The centralized server only stores and visualizes the DL model inference results following the data's filtering and processing at the edge AI unit.

For every scenario, the table presents three more metrics that are computed and presented: average bandwidth use, data transmission latency, and average duration for inference. The strain resulting from the transfer and processing of all the vibration and audio data collected from sensors to the central server is shown in the raw sensor data column. The outcomes of moving the data to the central server for archiving and visualization, however, are displayed in the "processed data at the edge AI unit" column. When numerous autonomous devices and human operators need to collaborate in the same space, real-time, safety-critical systems compute these performance indicators.

CONCLUSION

In addition to an in-house, specially designed middleware solution built on the FIWARE platform, this study demonstrated a comprehensive real-time defect detection and condition monitoring architecture utilizing edge AI. A wide range of use cases, such as logistics, agricultural, health applications, and security systems, can make use of the architecture's adaptability. As part of a smart manufacturing solution, we used an ATV condition monitoring use case to show off the system. The architecture is adaptable enough, even with ROS, to work with a range of IoT devices that can be used with various kinds of generic enablers and data adapters. Based on data from its numerous sensors, the ATV's edge AI gadget makes inferences.

When tested using offline ATV data, the deep learning method produces very high test accuracy, averaging 99.94% over 25 trials. The collected information and findings are then added to the data archive. The middleware system's customizing option for sensor subscriptions makes it simple to discontinue transferring raw sensor data over the network. After that, the data visualization tool generated dashboards and informed administrators as needed. To demonstrate the efficiency of the installed system, trials were conducted in a model industrial environment with a variety of operational irregularities, using ATVs to perform predetermined load-carrying activities.The developed system tracked and diagnosed ATV issues in real time using the data that was observed. Also, when the suggested edge AI solution is applied, the ATV use case requires 43 times less network traffic. Furthermore, the total data transfer time is reduced by a factor of 37 in comparison to conventional central server systems. The middleware layer may be readily expanded to handle other IoT devices, even if the experiments and system viability are shown with a single ATV. This is because different FIROS platform instances may be configured to run on different ATV units. However, there are trade-offs between the number of ATV units that are available and the cost of equipment. In order to improve operational efficiency in a smart manufacturing context, the system may therefore be integrated into actual problems. In subsequent work, we'll take a number of actions to extend the study's scope. We will provide a dynamic edge AI model training and update mechanism in the event that the DL model's performance deviates. Adding Auto ML integration and a dynamic model approach will help us improve our architecture. Model auto-update scenarios can be more easily created by integrating Auto ML techniques into the suggested architecture. Furthermore, since the suggested design incorporates a variety of equipment monitoring interface types, our upcoming work will also concentrate on extensive testing.

REFERENCES

Bae, I., Moon, J., & Seo, J. (2019). Toward a comfortable driving experience for a self-driving shuttle bus. *Electronics (Basel)*, *8*(9), 943. doi:10.3390/electronics8090943

Bakdi, A., Bounoua, W., Guichi, A., & Mekhilef, S. (2021). Real-time fault detection in PV systems under MPPT using PMU and high-frequency multi-sensor data through online PCA-KDE-based multivariate KL divergence. *International Journal of Electrical Power & Energy Systems*, *125*, 106457. doi:10.1016/j.ijepes.2020.106457

Cai, Y., Luan, T., Gao, H., Wang, H., Chen, L., Li, Y., Sotelo, M. A., & Li, Z. (2021). YOLOv4-5D: An effective and efficient object detector for autonomous driving. *IEEE Transactions on Instrumentation and Measurement*, *70*, 1–13. doi:10.1109/TIM.2021.3065438

Carranza-García, M., Torres-Mateo, J., Lara-Benítez, P., & García-Gutiérrez, J. (2021). On the performance of one-stage and two-stage object detectors in autonomous vehicles using camera data. *Remote Sensing (Basel)*, *13*(1), 89. doi:10.3390/rs13010089

Chen, Z., Liu, Q., & Lian, C. (2019, June). Pointlanenet: Efficient end-to-end cnns for accurate real-time lane detection. In 2019 IEEE intelligent vehicles symposium (IV) (pp. 2563-2568). IEEE.

Chiu, Y. C., Tsai, C. Y., Ruan, M. D., Shen, G. Y., & Lee, T. T. (2020, August). Mobilenet-SSDv2: An improved object detection model for embedded systems. In *2020 International conference on system science and engineering (ICSSE)* (pp. 1-5). IEEE. 10.1109/ICSSE50014.2020.9219319

Deng, W., & Wu, R. (2019). Real-time driver-drowsiness detection system using facial features. *IEEE Access : Practical Innovations, Open Solutions*, *7*, 118727–118738. doi:10.1109/ACCESS.2019.2936663

Guo, N., Lenzo, B., Zhang, X., Zou, Y., Zhai, R., & Zhang, T. (2020). A real-time nonlinear model predictive controller for yaw motion optimization of distributed drive electric vehicles. *IEEE Transactions on Vehicular Technology*, *69*(5), 4935–4946. doi:10.1109/TVT.2020.2980169

Gupta, A., Anpalagan, A., Guan, L., & Khwaja, A. S. (2021). Deep learning for object detection and scene perception in self-driving cars: Survey, challenges, and open issues. *Array (New York, N.Y.)*, *10*, 100057. doi:10.1016/j.array.2021.100057

Hong, Y., Pan, H., Sun, W., & Jia, Y. (2021). Deep dual-resolution networks for real-time and accurate semantic segmentation of road scenes. *arXiv preprint arXiv:2101.06085*.

Jabbar, R., Shinoy, M., Kharbeche, M., Al-Khalifa, K., Krichen, M., & Barkaoui, K. (2020, February). Driver drowsiness detection model using convolutional neural networks techniques for android application. In *2020 IEEE International Conference on Informatics, IoT, and Enabling Technologies (ICIoT)* (pp. 237-242). IEEE. 10.1109/ICIoT48696.2020.9089484

Langarica, S., Rüffelmacher, C., & Núñez, F. (2019). An industrial internet application for real-time fault diagnosis in industrial motors. *IEEE Transactions on Automation Science and Engineering*, *17*(1), 284–295. doi:10.1109/TASE.2019.2913628

Li, D., Zhang, Z., Liu, P., Wang, Z., & Zhang, L. (2020). Battery fault diagnosis for electric vehicles based on voltage abnormality by combining the long short-term memory neural network and the equivalent circuit model. *IEEE Transactions on Power Electronics*, *36*(2), 1303–1315. doi:10.1109/TPEL.2020.3008194

Liu, L., Lu, S., Zhong, R., Wu, B., Yao, Y., Zhang, Q., & Shi, W. (2020). Computing systems for autonomous driving: State of the art and challenges. *IEEE Internet of Things Journal*, *8*(8), 6469–6486. doi:10.1109/JIOT.2020.3043716

Liu, Z., Cai, Y., Wang, H., Chen, L., Gao, H., Jia, Y., & Li, Y. (2021). Robust target recognition and tracking of self-driving cars with radar and camera information fusion under severe weather conditions. *IEEE Transactions on Intelligent Transportation Systems*, *23*(7), 6640–6653. doi:10.1109/TITS.2021.3059674

Manoharan, D. S. (2019). An improved safety algorithm for artificial intelligence enabled processors in self driving cars. *Journal of Artificial Intelligence and Capsule Networks*, *1*(2), 95–104. doi:10.36548/jaicn.2019.2.005

Muhammad, K., Ullah, A., Lloret, J., Del Ser, J., & de Albuquerque, V. H. C. (2020). Deep learning for safe autonomous driving: Current challenges and future directions. *IEEE Transactions on Intelligent Transportation Systems*, *22*(7), 4316–4336. doi:10.1109/TITS.2020.3032227

Ouyang, Z., Niu, J., Liu, Y., & Guizani, M. (2019). Deep CNN-based real-time traffic light detector for self-driving vehicles. *IEEE Transactions on Mobile Computing*, *19*(2), 300–313. doi:10.1109/TMC.2019.2892451

Rohan, A., Rabah, M., & Kim, S. H. (2019). Convolutional neural network-based real-time object detection and tracking for parrot AR drone 2. *IEEE Access : Practical Innovations, Open Solutions*, *7*, 69575–69584. doi:10.1109/ACCESS.2019.2919332

Simon, M., Amende, K., Kraus, A., Honer, J., Samann, T., Kaulbersch, H., & Michael Gross, H. (2019). Complexer-yolo: Real-time 3d object detection and tracking on semantic point clouds. In *Proceedings of the IEEE/CVF Conference on Computer Vision and Pattern Recognition Workshops* (pp. 0-0). IEEE. 10.1109/CVPRW.2019.00158

Sun, L., Yang, K., Hu, X., Hu, W., & Wang, K. (2020). Real-time fusion network for RGB-D semantic segmentation incorporating unexpected obstacle detection for road-driving images. *IEEE Robotics and Automation Letters*, *5*(4), 5558–5565. doi:10.1109/LRA.2020.3007457

Van Wyk, F., Wang, Y., Khojandi, A., & Masoud, N. (2019). Real-time sensor anomaly detection and identification in automated vehicles. *IEEE Transactions on Intelligent Transportation Systems*, *21*(3), 1264–1276. doi:10.1109/TITS.2019.2906038

Verma, A. K., Nagpal, S., Desai, A., & Sudha, R. (2021). An efficient neural-network model for real-time fault detection in industrial machine. *Neural Computing & Applications*, *33*(4), 1297–1310. doi:10.1007/s00521-020-05033-z

Yao, L., Fang, Z., Xiao, Y., Hou, J., & Fu, Z. (2021). An intelligent fault diagnosis method for lithium battery systems based on grid search support vector machine. *Energy*, *214*, 118866. doi:10.1016/j.energy.2020.118866

Yao, L., Xiao, Y., Gong, X., Hou, J., & Chen, X. (2020). A novel intelligent method for fault diagnosis of electric vehicle battery system based on wavelet neural network. *Journal of Power Sources*, *453*, 227870. doi:10.1016/j.jpowsour.2020.227870

Yaqoob, I., Khan, L. U., Kazmi, S. A., Imran, M., Guizani, N., & Hong, C. S. (2019). Autonomous driving cars in smart cities: Recent advances, requirements, and challenges. *IEEE Network*, *34*(1), 174–181. doi:10.1109/MNET.2019.1900120

Yin, S., Rodriguez-Andina, J. J., & Jiang, Y. (2019). Real-time monitoring and control of industrial cyberphysical systems: With integrated plant-wide monitoring and control framework. *IEEE Industrial Electronics Magazine*, *13*(4), 38–47. doi:10.1109/MIE.2019.2938025

Chapter 5
IoT in Day-to-Day Life:
Vehicle Body Fatigue Analysis

Rajaram Vassudev Pai Kuchelkar
Liverpool John Moores University, India

Anupama Jawale
https://orcid.org/0000-0002-9982-5218
Narsee Monjee College of Commerce and Economics, India

ABSTRACT

One of the analyses is fatigue analysis which helps determine the durability of the product for its intended life. The fatigue analysis requires an input data from vehicle-road interaction obtained as vertical acceleration. The input data is then used in finite element analysis software or in a test rig instrumentation to perform a fatigue analysis. The focus of this study is about collecting the input data using a data acquisition device. OEM of the automobiles use proprietary data acquisition device for testing their vehicles for durability. However, the automotive body building companies or vehicle body builders which are 1-Tier companies below OEM have a limited budget for such an analysis and therefore they principally lag in advance analysis approach such as a fatigue analysis for the design of vehicle body. Instead, to sustain durability of vehicle, the design is made heavier. This research study is an effort to provide an affordable setup for data acquisition by making use of IoT and web technologies which are cost effective.

INTRODUCTION

The term fatigue in this context means the cyclic loading of material. The principle of fatigue is based on the material S-N curve, which is material strength (S) versus the number of load cycles (N). As the number of load cycles increases, the material strength weakens. A fatigue analysis is conducted to determine the durability and life of the component. A lab setup with test rigs and instrumentation is a conventional technique used for conducting fatigue analysis. In the modern world of digitization, finite element-based approaches are commonly used. The finite element approach is

DOI: 10.4018/979-8-3693-1962-8.ch005

considered the best approach due to savings in time and resources over the conventional approach. To conduct fatigue analysis for the vehicle body using a finite element approach, the inputs from vehicle-road interaction such as vertical acceleration and number of cycles are necessary. The vehicle is instrumented with data acquisition devices and the vehicle-road response is captured. The road surface quality equally influences fatigue behaviour. If the road is rough or undulated, the vertical acceleration is more intense, which results in reduced fatigue life. If the road is good, fatigue life is longer. Vehicles travel mostly on concrete or asphalt types of roads during their life span. The seasonal changes like monsoon rains and floods spoil the road surface, and this in turn influences fatigue. Therefore, study of fatigue is necessary on type of roads intended for the design of the vehicle. The fatigue analysis is used by Original Equipment Manufacturer (OEM) for fully built bodies like cars, utility vehicles, truck cabins, and chassis whereas hardly ever used by vehicle body builders for body building applications like buses, truck load bodies, containers etc. This is due to the lack of road load input data for benchmark vehicles, the high cost of proprietary data acquisition equipment to generate own data, and time constraints for body building development activity required by customers. In the Indian market, vehicle body builders are facing stiff competition from their peers in terms of product pricing, weight, mileage, and payload carrying efficiencies. Vehicle body builders must make constant efforts on light weighting of vehicle body to gain leverage over their competitors and therefore may consider a proper analytical approach such as fatigue analysis for durability confirmation. This research study focuses on the way of capturing and recording the vehicle-road response using low-cost data acquisition methods, making it affordable for the vehicle body builders to conduct vehicle body fatigue analysis.

The data acquisition instruments are required for recording the vehicle-road response. The proprietary data acquisition instruments available on the market are expensive as they are designed for robust usage like measuring the leaf springs or vehicle axle components which have high vertical acceleration values. The vehicle body has a lower operating range of vertical acceleration as compared to vehicle components below vehicle suspension which makes it feasible to use low-cost IoT components for data acquisition. The vehicle body builders cannot afford the cost of proprietary data acquisition instrumentation. As a result, as a first step, there is a need to suggest an affordable tool for data acquisition and a method for preparing input data for fatigue analysis, which are discussed in this research study.

The main aim of this research is to propose an affordable solution to the vehicle body builders on data acquisition for vehicle body fatigue analysis. With this affordable solution, the body builders will incur low expense on the data acquisition instrument and can focus on the analysis part. The research objectives are formulated based on the aim of this study which are as follows:

- To study the recent developments in the tools and techniques used for data acquisition of vehicle-road interaction.
- To propose a low-cost data acquisition device using IoT for acquiring the parameters for fatigue analysis.
- To acquire a dataset practically using the proposed data acquisition instrument.
- To analyse the acquired dataset and prepare input test cases for fatigue analysis.

The research questions are as follows.

- What are the modern tools and techniques used for vehicle-road data acquisition?
- How to build simple IoT based circuit and a tool for data acquisition for fatigue analysis?
- How to acquire a dataset for fatigue analysis using a simple data acquisition?
- How to analyse and post-process the acquired data and prepare input test cases for fatigue analysis?

Vehicle body fatigue analysis is conducted with inputs of vertical acceleration and number of cycles. The analysis is then performed using a finite element analysis approach. The scope of the study is about the data acquisition of vertical acceleration and devising methods for the determination of the number of cycles and test cases for finite element analysis. The scope excludes performing a finite element analysis. As the vehicle bodies operate in the lower vertical acceleration range described in the problem statement, the scope of the study is applicable to all types of vehicle bodies only. On the other hand, a vehicle chassis is subjected to a higher range of vertical acceleration from the vehicle-road response. As this needs a higher degree of accuracy and precision on data acquisition instruments fulfilled by proprietary instruments, the vehicle chassis is excluded from the scope of study. The scope of the study is to devise affordable data acquisition instrumentation using IoT technologies in the field of computer science.

This study will provide a cost-effective method for the vehicle body builder and guide them to create their own data acquisition device and road test database using low-cost electronics and open-source IDE platforms for programming microcontrollers. This study will also show how to process the collected data and prepare the input test cases for fatigue analysis.

BACKGROUND

Introduction

In this section, we will explore the literature on fatigue analysis studies conducted by researchers and the data acquisition methods using IoTs that can be adopted for capturing the vehicle-road response. The literature study is broadly classified based on Fatigue Analysis, Data Acquisition Process, Data Acquisition Devices and Data Analysis as tabulated in Table 1,2,3 and 4 respectively.

Inputs for Fatigue Analysis

From the above review, it can be summarised that the inputs principally needed for fatigue analysis are a data acquisition system for collection of vehicle-road response, sensors connected to vehicle body, and a data recording device such as a SD card or a computer system/smart phone. Optionally, GPS, time module, and Wi-Fi modules may be used to record the location of the vehicle, record the time of the capturing event, and transmit the data wirelessly.

Various Approaches to Collecting Inputs for Fatigue Analysis

From the above literature, the various approaches used for collecting input data are as below.

Table 1. Literature review based on fatigue analysis

Reference	Description
(Xuewen et al., 2020)	In this research study, the finite element method was used to find out the S-N curve of the structure. The load spectrum was determined based on empirical data. The load spectrum is fatigue load and the number of cycles per hour. With the above load spectrum and S-N curve of the structure, the linear Palmgren-Miner cumulative damage model was used to calculate the fatigue life. The topology optimization technique is used for the improvement of structural design.
(Polat, 2017)	In this research study, the CAD model of the subject frame was done in the Solid Works CAD software. The Ansys Static Programmer software was used for the finite element analysis model. A static analysis was performed using the on-vehicle weight method. The fatigue analysis is performed using Ncode Design Life software using the test track data and then the fatigue life of the parts on the model is examined. Three important steps are mentioned for fatigue analysis, such as selection of fatigue methods, selection of installation methods, and selection of a curve for material fatigue (S-N curve).
(Salokhe et al., 2016)	In this research study, the damage detection of a structure is studied, which is like fatigue. The technique used here is based on dynamic response collected from the structure under study. A technique called Singular Value Decomposition (SVD) is used to predict the damage of the structure.
(Ogunoiki, 2015)	This research study has discussed three methods of data acquisition: the empirical method, the analytical method, and the semi-analytical method. The empirical method consists of a customer survey and proving ground. The analytical method consists of finite element analysis and multi-body dynamics. The semi-analytical method consists of a combination of analytical and empirical methods. A lab setup is used here with a quarter vehicle test rig. Transducers such as Piezoelectric Accelerometer, load cell, and linear variable differential transformer are used in the equipment for measurement of signals. A hydraulic power unit is used to apply the road input excitations to the test rig. Different computer software such as Cubus, QanTiM, and LabVIEW are used to acquire the data from the test rig. The above experiment is also done in CAD and SIMPACK multi-body dynamics and the results are compared in this study. Machine Learning aspects are also studied in this paper.
(Shafiullah & Wu, 2013)	In this research study, an Accelerated Durability test is modelled. A conjugative approach consisting of finite element analysis and fatigue analysis for a specific durability life is used. Instead of a time-based technique, a frequency-based technique is used to accelerate the fatigue analysis. For fatigue analysis, Ncode Design Life software is used.
(Dongpo & Xuhui, 2011)	This research study follows a CAE-based approach. The data for the virtual proving ground is verified using MATLAB software. The durability analysis is carried out using explicit finite element code using LSDYNA, a proprietary CAE software package.
(Mo et al., 2000)	This research study uses a different technique for durability study (or fatigue analysis) using a virtual proving ground. A virtual proving ground is a laboratory test with finite element analysis. From the analysis, fatigue life is predicated on the stress magnitude and load history. Fatigue life results of the vehicle components from the computer simulation are compared with physical test results.
(Sener, n.d.)	This research study applies to the fatigue analysis of automatic transmission shafts. The design of shafts requires a service history or load history of the shaft. The process of capturing service history and load history is done with a data acquisition system. The amount of service history data needed is huge, so a 12-bit data acquisition system is used. A high sampling rate such as 4KHz is used, which generates a data rate of 20MB per hour. In this study, proprietary data acquisition systems are referred to as expensive equipment. Therefore, an innovative approach to cost-effective data acquisition technique with data compression is studied.
(Ilic et al., n.d.)	This research study was carried out in Turkey. Around fifty Turkish roads were studied to capture the characteristics of fatigue and define a load spectrum. A leaf spring of a test vehicle was studied for equivalent fatigue damage caused by an accelerated test method. Fatigue analysis and estimated fatigue life were calculated using a finite element analysis approach and verified by using the Palmgren-Miner rule. Two types of methods have been used in this study. One is the questionnaire method, and the second is the black box method. The questionnaire method is used to survey the types of roads and types of loads. After getting the questionnaire results, a road test is programmed with a vehicle equipped with transducers to run on several types of roads with different load cases. The measured data is analysed and processed for spikes and is filtered. Frequency analysis and arithmetic manipulations are done to generate inputs for fatigue analysis. As the road signals are formed from random vibrations, the data is processed with approaches such as range pair, rain flow level crossing, and counting methods to compare the signals in a meaningful manner.

- Collecting data on different types of road surfaces.
- Collecting data at different speeds.
- Collecting data on proving grounds.
- Using a CAE tool such as multi-body dynamics on virtual proving ground.

*Table 2. **Literature** review based on data acquisition process for fatigue analysis*

Reference	Description
(Xuewen et al., 2020)	In this research study, the load spectrum was determined using empirical data. The S-N curve of the structure was determined using finite element analysis.
(Wang et al., 2019)	In this research study, a pressure sensor is used in the air spring circuit to sense the vertical loads. The data acquisition of pressure signals is done using the Agilent Digital Signal Acquisition System.
(Polat, 2017)	In this research study, the acceleration data was collected from four different test tracks. The length of the test track is about 3 km. These acceleration values are converted to gravitational acceleration format, which is in terms of g.
(Salokhe et al., 2016)	In this research study, the data acquisition of acceleration is done by applying a heavy-duty sensor to the subject structure.
(Ogunoiki, 2015)	In this research study, data acquisition is done on a test rig setup using multi-body dynamics Finite Element Analysis tools.
(Shafiullah and Wu, 2013)	In this research study, data acquisition is done by a proprietary tool, somat eDAQ, and accelerations and strains are captured in this process. An extremely high sampling rate of 2500 Hz was used in this study.
(Mo et al., 2000)	In this research study, the data acquisition of the acting forces takes place within the finite element analysis. The vehicle model runs on the digital proving track developed in the form of finite elements. There is no physical equipment or device used for data acquisition in this study. The kind of data acquisition device used on a physical test vehicle is not described in this study.
(Sener, n.d.)	A stand-alone real-time data acquisition system was used in this research study. The system is attached to a single-board-based computer. This system is 12bit and offers a high sampling rate of 100KHz and has a storage capability of 1 GB.
(Ilic et al., n.d.)	In this research study, the data acquisition was done by deploying a load transducer and two strain gauges on the leaf spring of the suspension system, one strain gauge on the transmission shaft, and one strain gauge on the steering shaft. The vertical loads created by road roughness, manoeuvres, and acceleration were measured by two half-bridge strain gauges.

- Mounting sensors at different locations on the vehicle body.
- Using different types of chassis platforms for same type body application.

Conventional Methods versus Modern Methods of Data Acquisition

It has been discovered that conventional approaches use proprietary data acquisition devices, which are both expensive and accurate when used for high data sampling rates. The researchers have discovered low-cost data acquisition techniques that use smartphones, inexpensive sensors, and open platforms like Arduino to produce findings that are comparable to those of more traditional instruments. In situations when accuracy is crucial, like with dynamic machinery or high-speed applications, conventional approaches should be employed. While using modern methods, it is important to be aware of the appropriate sampling rate and measurement range.

Data Processing and Analysis Techniques

Fourier transforms and related approaches, including windowed Fourier transforms, are frequently used to process data for vibration signals with a sinusoidal component. However, additional methods, such as range pair, rain flow, and counting procedures, are used to process the random vibration signals from the vehicle-road (Ilic et al., n.d.).

Table 3. Literature review based on data acquisition devices

Reference	Description
(Ur Rehman et al., 2021)	This research study uses IoT for monitoring and control of substations to provide safe electricity to consumers. The equipment used is a Microcontroller ESP32 with the Cayenne IoT platform. It uses different types of sensors, namely current sensors, voltage sensors, frequency sensors, humidity sensors, temperature sensors, and oil level sensors. The Cayenne platform is useful for real-time monitoring, data logging, and control.
(Shuai et al., 2021)	In this research study, the data was acquired with sensors which include acceleration sensors, pull-type displacement sensors, strain flowers, and GPS with the help of a video auxiliary system. The collection equipment consisted of an eDAQ system and NCode Design Life analysis software.
(Pal Singh & Singh, 2021)	In this research study, a low-cost data acquisition system consisting of an Arduino Uno is used for vehicle-road interaction. The accelerometer ADXL345 is used for capturing acceleration factors, and the GPS receiver Ublox Neo 6M is used for capturing the location coordinates. The Arduino Uno uses an 8-bit ATmega328P microcontroller.
(Jamaludin, 2021)	The Arduino based board is used in this research study. The different modules used in the circuit are a current and voltage sensor unit, a real-time clock module, a power supply module, and a secure digital card module. The type of microcontroller on the Arduino board is the Atmega328P type.
(Moreno et al., 2020)	This research study states that the configuration is required for measurement of high-frequency-based vehicular applications. The sensors required for the study are displacement sensors, acceleration sensors, and speed sensors. The different kinds of sensors make the price of data acquisition equipment skyrocket. This study proposes a low-cost data acquisition, 'Arduino Due', with enhancement of its limitations on the speed of data transfer. By using some programming techniques, the 'Arduino Due' is made compatible with the requirements of high-frequency vehicular applications. The paper compares various Arduino boards and selects the 'Arduino Due' as the most relevant board for this application.
(Amestica et al., 2019)	The researcher studied different types of digital platforms and programmed using the Arduino IDE platform, which is used for data acquisition applications and performing digital controls. The three types of Arduino boards studied in this research paper were Arduino Uno, Arduino Mega, and Arduino Due. ESP-based boards, that is, ESP8266 and ESP32 boards, were also used in this study. Comparison is made for execution time with the above boards in this study. ESP boards can be connected to a Wi-Fi network.
(Wong et al., 2019)	In this research study, a data logger is prepared using Arduino. The 1Sheeld Android Application is used, which connects to the timestamp, accelerometer, and Bluetooth of the smartphone.
(González et al., 2018)	This research study uses an Arduino-based system along with low-cost accelerometers. The accelerometer ADXL345 and MPU6050 are used. Vehicle dynamics applications are being studied. A comparison of the performance of the Arduino-based data acquisition is made with professional data acquisition systems. A professional data acquisition system with 4 channels connected to piezoelectric accelerometers has been used. The study concludes that the low-cost system is suitable for data acquisition for vehicle dynamics applications.
(Chellaswamy et al., 2018)	This research study is about road surface monitoring. The road surface deformities, such as the potholes are studied. The data acquisition equipment used in this study consists of the following: Arduino Uno with ESP8266, Ultrasonic Sensor, MEMS Accelerometer, GPS Receiver, and Access Point. A layered architecture has been used, starting from top level to bottom level, as Cloud Server, Edge Server, Node IoT, and Sensors.
(Wali & Areeb, 2018)	In this research study, Proposed data acquisition equipment consists of an Arduino Mega 2560 and a Raspberry Pi. Different types of proprietary data acquisition are compared with Arduino board-based data acquisition in terms of cost.
(Wu et al., 2018)	This research study is about a massive data collection of driving events. An on-board device in the vehicle was used in this study. The on-board device is made of the MT662 chip, which contains a Wi-Fi/Bluetooth transceiver and an FM receiver. The sensing data on acceleration and angular velocity is generated by a built-in chip called the BMI120. The BMI120 chip is a low-powered 3-axis low-g accelerometer and 3-axis gyroscope. It produces 50 sets of data per second.
(Monterrey, 2018)	In this research study, road load data acquisition on heavy trucks is studied. A proprietary data acquisition device is used. Accelerometers and strain gauges are used for acquiring road signals. LabVIEW software is used to read, process, store, and analyse the information.
(Gonzalez et al., 2017)	The researcher studies road anomalies using a smart phone. The smart phone's inbuilt sensor accelerometer is used for the purpose of capturing acceleration in (m/s^2). The dataset collected in this study is made available to the public.
(Agrawal et al., 2016)	In this research study, the data acquisition system uses the Arduino Uno board, an MPU 6050 accelerometer, and a temperature sensor. The analysis is performed on MATLAB software.
(Du et al., 2016)	In this research study, a ZigBee module is used to collect accelerations. The acceleration is measured in the Z-axis. An embedded 3G/4G module is used to transmit the data to an FTP server. A micro controller (MCU) (type TC12C5608AD) is used. An accelerometer (type MMA8451Q) is used.

continued on following page

Table 3. Continued

Reference	Description
(Mohamed et al., 2014)	In this research study, smart phones are used for data acquisition. Accelerometer, Gyroscope and GPS inside the smart phone are used to collect data and the data is stored in the form of file. The file is dumped into micro-SD card in the smartphone. This paper studies capturing events for a speed bump. From the acceleration data, the features for speed bump are extracted using the conditions for acceleration in Z axis and X axis. This data is further classified and studied for machine learning.
(Of & Signal, 2013)	In this research study, various components of data acquisition are explained. Single channel, multi-channel and pc-based data acquisition are described. LabVIEW software is used for controlling of data acquisition. Noise filter is applied.
(Perttunen et al., 2011)	This research study is about road roughness measurement. Smart phones are used to collect data. (The smartphone model used is the Nokia N95 8GB). The sampling rate is 38Hz for the accelerometer and 1Hz for GPS readings. The data collection tool is written in Java. The position of the smart phone was on the rack of the windscreen. The test drive was 25km in 40 minutes. A camera was attached to capture the drive. Cobblestone data was filtered out. The data was classified based on types of road anomalies based on feature extraction and selection techniques, and the statistics of the findings were plotted.
(Chen, 2009)	In this research study, a remote-controlled signal analyser (HP3560A) is used in this study and controlled over the internet using LabVIEW.
(Junnila & Niittylahti, 2003)	This research study discusses various worldwide open standards for wireless communications. These include Bluetooth, ZigBee module, WLAN, Wireless USB, Spike, Nanonet, RF-232, UWB, RF Transceivers.

Table 4. Literature review based on data analysis or signal processing

References	Methodology used for processing of signals
(Men et al., 2018)	This research study discusses processing vibration signals. It says relative mean square (RMS) is an important index for the study of vibrations. Further, this paper discusses certain problems with the computation of RMS value and therefore proposes an improved method called the time domain filtering method in this study.
(P. Wang, 2017)	In this research study, the acceleration signal is studied. The signal consists of noise, which is removed using the Fourier transform. The wavelet analysis is applied, which achieves the noise removal. A wavelet analysis is a time-frequency analysis used for de-noising signals. A Fourier transform is used to convert the signal from the time domain to the frequency domain. The displacement signal is calculated in this study from the acceleration signal.
(U et al., 2015)	The limitations of Fast Fourier Transform (FFT) are discussed in this research study. The study discusses that if the measured signals have high sensitivity and many times invariant harmonics, then it is difficult to extract fault-related signals. Hence, a Windowed Fourier Transform (WFT) approach is used. This study was applied to the study of bearing vibrations and finding out the fault frequencies of the bearing.
(Baoquan, 2011)	A blasting vibration signal is studied in this research. The Fast Fourier Transform (FFT) is used in conjunction with the Short-time Fourier Transform (STFT).
(Hu et al., 2011)	In this research study, time domain sea maps with windowed Fourier transformation are examined.
(Q. Wang et al., 2009)	In this research study, MATLAB software is used to do frequency domain analysis to identify motor noise. Power spectrum analysis combined with holographic spectrum analysis is conducted to find the abnormal vibration of the motor more precisely.

Related Datasets for Fatigue Analysis

(Gonzalez et al., 2017) freely publishes data sets of the data collected on various types of roads for analysis. The acceleration data is collected in the X, Y, and Z axes. The researcher has collected the data on asphalt bump roads, potholed roads, metal bump roads, and regular roads. This data can be studied for fatigue analysis inputs.

Highlights of Literature Review

According to the literature evaluation, a vehicle's fatigue analysis can also be referred to as its durability analysis. There are several different approaches, including lab-based and finite element-based approaches. The cost of using lab-based procedures makes them unsuitable for our target users, Vehicle Body Builders (VBB), who operate as lean businesses with little resources for research and development. The optimum approach is a combination of realistic, low-cost data acquisition of vehicle-road on a chosen road track defined by the vehicle body designer and employing a finite element analysis tool for fatigue analysis. This ought to be the most affordable solution for vehicle fatigue analysis.

In the different reviews of literature on vehicle fatigue analysis, the open-source, low-cost data acquisition system known as Arduino has been offered and demonstrated as a low-cost alternative. For gathering data, sensors like the MCU6050 and the ADXL3XX series have been developed. For remotely acquiring or monitoring the signals, web technologies including cloud servers, client servers, and Wi-Fi modules like ESP8266 and ESP32 have also been employed. Road roughness has been measured using the aforementioned technologies. The built-in accelerometers and other sensors of smartphones have been extensively used by researchers to collect data. Researchers have utilised rain-flow or counting methods to study random vibrations, while Fourier transform techniques have been used to study dynamic machinery and vibration signals with cyclic patterns.

FATIGUE ANALYSIS

Introduction

The methodology for vehicle body fatigue analysis using IoT broadly covers the following steps as depicted in Figure 1.

- The vehicle-road interaction leads to vertical acceleration.
- The vertical acceleration is captured using a data acquisition instrument sampled for a short duration of about 3 to 5 minutes.
- The sampled data is analysed for a frequently occurring pattern of acceleration values.
- The frequency of occurrences is extrapolated to desired life span of the vehicle.
- Finally, the number of cycles is computed which serves as input test case for fatigue analysis.

Methodology for Data Acquisition

- A data acquisition device shall be deployed in the vehicle as per the vehicle designer's requirement.
- The accelerometer of the data acquisition device shall be attached to a hard point on the vehicle body, such as a seat mounting or over the front or rear wheels.
- The accelerometer's Z axis shall be aligned parallel to the vertical axis as shown in Figure 2, which is also perpendicular to the ground plane.

Figure 1. Flow chart for fatigue analysis study

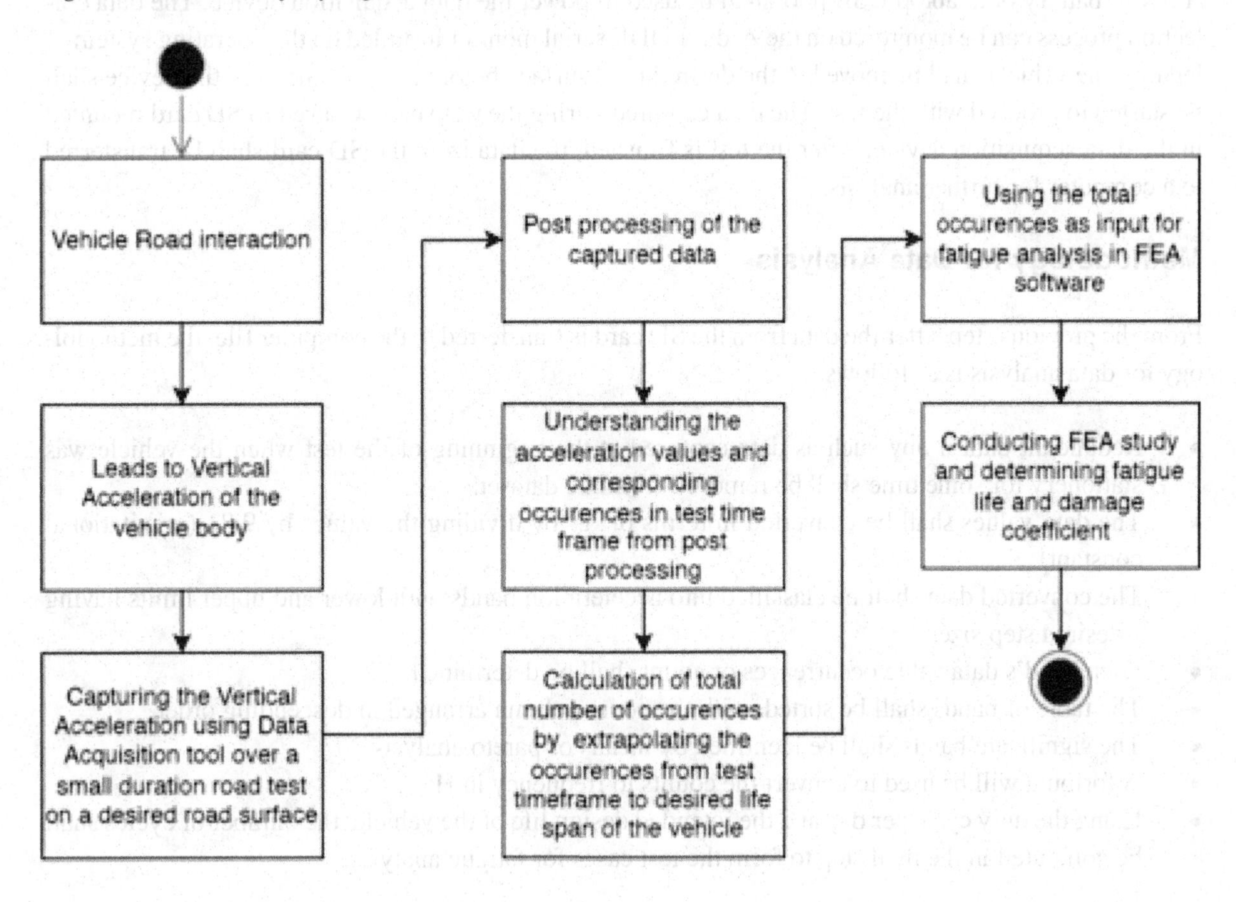

Figure 2. Vehicle layout showing ground plane and vertical axis

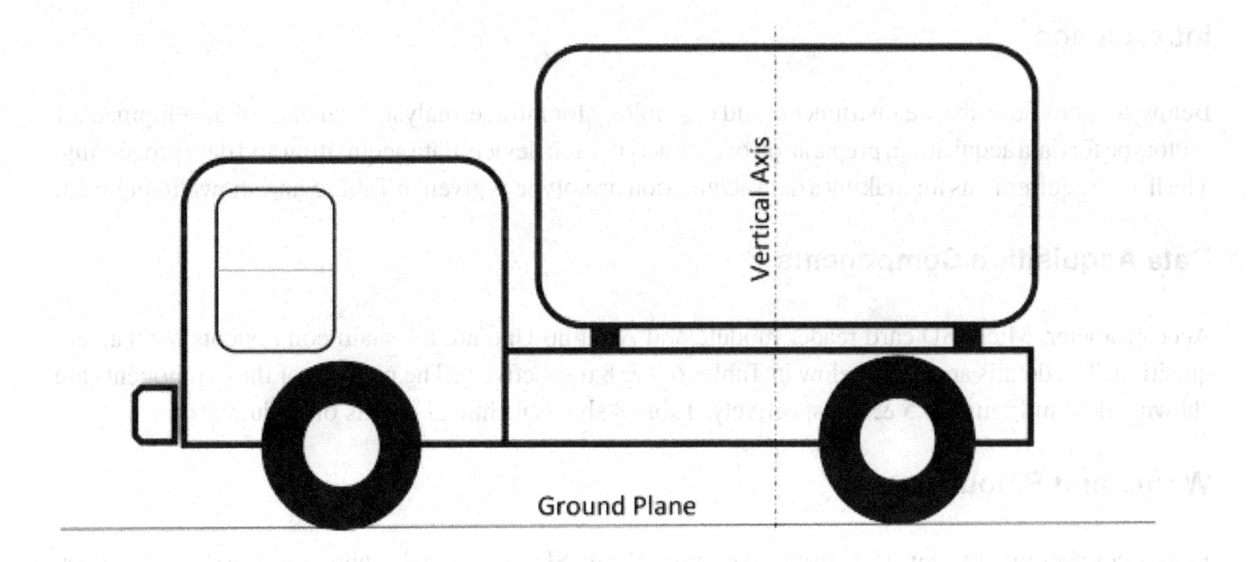

A 5-volt battery or a laptop USB port shall be used to power the data acquisition device. The data collection process can be monitored on the Arduino IDE serial monitor installed on the operating system of laptop. The vehicle shall be moved to the desired road surface before the test, and then the device shall be started to proceed with the test. The data captured during the test shall be saved to SD card mounted in the data acquisition device. After the test is finished, the data from the SD card shall be transferred to a computer for further analysis.

Methodology for Data Analysis

From the previous step, after the data from the SD card is transferred to the computer file, the methodology for data analysis is as follows.

- Redundant data if any such as data captured at the beginning of the test when the vehicle was stationery for some time shall be removed from the dataset.
- The data values shall be converted in terms of 'g' by dividing the values by 9.81 (gravitational constant).
- The converted data shall be classified into acceleration bands with lower and upper limits having a desired step size.
- Each band's data value occurrences or count shall be determined.
- The table of bands shall be sorted, with the count column arranged in descending order.
- The significant bands shall be identified by means of pareto analysis.
- A formula will be used to convert the counts to frequency in Hz.
- Using the duty cycle per day and the intended design life of the vehicle, the number of cycles shall be computed in the final step to form the test cases for fatigue analysis.

INSTRUMENTS AND TECHNOLOGY FOR FATIGUE ANALYSIS

Introduction

Below sections describe the instruments and technology for fatigue analysis by means of development of prototype for data acquisition, preparation of data acquisition device, data acquisition and data processing. The list of requirements for making a data acquisition prototype is given in Table 5 and shown in Figure 3.

Data Acquisition Components

Accelerometer, Micro SD card reader module and Arduino Uno are the main components for data acquisition. The details are given below in Tables 6,7 & 8 respectively. The pictures of the components are shown below in Figures 4,5 & 6 respectively. Table 8 shows technical details of Arduino.

Wiring and Pinout

Pinout connections for wiring between Arduino, Micro SD Card Reader and Accelerometer is given below in Table 9 and 10. Wiring layout is given in Figure 7.

Table 5. Requirements for Prototype data acquisition

Sn	Description	Qty
1	Arduino Uno Board	1
2	ADXL 345 Accelerometer	1
3	Micro SD Card Reader Module	1
4	Jumper Wires	10
5	USB Power Cable for Arduino Uno	1
6	Laptop installed with Arduino IDE	1
7	Object Vehicle	1
8	Adhesive Tape	20 centimetres
9	Micro SD Card	1

Figure 3. Layout of data acquisition device

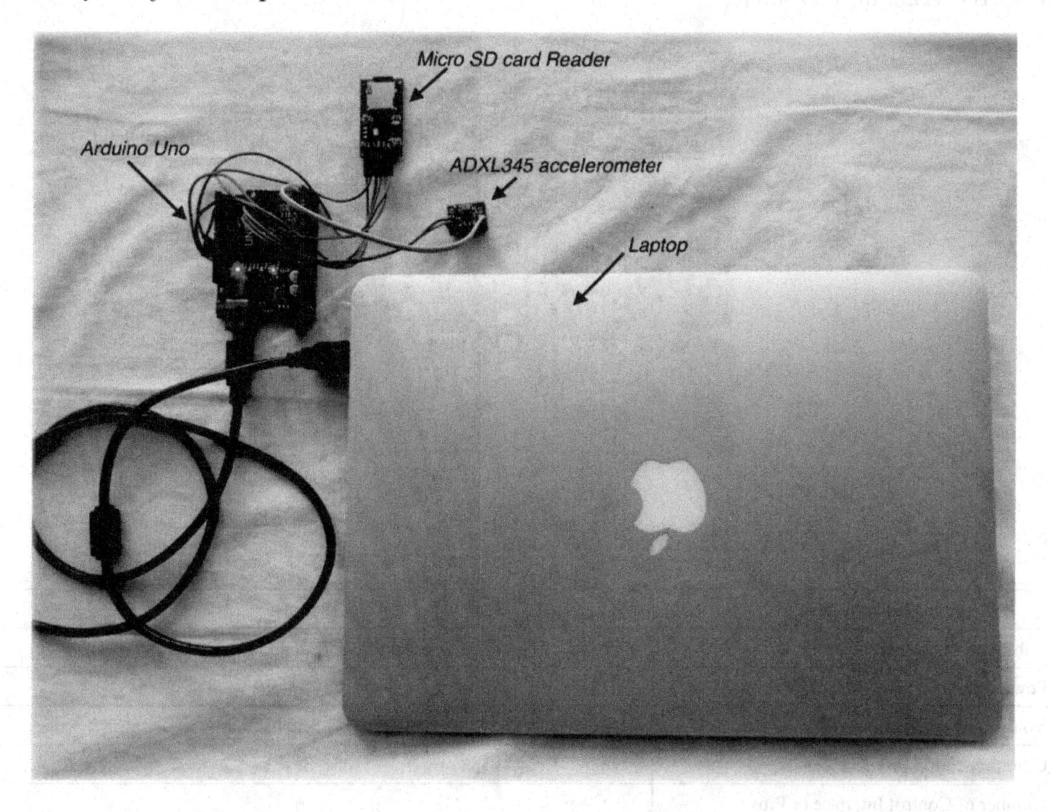

The data acquisition device preparation is done by uploading a code (also known as a sketch) to the Arduino Board microcontroller. The code for the data acquisition device is written in C++ using the Arduino IDE software. For uploading the code, the code is opened in the Arduino IDE on the laptop connected to the Arduino board of data acquisition, and the code is uploaded by the upload button on the IDE. The data acquisition device is ready for deployment. (LastMinuteEngineers.com, 2022), PiMyLife(2022), (Mechatronics, 2022; PiMyLife, 2022; toptechboy.com, n.d.) provide coding referrals.

Table 6. Technical data of ADXL345 accelerometer (Analog Devices Inc. and Inc., 2008)

Sn	Description	Values
1	Sensitivity range	+/- 2g to +/- 16g
2	Input voltage supported	5V / 3.3V
3	Output data rate (Measurement rate)	0.1 to 3200 Hz
4	Number of pins	8
5	Axes	3 (X-axis, Y-axis and Z-axis). X and Y axis are in plane of the board. Z axis is perpendicular to the plane of the board.
6	Pin configuration and description	GND, VCC, CS, INT1, INT2, SDO, SDA, SCL GND to ground, VCC to power supply, CS to chip select, INT 1: interrupt, INT2: interrupt, SDO: serial data output, SDA: serial data (I2C), SCL: serial communications clock

Note. In this study, we will use the Z axis to measure vertical acceleration.

Figure 4. ADXL354 accelerometer

Table 7. Technical data of micro-SD card reader module

Sn	Description	Values
1	Card type & capacity	Micro sd card (<=2GB) and micro sdhc card (<=32GB)
2	Power Supply or Input Supply Voltage (VDC)	4.5~5 Volts
3	Voltage regulator	3.3 Volts
4	Communication Protocol	SPI
5	Number of Control Interface or Pins	6
6	Control Interface	GND, VCC, MISO, MOSI, SCK, CS GND to ground, VCC is the power supply, MISO, MOSI, SCK is the SPI bus, CS is the chip select signal pin
7	PCB Size (L x W) mm	40 x 24

Figure 5. Micro SD card reader module
Courtesy: www.robu.in

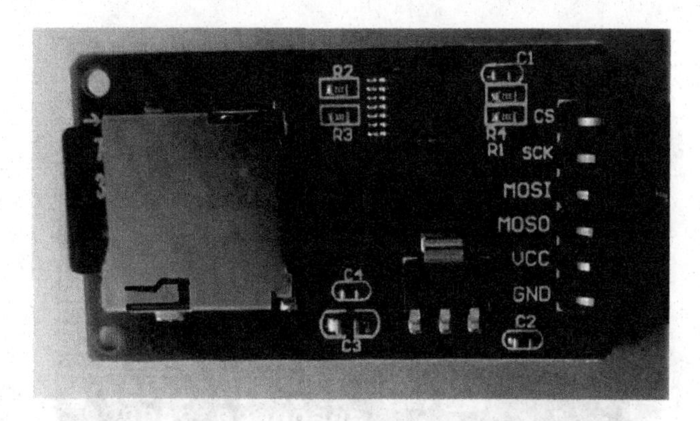

Table 8. Technical data of Arduino uno

Sn	Description	Values
1	Input Voltage (Recommended)	7-12V
2	Analog I/O Pins	6
3	Digital I/O Pins	14 (of which 6 provide PWM output)
4	PWM Digital I/O Pins	6
5	Clock Speed	16 MHz
6	Flash Memory	32 KB
7	SRAM (KB)	2
8	EEPROM	1 KB (ATmega328P)
9	DC Current for 3.3V Pin (mA)	50
10	DC Current per I/O Pin (mA)	20
11	On Board LEDs	Yes
12	Operating Temperature (C)	-10 to 60
13	Weight (gm)	25
14	Length (mm)	68.6
15	Width (mm)	53.4
16	Height (mm)	12.5

Data Acquisition

An asphalt road consisting of average roughness was selected as shown in Figure 8. The prototype data acquisition device was installed on the vehicle body. The accelerometer sensor of the data acquisition device is deployed at the co-driver seat mounting hard point as shown in Figures 9. The Z axis of the accelerometer is ensured to be perpendicular to the plane of the road surface. A road test was taken at vehicle speed 30 Kilometres per hour for 3-5 minutes duration. Vertical acceleration data was acquired

Figure 6. Arduino Uno
Courtesy: www.electronicscomp.com

Table 9. Pinout connection for Arduino uno and micro SD card reader

Micro SD Card Reader Pin	Arduino Uno Pin	Remarks
+5	5V	Power Supply
CS	4	Chip Select
MOSI	11	SPI Data
SCK	13	clock
MOSO	12	SPI Data
GND	GND	Ground

Table 10. Pinout connection for Arduino uno and ADXL 345 accelerometer

Arduino	ADXL345
A4 PIN(SDA)	SDA PIN
A5 PIN(SCL)	SCL PIN
GND PIN	GND
5V	VCC

during the test drive at a sampling rate of 2 Hertz (Hz). The test was monitored on a laptop by a co-driver using Arduino IDE serial monitor.

Data Processing and Results

The dataset collected in this study has 384 data points at a sampling rate of 2Hz. Higher sampling rates may also be selected by the designer, such as 5Hz, 10Hz etc. The accelerometer has a maximum capacity of 3200Hz. The total time duration of the test = number of data points/sampling rate

Figure 7. Writing diagram of data acquisition circuit preparation of data acquisition device

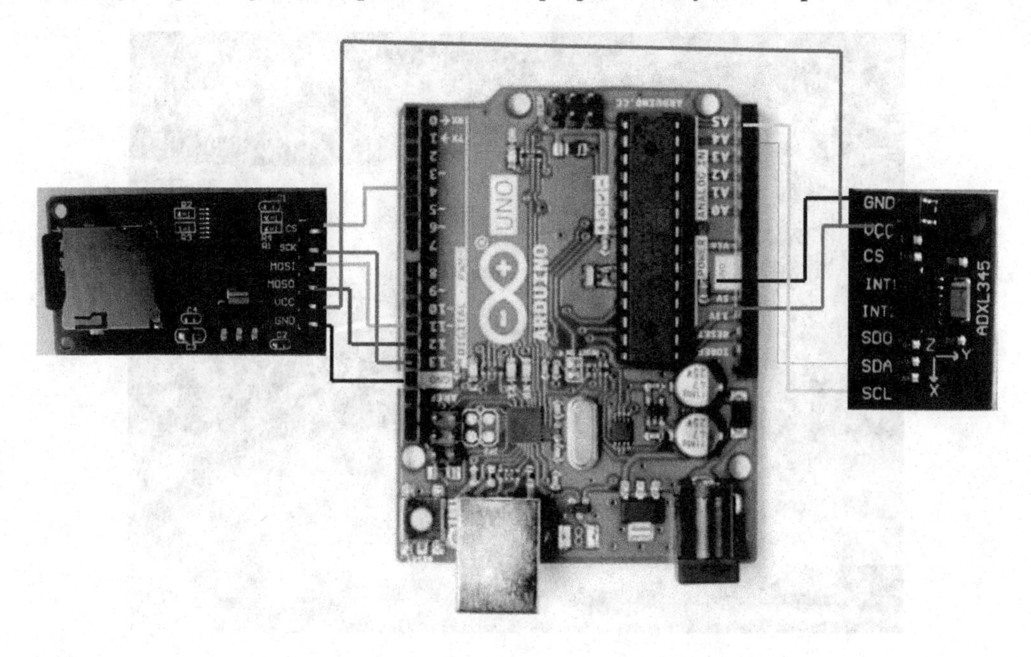

Figure 8. Road surface

= 384 / 2

= 192 seconds

= 3 minutes, 12 seconds

The dataset collected in this study can be downloaded from GitHub link https://github.com/rajaram-pai-kuchelkar/daq.git. File name: roadtest_24.05.2022.csv .

Figure 9. Deployment of data acquisition at vehicle seat mounting

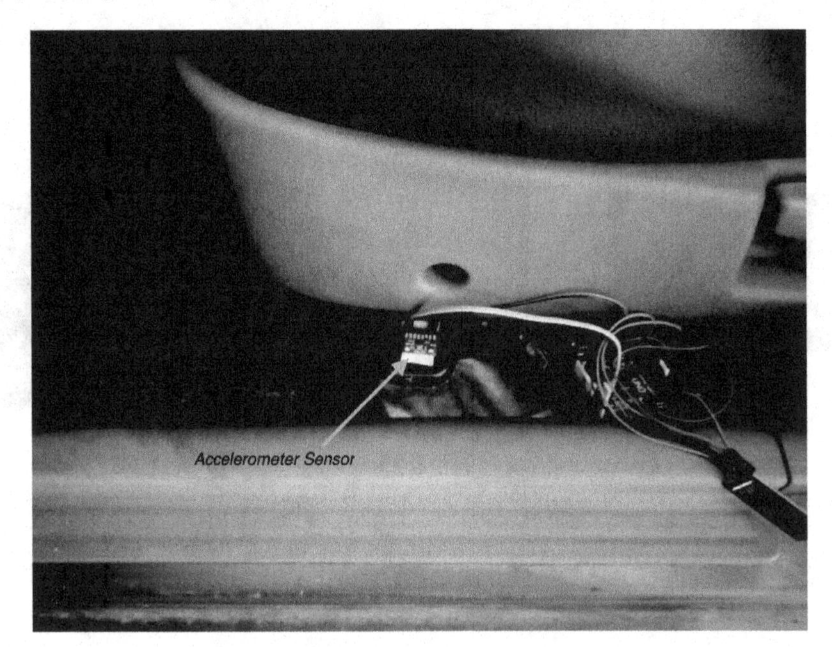

The redundant data at the beginning is removed and then converted in terms of g by dividing each value by 9.81. The converted data is shown in the file "data.txt" available via the GitHub link https://github.com/rajaram-pai-kuchelkar/daq.git. The graphical plot of the converted dataset against data points is shown in Figure 10 as below. The y-axis shows acceleration values in terms of g (gravitational constant) and the x-axis shows the data points.

Data Classification

The converted data file is classified in the form of bands of accelerations, with step size taken as 0.25g as shown in Table 11 below. A histogram is plotted as shown in Figure 11.

Data Sorting

The classified data is sorted based on count values field in descending order as shown in Table 12.

Figure 10. Dataset plot of road test

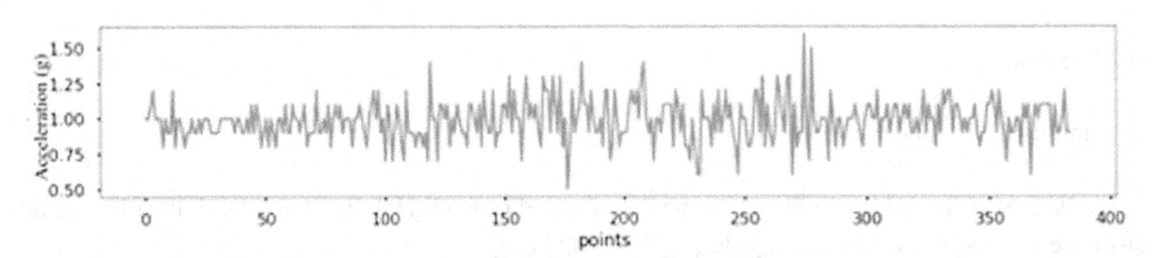

Table 11. Test sample classification of data

Acceleration Range (g)		count
lower limit	upper limit	
0	0.25	0
0.25	0.5	0
0.5	0.75	13
0.75	1	142
1	1.25	201
1.25	1.5	25
1.5	1.75	3
1.75	2	0

Figure 11. Histogram plot of test sample classification

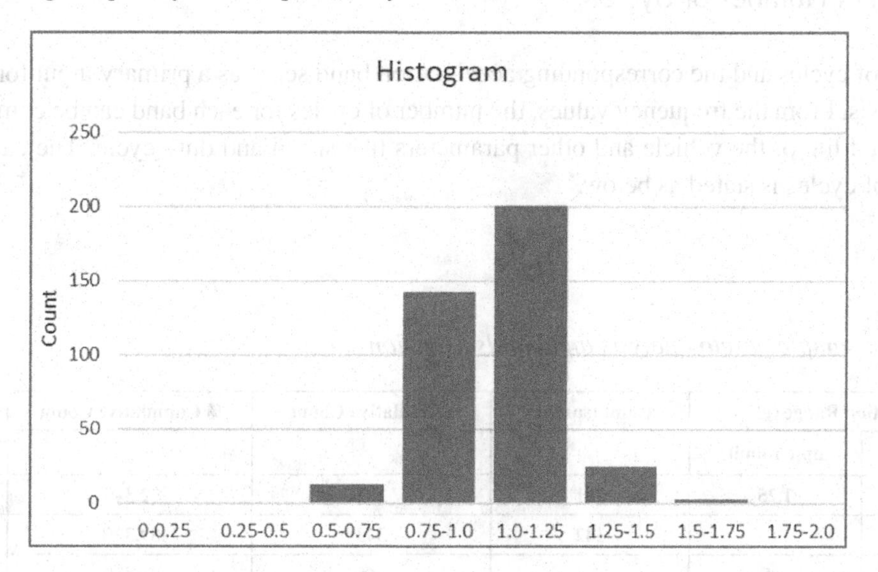

Table 12. Test sample sorted data

Acceleration Range		count (sorted)
lower limit	upper limit	
1	1.25	201
0.75	1	142
1.25	1.5	25
0.5	0.75	13
1.5	1.75	3
0	0.25	0
0.25	0.5	0
1.75	2	0

Pareto Analysis and Selection of Bands

The cumulative count and percentage cumulative count is determined as shown in Table 13. Pareto rule 80:20 is applied and the percentages up to 80% are selected.

Calculation of Frequency

The frequency is determined by dividing the number of occurrences (count) for each acceleration band by the total duration in seconds given by the below formula and shown in Table 14.

n is the total number of points
s is the sampling rate in Hz
o_i is the occurrence for i^{th} band
f is the frequency in Hz for i^{th} band, f = oxs/n

Calculation of Number of Cycles

The number of cycles and the corresponding acceleration band serve as a primary input for conducting fatigue analysis. From the frequency values, the number of cycles for each band can be computed based on the required life of the vehicle and other parameters like speed and duty cycle. The calculation for the number of cycles is stated as below.

Table 13. Test sample pareto analysis and bands selection

Acceleration Range (g)		count (sorted)	Cumulative Count	% Cumulative Count	Remarks
low limit	upper limit				
1	**1.25**	**201**	**201**	**52.34%**	**selected**
0.75	**1**	**142**	**343**	**89.32%**	**selected**
1.25	1.5	25	368	95.83%	discarded
0.5	0.75	13	381	99.22%	discarded
1.5	1.75	3	384	100.00%	discarded
0	0.25	0	384	100.00%	discarded
0.25	0.5	0	384	100.00%	discarded
1.75	2	0	384	100.00%	discarded

Table 14. Test sample frequency determination

Acceleration Range (g)		Count (o) (sorted)	Frequency (f) Hz
lower limit	upper limit		
1	1.25	201	1.05
0.75	1	142	0.74

f is the frequency of occurrence of an acceleration in a particular band in Hz.
D is the duty cycle in kilometres (kms) per day.
v is the average speed in kms per hour.
L is the life in years.
N is number of cycles, calculated as per below formula: N = (fxLxDx1.314e+6)/v

Where, the value 1.314e+6 is obtained by converting speed v in kms per second and life L in days i.e. 365x60x60 = 1.314e+6.

Considering, L = 15 years, D = 200 kms / day, v = 50 kms/ hour

The number of cycles calculated for sample data is shown in Table 15.

Results

From Table 15, the average acceleration range and number of cycles forms the input test cases for fatigue analysis as shown in Table 16.

SUMMARY

The findings of the research study showed that inexpensive Arduino-based IoT devices were the most recent breakthroughs in tools and approaches for data acquisition. The expense of proprietary data acquisition devices was the primary reason for this research study on utilising low-cost IoT for data acquisition. In this study, the authors constructed a prototype data acquisition device using open-source software, Arduino IDE, and low-cost IoT hardware, comprising Arduino Uno, an accelerometer ADXL345 and a MicroSD Card reader. The cost of components was within INR 2000. The prototype data acquisition device was used to collect real-world data of a vehicle driving through a road test. The vehicle-road

Table 15. Test sample number of cycles

Acceleration Range (g)		Value (sorted)	Frequency Hz	No. of Cycles
lower limit	upper limit			
1	1.25	201	1.05	8.25E+07
0.75	1	142	0.74	5.83E+07

Table 16. Results: Test cases for fatigue analysis

Test Case	Acceleration (g)	No of Cycles
1	1.125	8.25E+07
2	0.875	5.83E+07

response was acquired and analysed using dataset classification, sorting, pareto technique, and formulae to get the number of cycles for fatigue analysis.

REFERENCES

Agrawal, S., Kumar, V., Anand, N., Agarwal, V. K., & Islam, A. (2016). *Development of Data Acquisition System and Data Analysis Technique for Automotive Applications. 50*(Ic), 3–6.

Amestica, O. E., Melin, P. E., Duran-Faundez, C. R., & Lagos, G. R. (2019). An Experimental Comparison of Arduino IDE Compatible Platforms for Digital Control and Data Acquisition Applications. *IEEE CHILEAN Conference on Electrical, Electronics Engineering, Information and Communication Technologies, CHILECON 2019*, (pp. 1–6). IEEE. 10.1109/CHILECON47746.2019.8986865

Baoquan, G. (2011). *Time-domain Analysis and Research on blasting vibration signals Based on Fourier I* °. Research Gate.

Chellaswamy, C., Famitha, H., Anusuya, T., & Amirthavarshini, S. B. (2018). *IoT Based Humps and Pothole Detection on Roads and Information Sharing*. Research Gate.

Chen, E. N. (2009). Remote analysis of mechanical vibration based on client/server architecture. *ICEMI 2009 - Proceedings of 9th International Conference on Electronic Measurement and Instruments*, (pp. 10–13). IEEE. 10.1109/ICEMI.2009.5274399

Dongpo, L., & Xuhui, M. (2011). *Virtual Proving Ground.*

Du, Y., Liu, C., Wu, D., & Li, S. (2016). Application of Vehicle Mounted Accelerometers to Measure Pavement Roughness. *International Journal of Distributed Sensor Networks, 2016*(6), 8413146. doi:10.1155/2016/8413146

González, A., Olazagoitia, J. L., & Vinolas, J. (2018). A low-cost data acquisition system for automobile dynamics applications. *Sensors (Basel), 18*(2), 366. doi:10.3390/s18020366 PMID:29382039

Gonzalez, L. C., Moreno, R., Escalante, H. J., Martinez, F., & Carlos, M. R. (2017). Learning Roadway Surface Disruption Patterns Using the Bag of Words Representation. *IEEE Transactions on Intelligent Transportation Systems, 18*(11), 2916–2928. doi:10.1109/TITS.2017.2662483

Hu, F., He, Q., & Kong, F. (2011).. . *Time-Frequency Vibration Representation for Steel Mill Condition Monitoring., 1*, 1–5.

Ilic, S., Katupitiya, J., & Tordon, M. (n.d.). *In-vehicle data logging system for fatigue analysis of drive shaft.*

Jamaludin, W. A. W. (2021). *Photovoltaic-Thermoelectric Generator Monitoring System using Arduino Based Data Acquisition system Technique.*

Junnila, S., & Niittylahti, J. (2003). Wireless technologies for data acquisition systems. *Proceedings of the 1st International Symposium on Information and Communication Technologies*, (pp. 132–137). IEEE.

LastMinuteEngineers.com. (2022). *Interfacing Micro SD Card Module with Arduino*. Last Minute Engineers.

Men, X., Liu, H., Chen, N., & Li, F. (2018). *A new time domain filtering method for calculating the RMS value of vibration signals. 51605191*, 2016–2019.

Mo, K., Suh, K., & Hong, S. (2000). New Approach in Vehicle Durability Evaluation, Virtual Proving Ground. *Test*, (March), 1–5.

Mohamed, A., Fouad, M. M. M., & Elhariri, E. (2014). *RoadMonitor: An Intelligent Road Surface Condition Monitoring System RoadMonitor: An Intelligent Road Surface Condition Monitoring System*. Springer. doi:10.1007/978-3-319-11310-4

Monterrey, C. (2018). *Road Load Data Acquisition system with SAE-J1939 Communications Network: Integration and Laboratory Test*. [Thesis, Instituto Tecnológico y de Estudios Superiores de Monterrey].

Moreno, C., González, A., Olazagoitia, J. L., & Vinolas, J. (2020). The acquisition rate and soundness of a low-cost data acquisition system (LC-DAQ) for high frequency applications. *Sensors (Basel)*, *20*(2), 524. doi:10.3390/s20020524 PMID:31963552

Of, A., & Signal, V. (2013). *A LabVIEW BASED DATA ACQUISITION SYSTEM FOR MONITORING Sunita Mohanta Department of Electronics and Communication Engineering National Institute of Technology Rourkela, Odisha-769008 Sunita Mohanta Dr. Umesh Chandra Pati Department of Electronics and Co*.

Ogunoiki, A. O. (2015). *University of Birmingham Research Archive*. University of Birmingham.

Pal Singh, H., & Singh, R. (2021). Low Cost Data Acquisition System for Road-Vehicle Interaction Using Arduino Board. *Journal of Physics: Conference Series*, *1831*(1), 012031. doi:10.1088/1742-6596/1831/1/012031

Perttunen, M., Mazhelis, O., Cong, F., Ristaniemi, T., & Riekki, J. (2011). Distributed Road Surface Condition Monitoring. Lecture Notes in Computer Science. Springer. doi:10.1007/978-3-642-23641-9

PiMyLife. (2022). *Arduino Accelerometer using the ADXL345*.

Polat, F. (2017). *Journal of Engineering Research and Applied Science. Dece*.

Salokhe, N., Thakre, P., Awale, R. N., & Kambale, S. (2016). *Vibration Based Damage Detection using Overall Frequency Response and Time Domain*.

Sener, A. S. (n.d.). *Determination Of Vehicle Components Fatigue Life Based On Fea Method And Experimental Analysis Determination Of Vehicle Components Fatigue Life Based On Fea Method And*. *2*(Lcv), 133–146.

Shafiullah, A. K. M., & Wu, C. Q. (2013). Generation and validation of loading profiles for highly accelerated durability tests of ground vehicle components. *Engineering Failure Analysis*, *33*, 1–16. doi:10.1016/j.engfailanal.2013.04.008

Shuai, J., Heng, S., Jing, Y., Feng, Y., Han, N., & Mengqin, Y. (2021). *Load Spectrum Acquisition, Analysis and Application of The Electric Bus in Road Simulation Test Based on*. 479–483. toptechboy.com. (n.d.). *ARDUINO LESSON 21: LOG SENSOR DATA TO AN SD CARD.*

Ur Rehman, S., Mustafa, H., & Larik, A. R. (2021). IoT Based Substation Monitoring Control System Using Arduino with Data Logging. *Proceedings - 2021 IEEE 4th International Conference on Computing and Information Sciences, ICCIS 2021*. IEEE. 10.1109/ICCIS54243.2021.9676384

Wali, S., & Areeb, M. (2018). Development of Low-Cost DAQ for Power System Signals Using Arduino. *2018 IEEE 21st International Multi-Topic Conference (INMIC)*, (pp. 1–5). IEEE. 10.1109/INMIC.2018.8595519

Wang, J., Zhao, Y., Yang, Y., & Yang, J. (2019). *Fatigue Analysis for Bogie Frame of Urban Transit Rail Vehicle under Overload Situation. Qr2mse*. Research Gate.

Wang, Q., Yan, K., & Li, H. (2009). *Motor Noise Source Identification Based on Frequency Domain Analysis*. Research Gate.

Chapter 6
Design and Fabrication of a Softrobotic Gripper for Involving Underwater Vehicles in Seaweed Farming

Prabhakar Gunasekaran
Thiagarajar College of Engineering, India

Meenakshi S.
Thiagarajar College of Engineering, India

Jainulafdeen A.
K. Ramakrishnan College of Engineering, India

Ayyanar N.
https://orcid.org/0000-0002-4452-5293
Thiagarajar College of Engineering, India

Rajalakshmi Murugesan
https://orcid.org/0000-0001-8532-2452
Thiagarajar College of Engineering, India

ABSTRACT

Seaweeds, crucial components of marine ecosystems, thrive in marine and coastal waters, notably in the Gulf of Mannar Biosphere. Despite their ecological importance, certain seaweed species pose a threat to coral reefs due to the release of hydrophobic allelochemicals. Fisherwomen, risking their lives in collecting seaweed seed stock from littoral zones, rocky shores, and deep-sea waters, encounter hazards such as sharp rocks, poisonous algae, cyanobacteria, and coral reefs. Societal research pinpoints risks, overexploitation, unorganized harvesting, and algal blooms as significant issues. The proposed solution involves integrating intelligent Soft Robotic Grippers into underwater vehicles for precise seaweed farming without harming coral reefs. Abacus FEA software aids in deformation analysis, guiding the gripper's design to safeguard fisherwomen and preserve coral reefs.

DOI: 10.4018/979-8-3693-1962-8.ch006

INTRODUCTION

Underwater robots has advanced significantly, opening up new avenues for industrial applications, research, and exploration in aquatic environments. Rigid grippers are widely used in traditional underwater robotic systems. Although they work well in some situations, they have significant drawbacks. Underwater environments present significant challenges for rigid grippers due to their severe conditions, which include changing pressure, unexpected currents, and complex topography (Wei et al., 2016). A study conducted in (Bao et al., 2020; Fischell et al., 1806; Stenius et al., 2022) looked at the opportunities and difficulties of using autonomous underwater vehicles (AUVs) for seaweed farming. Prior research has brought to light the challenges associated with implementing autonomous systems in maritime settings, especially when it comes to tasks like navigation and farm localization. Scholars have underscored the significance of creating resilient technologies to tackle these obstacles and maximize AUV functioning in seaweed farms. Primary focus areas in the literature are sidescan sonar data collection, dead-reckoning navigation techniques, and early farm localization methods. The development and application of AUV systems for efficient seaweed farming techniques has benefited greatly from the insightful information these studies have offered. These obstacles make it harder to manage delicate and accurate interactions with marine creatures, handle breakable items, and maneuver in small areas. Furthermore, because of their limited adaptability and chance of unexpected accidents, inflexible grippers may unintentionally disrupt the surrounding marine habitat (Hughes et al., 2016). These soft grippers present a viable path for developing the capabilities of underwater robotic systems because they provide improved compliance, adaptability, and the capacity to negotiate challenging underwater terrains more effectively (Soft Robotics Toolkit, n.d.). Soft grippers' increased versatility makes them ideal for a variety of applications, including environmental monitoring, marine exploration, and conservation. It also allows for effective navigation through intricate underwater environments. The effective use of soft robotic grippers as we explore the ocean's depths is a sign of our dedication to comprehending and protecting the delicate balance of underwater ecosystems, in addition to being a technological accomplishment (Chen & Xu, 2023). The structural behavior of soft robotic grippers is evaluated with the use of Finite Element Analysis (FEA). This simulation method helps with gripper design optimization by enabling a thorough analysis of deformations, stresses, and performance under varied circumstances (Huang, 2021). To ensure the effectiveness and dependability of soft robotic grippers in their intended applications, FEM plays a critical role in bridging theoretical design concepts with practical implementation (Nordin et al., 2013; Soft Robotics Toolkit, n.d.; Xavier et al., 2021).

PROBLEM STATEMENT

Red seaweed called Kappaphycus alvarezii is also referred to as "cottonii" or "cottonii seaweed as shown in figure 1(c). " It is widely grown for commercial purposes and is a member of the Solieriaceae family. Carrageenan is a type of hydrocolloid that is used in many different industries. Seaweeds are macroscopic algae growing in the marine and shallow coastal waters and on rocky shores. It plays a major role in marine ecosystems and abundantly available in the region of Gulf of Mannar Biosphere. However, some species of seaweed clash coral reefs and damage them severely due to the release of hydrophobic allelochemicals. Therefore, the ecological balance is required in seaweed growth, otherwise it results in algal bloom, which is shown in figure 1(e). This factor induces the Indian government to promote the seaweed cultivation to a greater extent. Over the years, fisherwomen in Ramanathapuram district of Tamilnadu are engaged in collecting the Seed stock of seaweeds from littoral zones, rocky shores and deep-sea waters for economic gain. It endangers their lives due to presence of sharp rocks, poisonous algae's, cyanobacteria, coral reefs which is shown in figure 1(f) and other species in the ocean. They are staying in the seabed lined with slippery coral for more than 7 hours to search and collect the considerable number of seaweeds without any proper safety materials.

The risks associated with seaweed collection, human seeking effort, overexploitation of sea resources (continuous, indiscriminate, and unorganized harvesting of seaweeds), and algal bloom have been identified as significant problems by this societal research. The figure 1(d) depicts fisherwomen engaged in the traditional method of seaweed harvesting. In the illustration, fisherwomen are shown wading in shallow waters, meticulously handpicking seaweed from underwater vegetation. Traditional harvesting techniques are labor-intensive because they frequently require manual labor. Harvesting can be labor-intensive and require specialized personnel. The underwater robot currently employing the Newton Sea Gripper, as depicted in Figure 1a and 1b is specifically designed for seaweed farming, representing a rigid approach. When using hard grippers or any equipment in delicate marine environments, there is a serious risk of damaging coral reefs. Coral reefs are vulnerable to physical harm because they are delicate ecosystems that offer vital habitats for a wide variety of marine species. Current R&D projects are investigating novel farming methods, cutting-edge monitoring systems, and environmentally friendly seaweed farming techniques in an attempt to solve some of these problems. Resilience and efficiency in seaweed farming systems can be increased by taking a more comprehensive and technologically-aware approach. Underwater operations can be carried out with less risk to coral reefs and other delicate marine ecosystems by combining these factors with developments in soft robotics and responsible operational procedures. Underwater robotics must prioritize environmental conservation if our oceans are to remain healthy in the long run.

An important step toward environmental sustainability in the seaweed farming sector has been taken with the introduction of soft robotic grippers. These grippers minimize disturbances to fragile marine habitats, like coral reefs, while preserving biodiversity and reducing environmental risks because they provide a gentler alternative to traditional harvesting methods. Soft robotic grippers ensure the long-term health and resilience of marine ecosystems by utilizing sustainable harvesting practices. This helps the ecosystems adapt to climate change. By giving ecological factors top priority, this proactive strategy promotes a more sustainable future for both coastal communities and marine life.

The Indian government took many initiatives to enhance the livelihood of fisherwomen.

Among them, seaweed cultivation is one of the initiatives to gain marginal income. Coastal fisher families, especially fisherwomen in Self Help Groups (SHG) are actively involved in collecting the seaweeds from the deep-sea waters and rocky shores. It seems to be risking their lives for occupation. Therefore, the difficulties are faced by the fisherwomen in coastal areas of Tamilnadu, where the seaweeds are abundantly available. So, it is important to save the lives of fisherwomen involved in seaweed collection.

Figure 1. a. Underwater robot, b. Hard gripper, c. Kappaphycus alvarezii (seaweed), d. seaweed farming, e. Algal bloom f. Coral reef

ACTUATION TECHNOLOGIES FOR SOFT ROBOTIC GRIPPERS AND MANIPULATORS

Pneumatic Actuation

Pneumatic systems and compressed air enable controlled and gentle manipulation in an unpressurized soft robotic gripper intended for seaweed grabbing. Because it is made of materials that are both flexible and water-resistant, the gripper can flex and conform to the irregular shapes of seaweed, deforming compliantly under pressure and fitting the curves of the underwater environment. Using soft pneumatic actuators, like bellows or chambers, the soft robotic gripper gently and non-intrusively grasps seaweed without causing damage. The actuators expand or contract in response to air pressure. Control valves precisely adjust the grip force by controlling the flow of compressed air. A pressure regulation system keeps the pressure at just the right levels to allow for efficient seaweed harvesting without endangering

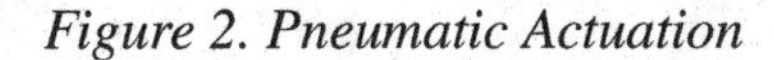

Figure 2. Pneumatic Actuation

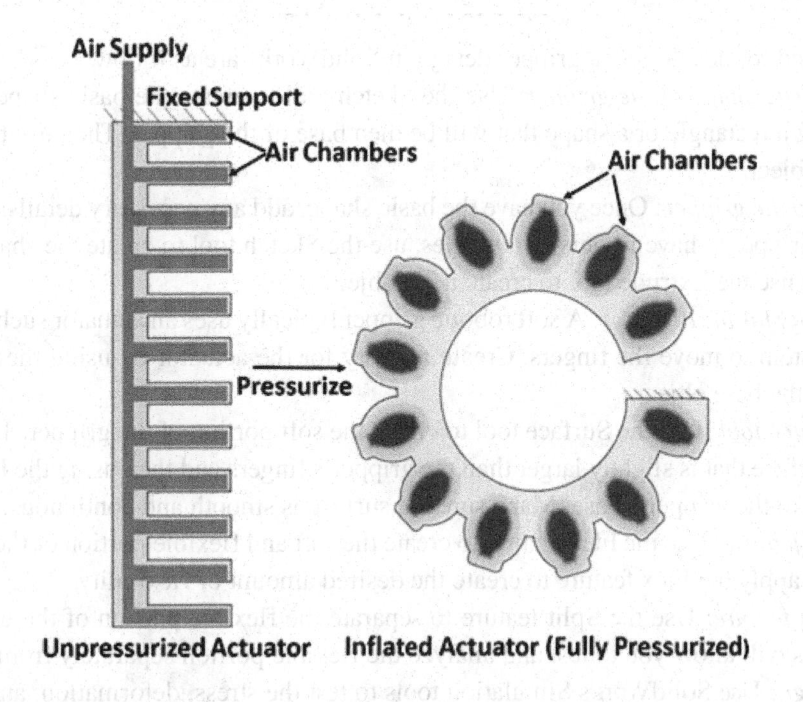

the delicate marine flora. In underwater applications, safety features such as pressure relief valves add to the dependability of the gripper.

A control unit incorporates automation into the soft robotic gripper, coordinating the deflation and inflation cycles to manipulate seaweed with grace. This technology is especially useful for jobs requiring a soft and adaptable grip to prevent harming fragile marine ecosystems, such as environmental monitoring or underwater harvesting. The ability of soft robotics to facilitate sustainable and non-intrusive interactions with marine environments is exemplified by the seaweed-grabbing gripper. The soft robotic gripper for seaweed grabbing operates through a carefully designed process, as illustrated in the accompanying Figure 2.

Hydraulic Actuation

In soft robotics, hydraulic actuation is a ground-breaking method that offers robotic systems unparalleled accuracy, adherence, and flexibility (Chen & Xu, 2023). This technology can be used to create soft robotic structures that can easily interact with delicate objects, navigate through complex environments, and dynamically adjust their stiffness for optimal performance. Because of its high force transmission, modular design, and capacity for complex deformations, hydraulic actuation is an essential element in the development of bioinspired and flexible soft robotic solutions. Applications for hydraulic actuation can be found in many different domains, such as enhancing dexterity in manipulating objects and ensuring safety in human-robot interactions. This emphasizes how important hydraulic actuation will be in determining how soft robotics develops in the future.

SOLIDWORKS DESIGN

The steps outlined for developing a gripper design in SolidWorks are as follows:

Sketch the basic shape of the gripper: Use the Sketch tool to create the basic shape of the gripper. Start by drawing a rectangle or a shape that will be then base of the gripper. Then use the Extrude tool to create a 3D object.

Add details to the gripper: Once you have the basic shape, add any necessary details. For example, if you want your gripper to have fingers or tentacles, use the Sketch tool to create the shape of the finger or tentacle, then use the Extrude tool to create a 3D object.

Create a cavity for the actuator: A soft robotic gripper typically uses an actuator such as a pneumatic or hydraulic system to move the fingers. Create a cavity for the actuator by using the Extrude tool to create a void in the base object.

Use the Surface tool: Use the Surface tool to create the soft portion of the gripper. This can be done by creating a surface that is slightly larger than the gripper's fingers and then using the Loft tool to connect the surface to the gripper's base. Make sure the surface is smooth and continuous.

Use the Flex feature: Use the Flex feature to create the soft and flexible portion of the gripper. Select the Surface and apply the Flex feature to create the desired amount of flexibility.

Use the Split feature: Use the Split feature to separate the flexible portion of the gripper from the base object. This will allow you to test and analyze the flexible portion separately from the base.

Test the design: Use SolidWorks Simulation tools to test the stress, deformation, and motion of the flexible portion of the gripper. You can apply various loads and boundary conditions to test the gripper's behavior.

Make any necessary changes: Based on the results of your testing, make any 21 necessary changes to the design until you are satisfied with the functionality of the gripper.

Finalize the design: Once you have made all necessary changes, finalize the design and save it as a 3D file that can be used for manufacturing.

Figure 2a, 2b,2c and 2d displays the developed gripper model in SolidWorks with different design views. These illustrations provide a thorough examination of the gripper's design and attributes from several perspectives.

FINITE ELEMENT MODELING

Finite element modeling simulation in Abaqus involves creating a digital model of a physical structure or system, dividing it into small, finite elements, and then applying boundary and loading conditions to simulate real-world scenarios. The software then uses numerical methods to solve the equations governing the behavior of the system, and provides information on stress, deformation, temperature distribution, fluid flow, and other variables of interest. The following steps are used to simulate the gripper in Abaqus using finite element modelling:

Create the model: Import the 3D model of the soft robotic gripper that was designed in SolidWorks into Abaqus. Make sure the model is properly meshed.

Create a step: In Abaqus, create a new step and name it something like "Pressure Load." This step will be used to apply the pressure to the gripper.

Figure 3. 3D view

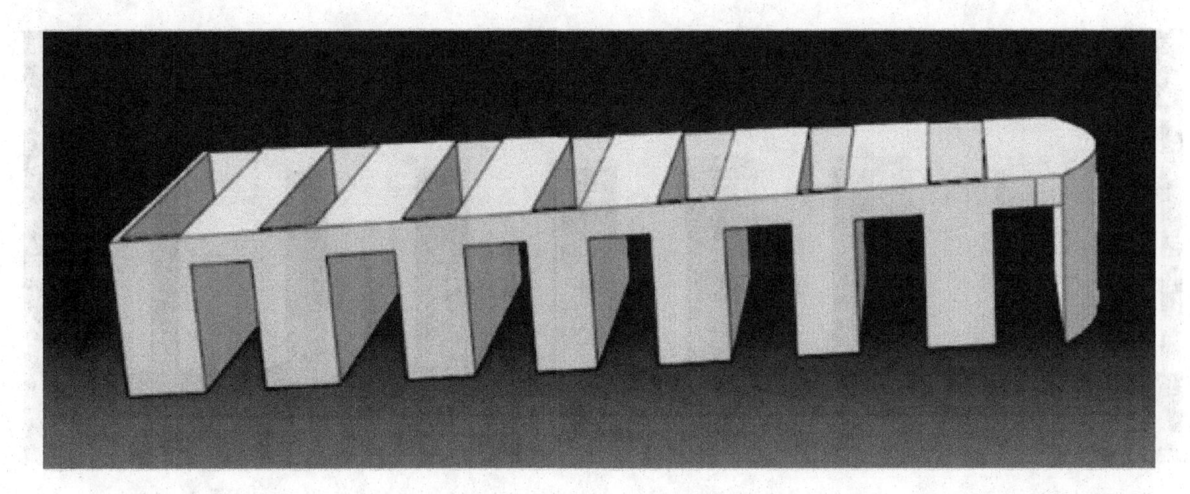

Figure 4. Top view

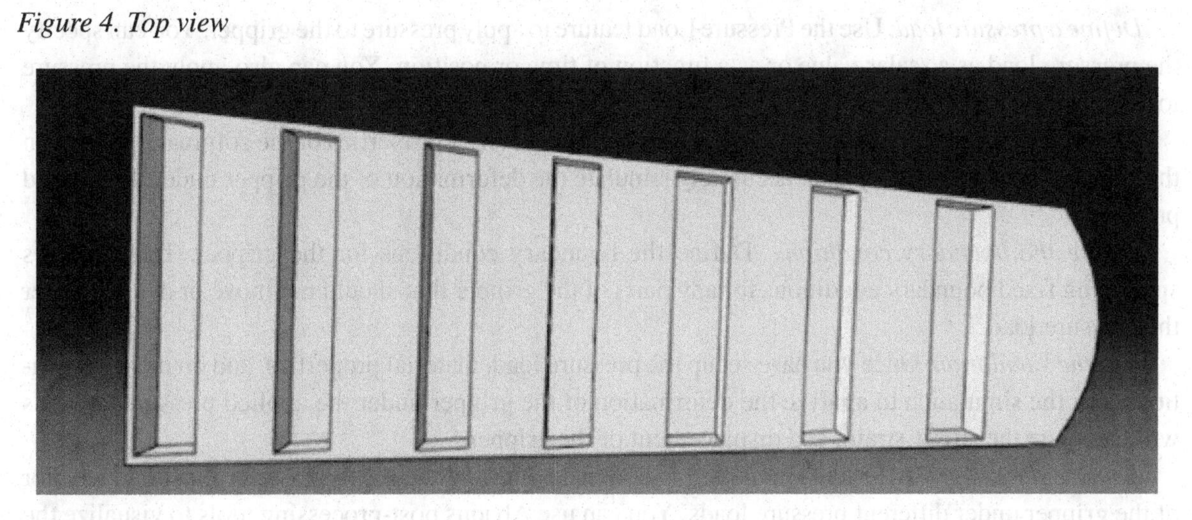

Figure 5. Bottom view

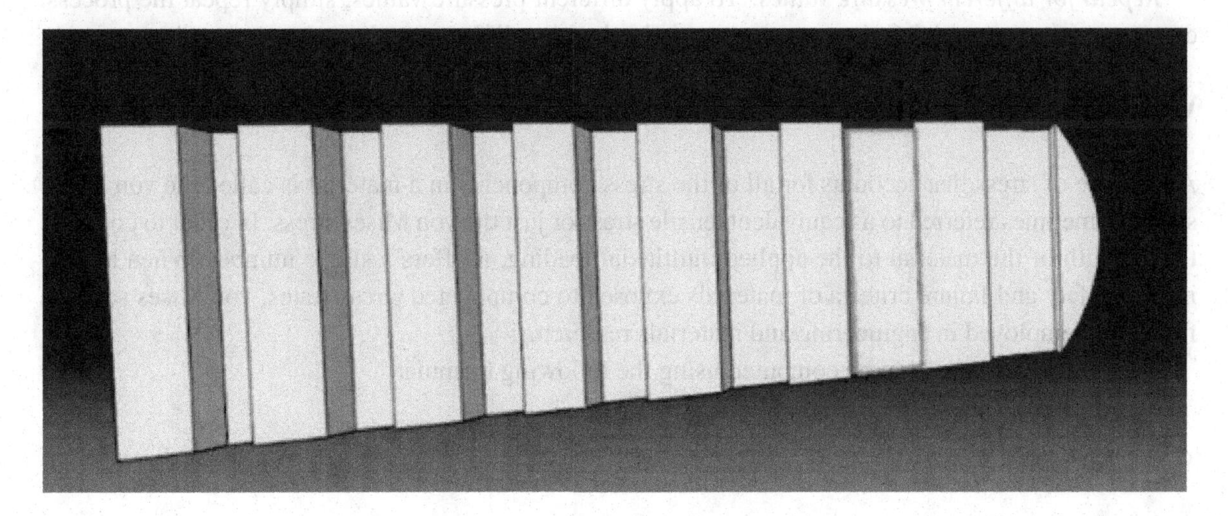

Figure 6. Side view

Define a pressure load: Use the Pressure Load feature to apply pressure to the gripper. You can specify the pressure load as a scalar value or as a function of time or position. You can also apply the pressure load to a specific set of nodes or to a region of the gripper.

Define the material properties: In Abaqus, define the material properties of the soft material used in the gripper. This will allow you to accurately simulate the deformation of the gripper under the applied pressure.

Define the boundary conditions: Define the boundary conditions for the gripper. This includes specifying fixed boundary conditions for any parts of the gripper that should not move or deform under the pressure load.

Run the simulation: Once you have set up the pressure load, material properties, and boundary conditions, run the simulation to analyze the deformation of the gripper under the applied pressure. Abaqus will calculate the stress, strain, and displacement of the gripper.

Analyze the results: After the simulation is complete, analyze the results to determine the behavior of the gripper under different pressure loads. You can use Abaqus post-processing tools to visualize the results, including deformation and stress contours.

Repeat for different pressure values: To apply different pressure values, simply repeat the process, creating a new step and defining a new pressure load with a different scalar value or function.

Von-Mises Stress Analysis

A measure of stress that accounts for all of the stress components in a material is called the von Mises stress, sometimes referred to as equivalent tensile stress or just the von Mises stress. In order to compare the strength of the material to the applied multiaxial loading, it offers a single number. When assessing the safety and failure criteria of materials exposed to complicated stress states, von Mises stress is frequently employed in engineering and materials research.

The von Mises stress can be computed using the following formula:

$$\sigma_{von-mises} = \sqrt{(\tilde{A}_x^2 + \tilde{A}_y^2 + \tilde{A}_z^2 - \sigma_x\sigma_y - \sigma_y\sigma_z - \sigma_z\sigma_x + 3\left(\tau_{xy}^2 + \tau_{yz}^2 + \tau_{zx}^2\right))}$$

where,

$\sigma v_{on\text{-}mises}$ is the von Mises stress.

σx, σy, and σz are the normal stresses in the x, y, and z directions, respectively.

τxY, τyz and τzx are the shear stresses on the xy, yz and zx planes, respectively.

MOULDING PREPARATION

Ecoflex Silicone 00-30mm is a type of two-part platinum-cure silicone rubber that is commonly used for mold making and casting applications. It has a Shore A hardness of 00, which means it is extremely soft and flexible, making it ideal for creating molds of delicate or intricate object.

De*sign the mould:* Create a 3D model of the mould using CAD software, such as SolidWorks or AutoCAD. The mould should be designed to the exact specifications of the desired soft robotic gripper, with the correct size, shape, and features.

C*hoose the moulding material:* Select a suitable material for the mould, such as silicone or polyurethane. The material should be able to withstand the heat and pressure of the moulding process without degrading or deforming.

Prepare the moulding surface: Prepare the surface of the mould by applying a release agent, such as silicone spray or wax, to prevent the moulding material from sticking to the surface.

Mix the moulding material: Mix the chosen moulding material according to the manufacturer's instructions. This typically involves mixing a base material with a curing agent in a specific ratio.

Pour the moulding material: Pour the mixed moulding material into the mould cavity, making sure to fill it completely and evenly. Cure the moulding material: Allow the moulding material to cure for the recommended amount of time. This may involve leaving it at room temperature or applying heat to accelerate the curing process.

Demould the soft robotic gripper: Once the moulding material is fully cured, demould the soft robotic gripper by carefully removing it from the mould. Be careful not to damage the gripper during the demoulding process. Trim and finish the soft robotic gripper: Trim any excess material from the gripper and finish the surface as desired. This may involve sanding, polishing, or applying a coating or paint.

RESULTS AND DISCUSSION

ABAQUS software was used to perform a Finite Element Analysis (FEA) to evaluate a modeled soft robotic gripper's performance at different pressure values. The 3D model considered the hyperelastic properties of soft materials, and a finely detailed mesh accurately captured deformation. Pneumatic actuation was simulated by applying pressure loads, and boundary conditions were designed to mimic actual operating conditions. The selection of nonlinear static or dynamic analyses was based on the magnitude of deformation. Post-processing provided information about pressure distribution, displacement, strain, and stress. The goal of the simulation was to gain insight into the gripper's behavior and deformation under various pressure conditions, which would aid in design optimization and practical application.

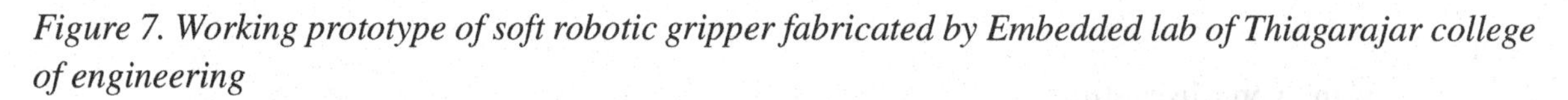

Figure 7. Working prototype of soft robotic gripper fabricated by Embedded lab of Thiagarajar college of engineering

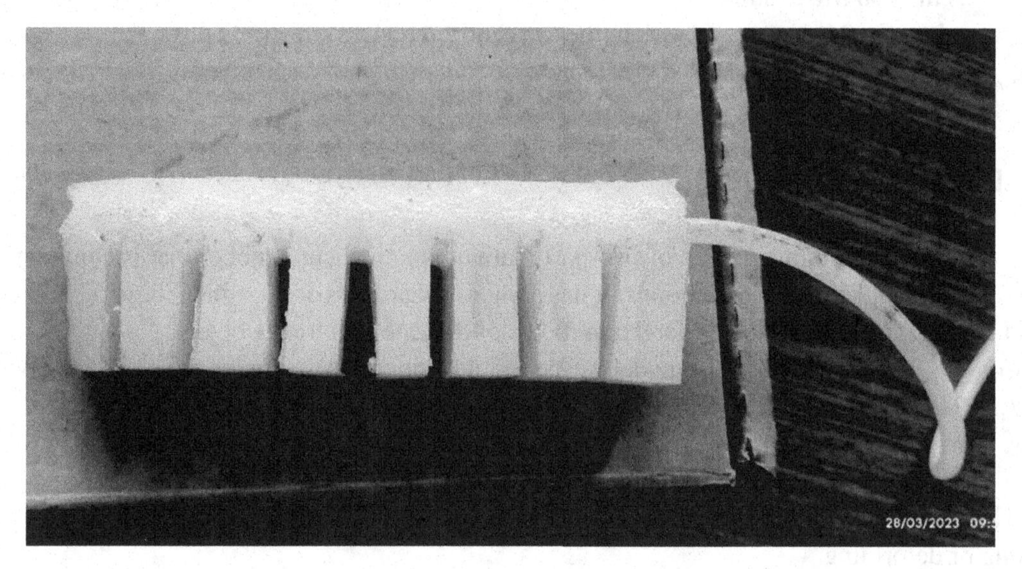

The von Mises stress distribution and gripper deformation were calculated for pressure values of 30 kPa, 40 kPa, and 50 kPa as illustrated in figures 5a, 5b, 5c and corresponding deformation magnitudes were tabulated in table1. The maximum magnitude of bending deformation of 57.4mm was found to occur at 50 kPa. This result suggests a relationship in which the gripper's ability to bend in the desired direction increases with pressure. The findings imply that manipulating the pressure levels may be a useful strategy for managing and enhancing the gripper's bending behavior in certain situations.

CONCLUSION

Soft robotic grippers have emerged as a promising technology for seaweed farming. Unlike traditional hard grippers, soft grippers can manipulate delicate and irregularly shaped seaweed without damaging it. This makes them ideal for tasks such as harvesting, sorting, and transporting seaweed. It can be designed to mimic the way that natural organisms manipulate objects. For example, researchers have developed grippers that use suction to attach to the seaweed, or that wrap around the seaweed like a vine. These grippers are perfect for use in underwater environments because they can be controlled remotely. Soft robotic grippers have the potential to revolutionize seaweed farming by making it more efficient, cost-effective, and environmentally friendly. They could enable farmers to harvest more seaweed with less labor, and to process the seaweed more quickly and efficiently. This could make seaweed farming more economically viable and help to meet the growing demand for sustainable food and fuel sources.

This initial study of the soft robotic gripper's bending behavior represents a significant step toward understanding its deformable properties and flexibility. The study's sophistication is highlighted by the use of ABAQUS software for Finite Element Analysis (FEA).

Figure 8. For 30KPa

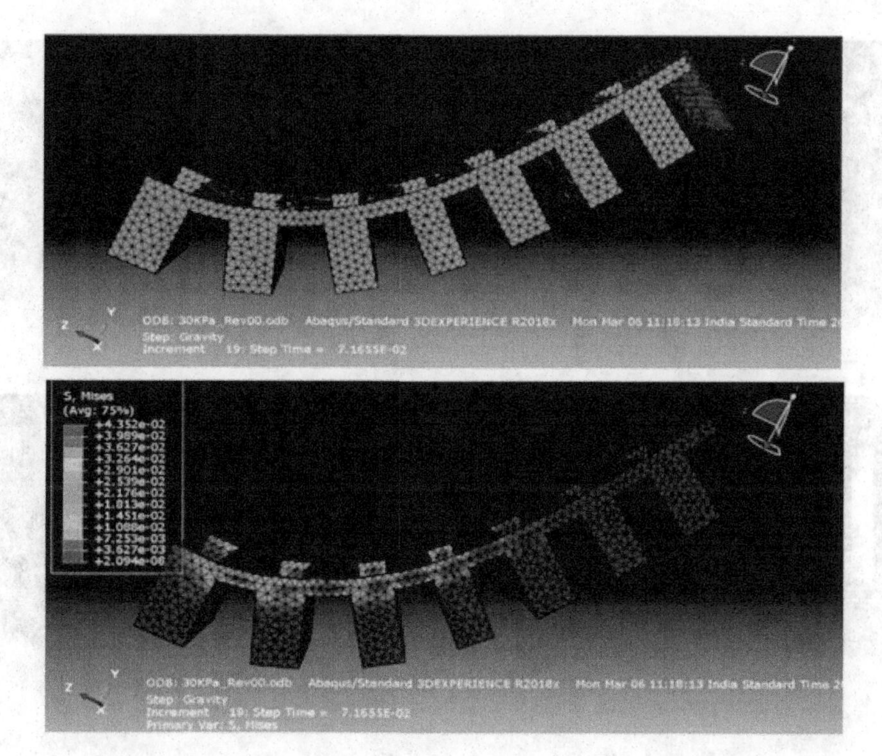

Figure 9. For 40KPa

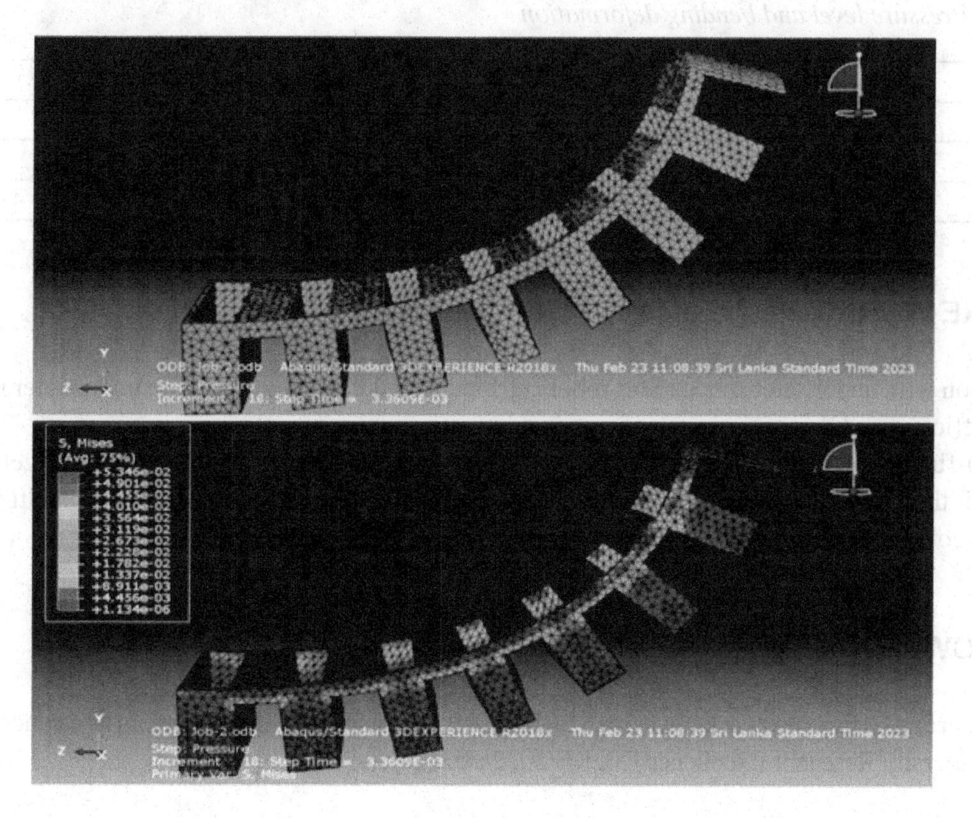

Figure 10. For 50KPa

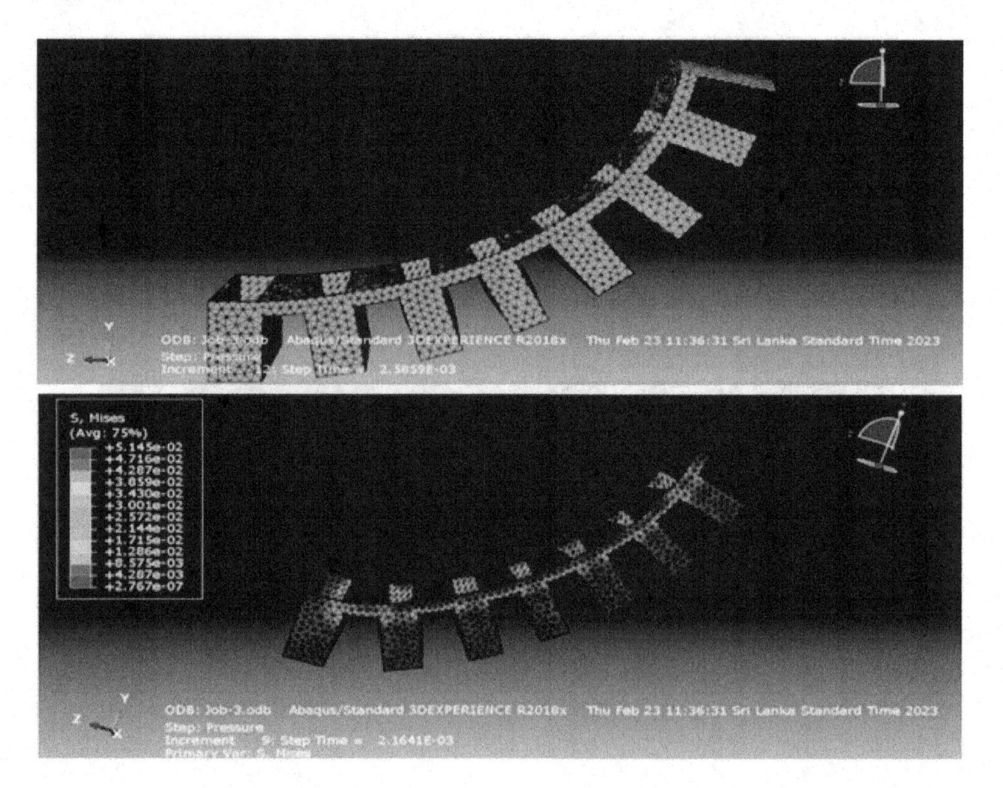

Table 1. Pressure level and bending deformation

Pressure Level (KPa)	Bending deformation(mm)
30	36.2
40	46.5
50	57.4

FUTURE WORK

Integration with underwater robots: Soft robotic grippers could be integrated with underwater robots to enable efficient and precise grasping of seaweed in marine environments.

Multi-fingered grippers: Current soft robotic grippers typically have only one or two fingers, which may limit their ability to grasp irregularly shaped objects such as seaweed. Developing multi-fingered grippers could increase the versatility and efficiency of seaweed grasping.

ACKNOWLEDGMENT

N. Ayyanar and G. Prabhakar acknowledge SERB SURE, India, for providing financial assistance through State University Research Excellence (SURE/2022/003424).

S. Meenakshi expresses her gratitude to Thiagarajar College of Engineering (TCE) for their support in conducting this research work. Additionally, she acknowledges the financial assistance provided by TCE under the Thiagarajar Research Fellowship scheme (File.no: TRF/Jul-2023/4) with heartfelt appreciation.

REFERENCES

Bao, J., Li, D., & Qiao, X. (2020). Integrated navigation for autonomous underwater vehicles in aquaculture: A review. *Information Processing in Agriculture, 7*(1).

Chen, S., & Xu, H. (2023). A pneumatic–hydraulic hybrid actuator for underwater soft robot swimming and crawling. *Sensors and Actuators A: Physical, 356.* doi:10.1016/j.sna.2023.114284

Fischell, E., Stanton, T. K., Kukulya, A., & Andone, C. (1806). Lavery, Monitoring of macroalgae (kelp) farms with autonomous underwater vehicle-based split-beam sonar, September 2018. *The Journal of the Acoustical Society of America, 144*(3).

Huang, W. (2021). Modeling soft swimming robots using discrete elastic rod method. *Proc. Bioinspired Sensing, Actuation, Control Underwater Soft Robotic Syst.*

Hughes, J., Culha, U., Giardina, F., Guenther, F., Rosendo, A., & Iida, F. (2016). Soft manipulators and grippers: A review. *Frontiers in Robotics and AI, 3*, 1–12. doi:10.3389/frobt.2016.00069

Nordin, I. N. A. M., Razif, M. R. M., & Natarajan, E. (2013). 3-D finite-element analysis of fiber-reinforced soft bending actuator for finger flexion. IEEE/ASME international conference on advanced intelligent mechatronic, Wollongong, Australia.

Soft Robotics Toolkit. (n.d.). Modeling and Design Tool for Soft Pneumatic Actuators. Soft Robotics Toolkit. https:// softroboticstoolkit.com/book/modeling-soft-pneumati c-actuators

Stenius, I., Folkesson, J., Bhat, S., Sprague, C. I., Ling, L., Özkahraman, Ö., Bore, N., Cong, Z., Severholt, J., Ljung, C., Arnwald, A., Torroba, I., Gröndahl, F., & Thomas, J.-B. (2022). A System for Autonomous Seaweed Farm Inspection with an Underwater Robot. *Sensors (Basel), 22*(13), 5064. doi:10.3390/s22135064 PMID:35808560

Wei, Y., Chen, Y., Ren, T., Chen, Q., Yan, C., Yang, Y., & Li, Y. (2016). A novel, variable stiffness robotic gripper based on integrated soft actuating and particle jamming. *Soft Robotics*, *3*(3), 134143. doi:10.1089/soro.2016.0027

Xavier, M. S., Fleming, A. J., & Yong, Y. K. (2021, February). Finite element modeling of soft uidic actuators: Overview and recent developments. *Advanced Intelligent Systems*, *3*(2), 2000187. doi:10.1002/aisy.202000187

APPENDIX

Solid Works Design Parameters

By using the following values, gripper is designed in solidworks software.

Number of Chambers: 8
Total height:115mm
Height of the chamber: 30mm
Gap between two chambers: 10mm
Thickness of the gripper: 5mm
Commercial tools for FEM analysis:

- ABAQUS
- ANSYS
- COMSOL
- MARC etc.

Chapter 7
Secure VANET Routing Protocols for Improved Vehicular Communication in Autonomous Systems

Abarna S.
https://orcid.org/0000-0001-6574-6821
National Engineering College, India

Naskath J.
National Engineering College, India

Rathi Pathi R.
National Engineering College, India

Jeyalakshmi C.
Mohamed Sathak Engineering College, India

ABSTRACT

Vehicular ad hoc networks (VANETs) represent an advanced iteration of mobile ad hoc networks (MANETs) designed specifically for internet communication within vehicles. VANETs aim to enable vehicle-to-vehicle (V2V) communication, enhancing safety and convenience for drivers and passengers. However, the open nature of ad hoc networks and the absence of a well-defined line of defense make security a crucial concern for VANETs. Prior to deploying mobile ad hoc networks in hostile or sensitive areas, it becomes imperative to establish robust security services. This study addresses the need for a trusted VANET routing protocol that incorporates a diverse range of security services. The proposed approach implements a secure routing protocol based on the Dijkstra Algorithm to identify the secure and shortest path. For ensuring secure routing, the protocol employs route request (RREQ) and route reply (RREP) mechanisms to identify trustworthy nodes. Additionally, message authentication is utilized to provide end-to-end, hop-to-hop, and entire-route authentication. To transmit messages securely, the Diffie-Hellman Key Exchange Protocol is employed for message encryption, ensuring safe delivery to the intended destination. To assess the performance of the suggested protocol, the authors conducted simulations using NS2. These simulation results demonstrate that the proposed routing protocol outperforms existing methods, affirming its effectiveness in VANET environments.

DOI: 10.4018/979-8-3693-1962-8.ch007

INTRODUCTION

Today's world has seen a rise in the use of wireless technology, which may be applied in a variety of contexts at any time. Wireless networks provide communication between nodes via wireless data links, whereas ad hoc networks rely on nodes banding together to route and forward messages. Vehicle ad hoc networks (VANETs) are the most popular and pertinent uses of ad hoc mobile networks. VANETs connect nodes automatically and don't require any pre-existing infrastructure. According to (Nidhal M et al, 2015), the main goals of VANET networks are to improve road safety and offer convenience services to drivers. Because it directly affects passenger lives, protecting VANETs from potential threats is essential. Vehicle connectivity to the internet and service access is made possible by VANETs (Lai W K et al, 2015). Vehicular Ad hoc Networks allow for communication between stirring cars in a limited region. Vehicles can interact directly with one another in Vehicle-to-Vehicle (V2V) communication, but they can also communicate with infrastructure components like roadside units (RSUs) in vehicle-to-infrastructure (V2I) communication (Rehman et al, 2013). Researchers have recently concentrated on a variety of VANET-related issues, such as broadcasting, routing, security, architectures, applications, and protocols; as well as Quality of Service (QoS). Cities are experiencing an increase in accidents and traffic congestion as a result of the growing number of automobiles on the road. To solve these problems, vehicles must communicate effectively and securely (Rehman et al, 2013). VANETs are networks that use a variety of routing protocols, which can be broadly classified into five categories: topology-based, position-based, cluster-based, geocast, and broadcast (Kamboj S et al, 2014).

When compared to conventional wireless networks, Vehicle Ad-Hoc Networks (VANETs) include a number of distinctive features. Above all, VANETs are extremely dynamic and fast-changing ecosystems. The network topology is always changing, as vehicles travel in different directions and at different speeds. VANET protocols must be flexible and able to handle sudden changes in network connectivity due to this dynamism. Fig 1 depicts the different characteristics of VANET which are most commonly known. The fact that VANETs depend on wireless communication is another important feature. Since cars interact wirelessly with other cars and with the infrastructure along the road, they are vulnerable to a number of problems with wireless communication, including fading, attenuation, and interference from other signals (Saini M et al, 2016). Time-sensitive applications such as traffic signal coordination and collision avoidance necessitate low-latency answers for communication in VANETs (Kumar V et al, 2013). Furthermore, because of the constant data interchange between vehicles, VANETs produce enormous amounts of data, which calls for effective data processing and management systems.

Commercial Applications of VANETs Commercial applications have a plethora of potential thanks to Vehicle Ad-Hoc Networks (VANETs). Among the most well-known is location-based marketing and advertising. Businesses are able to target potential customers with relevant adverts depending on their location and interests thanks to VANETs' connectivity and data exchange capability (Hamdi M M et al, 2020). For example, a coffee shop could entice passing vehicles to stop for a cup of coffee by sending them a special discount voucher. Furthermore, fleet management systems are made possible by VANETs, which help businesses save a lot of money by monitoring vehicle performance, optimizing routes, and enhancing fuel efficiency. Figure 2 Describes the application of VANET in different areas such as safety purpose, commercial and convenience purpose.

Convenient Services with VANETs Living in an urban environment has never been more convenient thanks to VANETs. Smart parking is among the most concrete instances. By having access to real-time information regarding parking spots that are accessible nearby, drivers can spend less time and get less

Figure 1. VANET characteristics

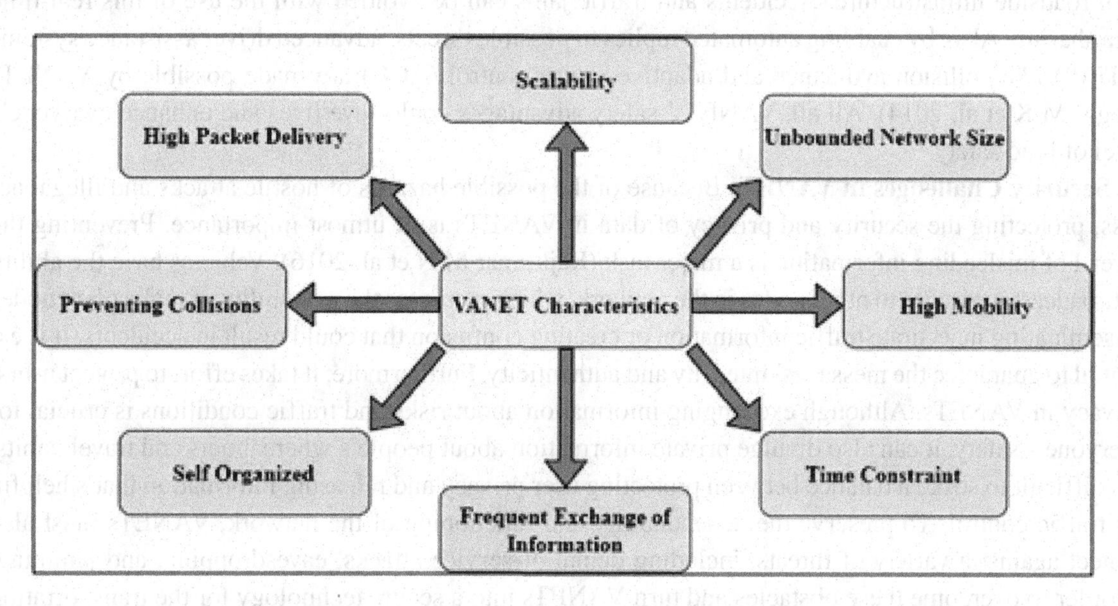

Figure 2. VANET applications

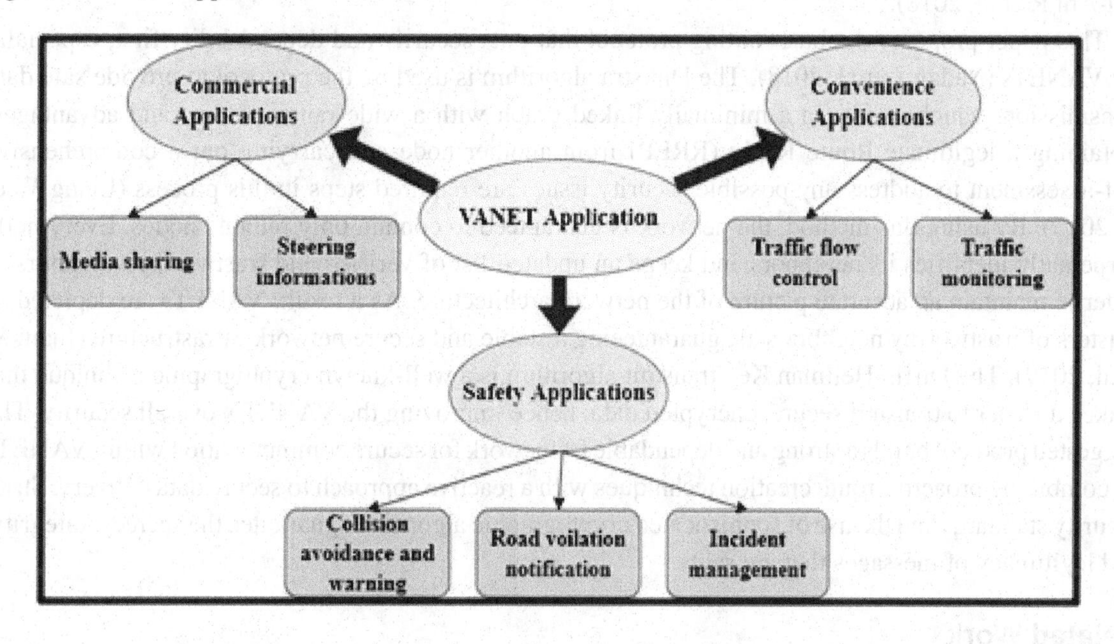

frustrated looking for a place to park. Additionally, ride-hailing and car-sharing services can be made easier for consumers to utilize without the inconvenience of owning a car thanks to VANETs. Dynamic routing and traffic congestion alerts are further advantages for commuters, since they enable them to bypass traffic bottlenecks and get at their destinations more quickly.

Safety Enhancements through VANETs The fact that VANETs improve road safety is among its most important features. In order to share information on traffic patterns, potential hazards on the road, and emergency scenarios, vehicles outfitted with VANET technology can connect with one another and with roadside infrastructure. Accidents and traffic jams can be avoided with the use of this real-time data sharing. Also, by enabling automated replies to possible threats, advanced driver assistance systems (ADAS) like collision avoidance and adaptive cruise control (ACC) are made possible by VANETs (Saggi M K et al, 2014). All all, VANETs' safety advantages could save lives and enhance everyone's level of road safety.

Security Challenges in VANETs Because of the possible hazards of hostile attacks and illegal access, protecting the security and privacy of data in VANETs is of utmost importance. Preventing the spread of misleading information is a major task (Rajkumar M N et al, 2016). Vehicles have the ability to broadcast messages to other nodes in the network, which increases the possibility of malevolent nodes disseminating inaccurate traffic information or creating confusion that could result in accidents. It is essential to guarantee the messages' integrity and authenticity. Furthermore, it takes effort to protect users' privacy in VANETs. Although exchanging information about risks and traffic conditions is crucial for everyone's safety, it can also divulge private information about people's whereabouts and travel habits. It's difficult to strike a balance between protecting user privacy and releasing information that's helpful for traffic control. To preserve the dependability and functioning of the network, VANETs must also protect against a variety of threats, including denial-of-service attacks, eavesdropping, and jamming. In order to overcome these obstacles and turn VANETs into a secure technology for the transportation of the future, it is imperative that strong security measures and encryption protocols be implemented (Al-Ani R et al, 2018).

This paper proposes a robust routing protocol that puts security and dependability first, especially for VANETs (Yadav S et al, 2018). The Dijkstra algorithm is used by the protocol to provide safe data transmission, which results in a minimally linked graph with a wide range of uses and advantages. Obtaining a legitimate Route Reply (RREP) from another node and carrying out a comprehensive self-assessment to address any possible security issues are required steps in this process (Liang W et al, 2015). By using this method, the network is guaranteed to contain only reliable nodes. Every node perpetually identifies its neighbors and keeps an updated list of verified and trustworthy neighbors in order to maintain an accurate picture of the network architecture. As a result, VANETs are depicted as clusters of trustworthy neighbor sets, guaranteeing a stable and secure network infrastructure (Inam M et al, 2019). The Diffie-Hellman Key transmit algorithm is a well-known cryptographic technique that makes it easier to transmit secure, encrypted data, hence improving the VANET's overall security. The suggested protocol builds a strong and dependable framework for secure communication within VANETs by combining proactive route creation techniques with a reactive approach to secure data delivery. Strict security standards and the use of sophisticated cryptographic algorithms guarantee the secrecy, integrity, and legitimacy of messages that are sent.

Related Work

Numerous researchers have dedicated their efforts to developing secure and efficient routing algorithms for VANETs. The aim is to enhance energy utilization and ensure secure routing, thereby eliminating network vulnerabilities and improving overall network performance. Routing protocols in VANETs can be categorized as topology-based and operation-based protocols (Singh R et al, 2018). Various secure routing protocols have been proposed to counteract attacks and enhance security. One such protocol is based on a game theory algorithm (Paramasivan B et al, 2015), which utilizes perfect Bayesian equilibrium and signaling to establish a secure routing protocol. This approach effectively reduces attacks from malicious nodes and incorporates neighbor reporting to enhance security. Another protocol, known as Secure B-MFR (Patil C et al, 2014), focuses on finding efficient routing paths with minimal hops while encrypting messages using secret keys.

The Trust Allocation Certificate (TAC) plays a significant role in establishing a trusted communication path (Charusheela M et al, 2014). TAC helps identify malicious behavior such as trust falsification and packet dropping, enhancing security. Additionally, the use of the elliptic curve digital signature algorithm (ECDSA) and digital verification with a certificate authority (Sarma A H K D et al, 2011) further improves performance. A hybrid routing protocol based on minimal spanning trees has been proposed, incorporating the Diffie-Hellman key exchange and certificate authority (Ghosh U et al, 2014). This protocol employs secure techniques for authentication and message verification. LETSRP (Biswas A K et al, 2020) utilizes the Winternitz One-Time Signature Scheme to authenticate transmitted data, ensuring secure communication in wireless sensor networks.

Position-Based Secure Routing Protocol (Singh G et al, 2019) combines elements from Most Forward within Radius protocols, while incorporating a security module using station-to-station key agreement. Machine learning and deep learning algorithms (Zhao L et al, 2016) have also been applied to derive routing protocols based on packet flow datasets, optimizing the routing environment to satisfy QoS and trust requirements.

Misbehavior detection is crucial for secure routing protocols, and a stacking approach utilizing classification techniques (Sonker A et al, 2020) can be employed. Secure routing protocols such as SE-AODV (Rajdeep S et al, 2014) and Energy-optimized Secure Routing (EOSR) (Yang T et al, 2018) address the identification and isolation of malicious nodes, taking into account factors such as trust level, remaining energy, and path length to optimize routing decisions. In summary, secure VANET routing protocols focus on overcoming attacks and ensuring trustworthiness, while considering parameters such as packet loss and malicious node detection. These protocols employ various techniques, including game theory, encryption, certificate authorities, and machine learning, to establish secure and efficient routing paths in VANETs.

In the area of VANETs, researchers have made significant strides in developing secure and efficient routing algorithms to enhance energy utilization and ensure secure routing. These efforts are aimed at eliminating network vulnerabilities and improving overall network performance. Routing protocols in VANETs can be broadly categorized as topology-based and operation-based protocols. Several secure routing protocols have been introduced to combat attacks and bolster security. For instance, a game theory-based protocol employs perfect Bayesian equilibrium and signaling to establish secure routing, effectively mitigating malicious node attacks. Secure B-MFR focuses on finding efficient routing paths with minimal hops while encrypting messages using secret keys. Trust Allocation Certificate (TAC) plays a vital role in identifying malicious behavior and enhancing security, further strengthened by techniques like the elliptic curve digital signature algorithm (ECDSA) and digital verification with a certificate authority. Additionally, a hybrid routing protocol utilizes minimal spanning trees, the Diffie-Hellman key exchange, and certificate authorities to ensure secure authentication and message verification. The use of the Winternitz One-time Signature Scheme in LETSRP guarantees secure data authentication in wireless sensor networks (Bangotra D K et al, 2018). The Position-Based Secure Routing Protocol combines various routing elements and security modules, including station-to-station key agreements, while machine learning and deep learning algorithms optimize routing based on packet flow datasets to meet Quality of Service (QoS) and trust requirements. Misbehavior detection, essential for secure routing, can be implemented using stacking approaches and classification techniques (Mejri M N et al, 2016). Secure routing protocols such as SE-AODV and Energy-optimized Secure Routing (EOSR) identify and isolate malicious nodes, considering factors such as trust levels, path length, and remaining energy to optimize routing decisions. In conclusion, secure VANET routing protocols aim to address attacks, ensure trustworthiness, and consider parameters like packet loss and malicious node detection. These protocols utilize a range of techniques, including game theory, encryption, certificate authorities, and machine learning, to establish secure and efficient routing paths in VANETs.

To address these issues, the protocol employs the Dijkstra algorithm to establish a minimally connected graph, ensuring efficient and reliable routing paths. This (Gadkari M. Y et al, 2012) introduces a stringent security process where nodes must obtain a valid Route Reply (RREP) and undergo self-assessment, guaranteeing that only trustworthy nodes are allowed into the network. Continuous neighbor detection and verified neighbor lists are used to create reliable clusters, enhancing network integrity and stability.

This work presents a robust VANET-specific routing protocol emphasizing security and reliability by employing the Dijkstra algorithm to establish a minimally connected graph. Nodes must acquire a valid RREP and undergo self-assessment for security, ensuring only trustworthy nodes join the network. Continuous neighbor detection and verified neighbor lists create reliable clusters, maintaining network integrity. The Diffie-Hellman Key Exchange enhances message security. The protocol blends proactive route construction with reactive data transmission, forming a dependable communication framework in VANETs, guaranteeing message confidentiality, integrity, and authenticity through advanced cryptographic techniques and stringent security protocols.

Proposed Work

In this section, we present a comprehensive description of a robust routing protocol designed specifically for Vehicular Ad Hoc Networks (VANETs). The protocol ensures a high level of security by allowing nodes to autonomously construct routes and transmit data securely using the Dijkstra algorithm. This innovative approach greatly contributes to the establishment of a minimally connected graph, opening up various applications and benefits. Prior to joining a VANET, each participating node undergoes a critical validation process. This involves acquiring a valid Route Reply (RREP) from another node and conducting a thorough self-assessment to address any potential security concerns. By adhering to these stringent security measures, the protocol guarantees that only trustworthy nodes become part of the network. Figure 3 depicts the formulation of VANET architecture.

To maintain an accurate understanding of the network topology, each node continually detects its neighboring nodes and maintains an up-to-date list of verified and trusted neighbors. Consequently, a

Figure 3. Formulation of VANET

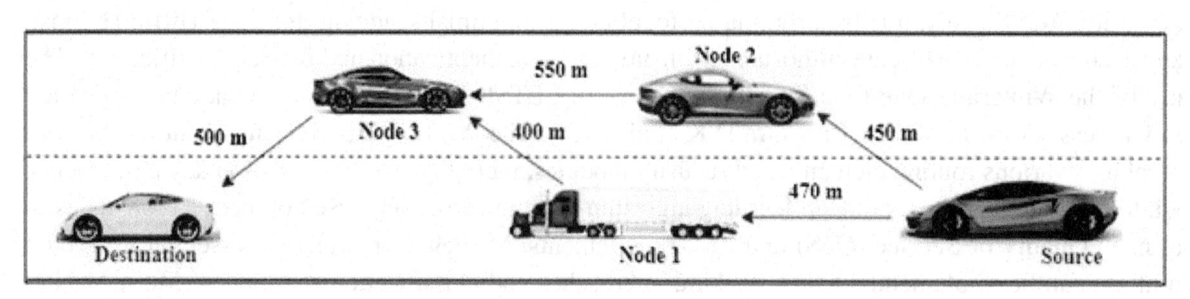

VANET can be represented as an amalgamation of multiple sets of reliable neighbor clusters, ensuring a dependable and secure network environment. To facilitate the transmission of secure messages, the protocol employs the renowned Diffie-Hellman Key Exchange algorithm. This cryptographic technique enables the exchange of trustworthy and encrypted messages, enhancing the overall security of the VANET. By combining the proactive measures of route construction and the reactive approach to secure data transmission, our protocol establishes a robust and reliable framework for secure communication within VANETs. Its utilization of advanced cryptographic algorithms and adherence to strict security protocols further ensures the confidentiality, integrity, and authenticity of the exchanged messages (Sampoornam K P et al, 2020). Figure 4 is a system diagram of VANET.

Identifying Trustworthy Neighbors and Forming of Safe Route

The process of identifying trustworthy neighbors and establishing safe routes within a network involves the utilization of Dijkstra's algorithm. Originally developed to address the single-source shortest route problem in static graphs, this approach offers an effective solution. It initiates from the source node and systematically explores the entire network to determine the shortest path between nodes. It's important to note that for a given transportation network, an upper bound on the distance between any two nodes can be precomputed. The proposed Dijkstra algorithm incorporates an Efficient Graph (EG) array, which contains a comprehensive list of vehicles and their corresponding reliability values. To commence the algorithm, the reliability value of the source vehicle, denoted as EG(source), is set to 0, while the reliability values of other unvisited vehicles are initialized as null (EG(unvisited)). Subsequently, unvisited vehicles stemming from the source node are evaluated to determine their journey reliability values. Overall workflow of proposed algorithm is shown in figure 5.

Once all the neighbors of a particular vehicle have been considered, that vehicle is marked as visited, and its journey reliability value is set as final. This iterative process continues until all nodes have been visited, and their final reliability values have been established. The accompanying pseudocode for the Dijkstra algorithm is presented below. Figure 4 illustrates the step-by-step execution of the Dijkstra algorithm, leading to the identification of the shortest and most efficient path. The algorithm generates a formulated graph that meticulously tracks the records of each node, facilitating the determination of the optimal route.

The provided pseudocode represents the Dijkstra algorithm, which is commonly used to get the shortest route in a graph. In the context of a VANET (Vehicular Ad Hoc Network), this algorithm can be applied to identify an efficient path between a given source vehicle and its destination. Digikstra work flow depicted in figure 6. Let's walk through the steps of the algorithm:

Initialization:

Set the distance of each node in the VANET formulated graph to infinity.

Mark all nodes as unvisited.

Set the distance of the source vehicle to 0.

Main Loop:

Repeat the following steps until the algorithm terminates.

Select the node (curr) with the lowest distance among the unvisited nodes in the graph.

Mark curr as visited.

Neighbor Evaluation:

Iterate through each neighbor (nxt) of the current node (curr).

Figure 4. Overall Process Flow Diagram

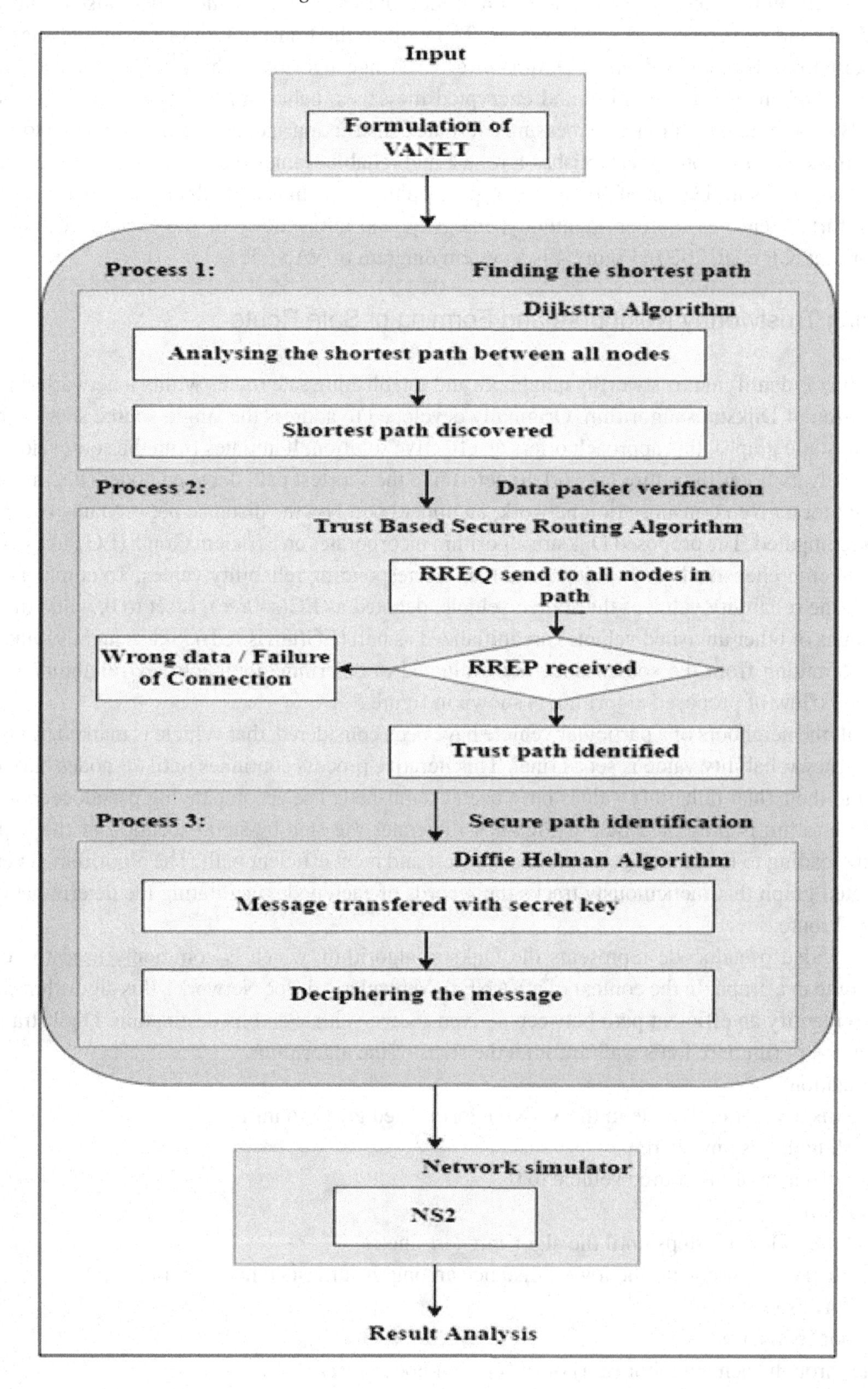

Figure 5. Algorithm flow diagram

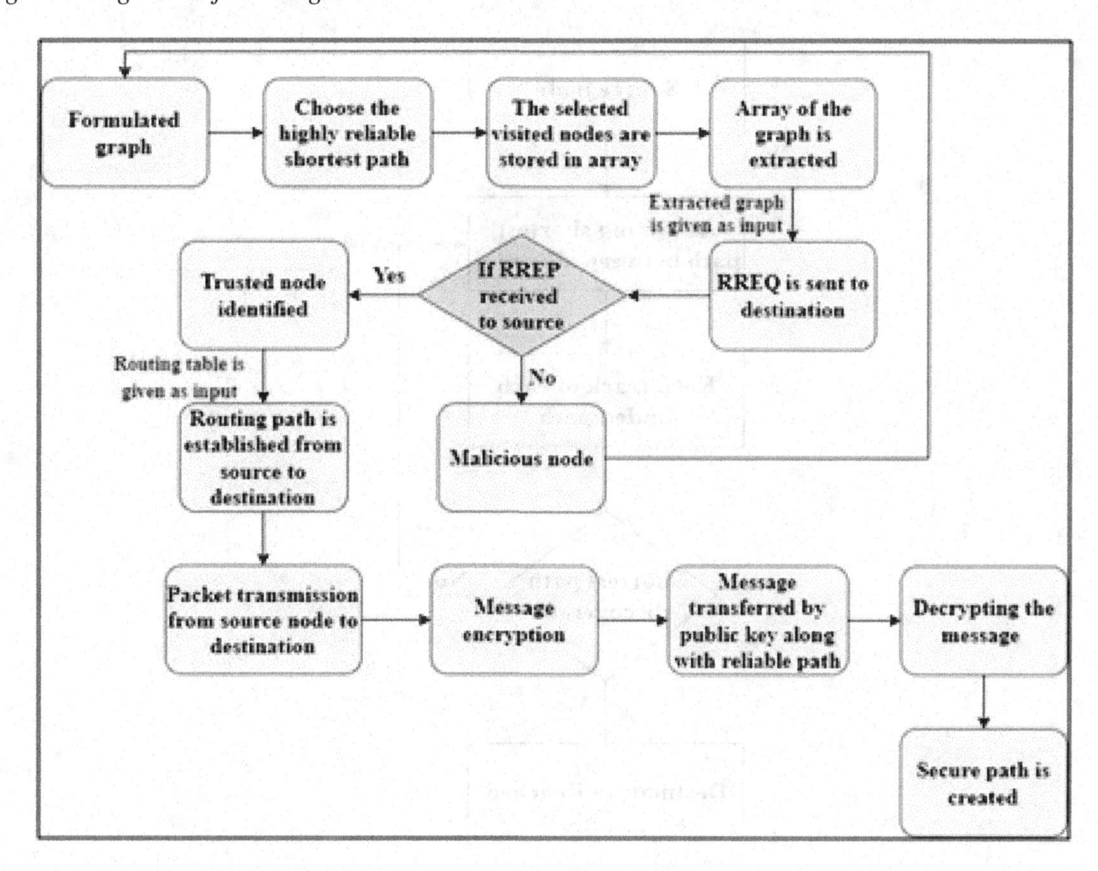

If nxt has not been visited:
Calculate the new distance (newdist) from the source node (curr) to nxt.
If the new distance is smaller than the current distance of nxt:
Update nxt's distance to newdist.
Set nxt's routeToNode as curr (the node that leads to nxt with the shortest distance).
Destination Reached:
If the current node (curr) is the destination (dest), return the path from the resource to the end.
No Valid Path:
If the node with the lowest score (distance) in the graph has a distance of infinity, it indicates that there is no valid path from the source to the destination.

The algorithm continues to iterate until either the destination is reached, in which case the optimal path is returned, or it determines that there is no valid path between the source and destination. The algorithm's efficiency lies in the fact that it progressively explores nodes with lower distances, updating their distances and routes as necessary. Note that the "calc(curr, nxt)" function in the pseudocode represents a calculation to determine the distance between the current node (curr) and its neighboring node (nxt). The specific calculation will depend on the characteristics of the VANET and the metrics used for evaluating distances.

Figure 6. Dijkstra algorithm working

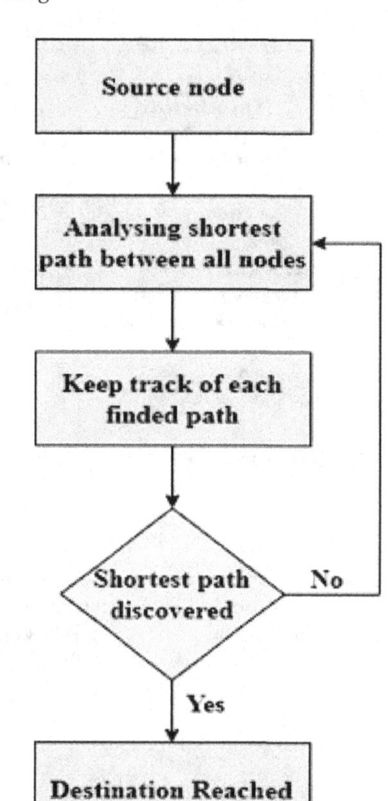

Trusted Node Identification

The process of identifying trusted nodes within the VANET involves a series of steps to ensure the reliability and security of communication along the efficient route. Let's elaborate on these steps:

Recording Heard Vehicles: As the efficient route is discovered, each vehicle encountered along the route receives a Route Request (RREQ) message. The receiving vehicle records information about the transmitting vehicle, specifically which vehicle is heard. This step allows for the establishment of a comprehensive record of the vehicles encountered during the routing process.

Forwarding RREQs: Upon receiving an RREQ, the vehicle determines the next hop based on the data updated by the RREQ extension. The RREQ message is then forwarded to the next hop along the route. It's important to note that intermediate vehicles along the route are not authorized to send a Routing Reply (RREP) message back to the source vehicle, even if they possess a valid route to the destination. This limitation ensures that only trusted nodes, specifically the destination vehicle, can initiate the response.

Dynamic Reliability Values: Due to the dynamic nature of the VANET, where vehicles are constantly moving, the reliability values at intermediate vehicles may become outdated over time. This is particularly relevant as the routing process incorporates the time domain, taking into account the changing positions and movements of vehicles. Therefore, the reliability values associated with intermediate vehicles need to be regularly updated and validated to maintain accurate trustworthiness assessments.

Algorithm

Algorithm:Dijkstra

```
Input: A Vanet formulated graph and a source vehicle (source),Destination of vehicle(des).
Output:An efficient path.
for each node in graph do:
node.dist=infinity
node.visited=false
end
source.dist=0
while true do:
curr=nodeWithLowestDistance(graph)
curr.visited=true
for nxt in curr.n do:
if nxt.visited==false then:
newdist=calc(curr,nxt)
if new.dist<nxt.dist then:
nxt.dist=newdist
nxt.routeToNode=curr
end
end
end
if curr==dest then:
return path(dest)
if nodeWithLowestScore(graph).dist==infinity then:
not correct path
end
end
```

Destination Vehicle Response: Once the RREQ reaches the destination vehicle, an appropriate response, namely the Routing Reply (RREP), is generated and sent back to the source vehicle. This RREP serves as the starting point for data transfer between the source and destination vehicles, initiating a secure and reliable communication session (Hu F et al, 2020).

Figure 7 visually represents the process of trusted node identification. By sending RREQ messages to each node encountered along the efficient route and receiving RREP responses, it becomes possible to assess whether a particular node can be considered trusted or potentially malicious (Patil C et al, 2014). This method ensures that only reliable and secure nodes are involved in the communication process, enhancing the overall security of the VANET. The provided pseudocode outlines the Trust Node Identification algorithm, which aims to determine the most reliable path from a source to a target vehicle in a given graph. Let's go through the steps of the algorithm:

Broadcast RREQ Packet: The algorithm starts by broadcasting a Route Request (RREQ) packet from the source vehicle, initiating the process of finding a reliable path to the destination.

RREP Reception Loop: A loop is initiated with the variable i set to 0. The loop continues until i reaches a value greater than or equal to 3 or until no Routing Reply (RREP) packet is received.

Path Comparison: If the ID of the source vehicle is less than the ID of the next node along the path, a break statement is executed, indicating that the current path is the most reliable and no further iterations are necessary. However, if the source vehicle's ID is greater than or equal to the next node's ID, the path is updated.

Path Update and RREQ Broadcast: If a node finds an updated path, it sends an RREQ packet to continue the process of identifying the most reliable path. Data packets are then sent from the source vehicle to the destination vehicle using the updated path.

Figure 7. Trust node identification protocol

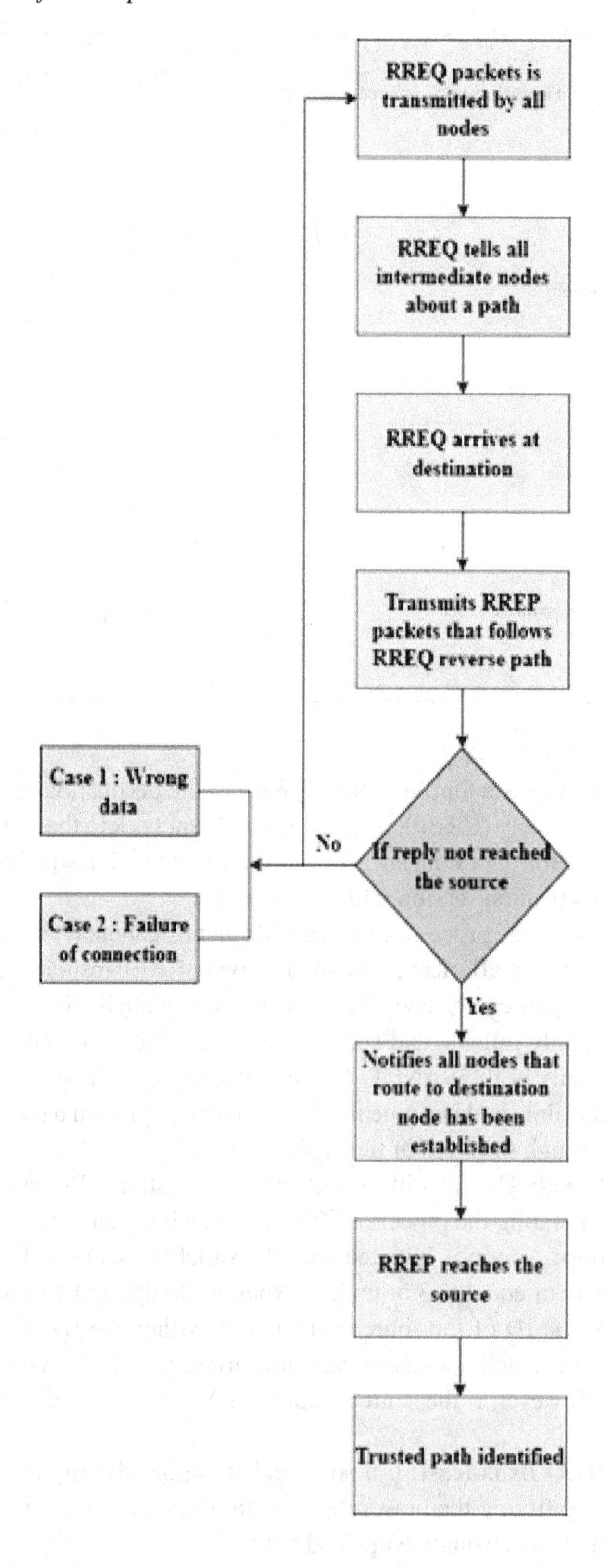

RREQ Update and Broadcast: If a node does not find an updated path, the previous ID is updated, and a check is performed. If the ID of the source vehicle's RREQ is greater than the next node's RREQ ID, the RREQ is updated, and a new RREQ packet is broadcast from the destination vehicle to the source vehicle. This step allows for further exploration of alternative paths.

RREP Reception and Data Transmission: If an RREP packet is received, it is used to update the routing table. Subsequently, data packets can be sent from the source to the target vehicle using the established reliable path.

Destination Not Reached: If an RREP packet is not received within the loop or the destination is not reached, it indicates that a reliable path to the destination could not be found.

The Trust Node Identification algorithm combines the use of RREQ and RREP packets to iteratively explore and update the path towards the destination vehicle, ensuring that the most reliable and secure route is identified.

Data Transfer Security

Ensuring the security of data transfer is paramount in any communication system. In this context, the proposed approach employs symmetric encryption to safeguard the data packets during transmission. The source node takes the data packets intended for transfer and encrypts them symmetrically using a secret key, denoted as K. Symmetric encryption algorithms, such as AES (Advanced Encryption Standard), are commonly used for this purpose. By applying the same secret key for encryption and decryption, the confidentiality and integrity of the data are preserved. In step 3 of the process (not explicitly mentioned in the provided context), a route is established for data transmission. (Lazrag H et al, 2018) This could involve the use of routing protocols, such as the Dijkstra algorithm, to get the optimal path from the source node to the target node. The route ensures that the encrypted data packets are sent along a reliable and efficient path. Upon receiving the encrypted data packets, the destination node decrypts them using the same secret key, K. By employing the corresponding decryption algorithm, the target node successfully retrieves the original valid message contained within the data packets.

Figure 8, which is not explicitly described in detail, illustrates the operation of the Diffie-Hellman algorithm in the context of message transfer from the source to the target node. The Diffie-Hellman algorithm allows two ends to share a secret key over an insecure communication channel (Pandit C M et al, 2014). While the algorithm itself is not directly mentioned in the provided context, it likely plays a crucial role in the secure exchange of the secret key K, ensuring that only authorized parties can decrypt the data packets.By combining symmetric encryption for data confidentiality and integrity and possibly incorporating the Diffie-Hellman algorithm for secure key exchange, the proposed approach ensures the security of data transfer from the source node to the destination node.

The provided pseudocode represents the Diffie-Hellman algorithm, which is a key exchange protocol used to produce a shared secret key between two end nodes over an insecure communication channel. Let's go through the steps of the algorithm:

- Global Public Elements: The algorithm begins by initializing the global public elements (q, a), which are known to both the source and destination nodes. These elements are typically prime numbers and generator values chosen in advance.
- Key Generation for the Source Node: The source node generates its private key, Xsrc, which is a randomly selected integer less than the value of q. Then, it computes its public key, Ysrc, using the

formula Ysrc = a^Xsrc mod q. Here, '^' denotes exponentiation and 'mod' represents the modulo operation.

- Key Generation for the Destination Node: Similarly, the destination node generates its private key, Xdst, and computes its public key, Ydst, using the same formulas as the source node.
- Calculation of Secret Key by the Source: The source node calculates the secret key by raising the destination node's public key, Ydst, to the power of its own private key, Xsrc. This is done modulo q. The resulting value represents the shared secret key.
- Calculation of Secret Key by the Destination: The destination node performs the same calculation as the source node, but using its private key, Xdst, and the source node's public key, Ysrc. Again, the calculation is performed modulo q.
- Shared Secret Key: The final step results in both the source and destination nodes having computed the same shared secret key. This shared secret key can now be used for various purposes, such as symmetric encryption or message authentication.

This algorithm helps two people share secret information safely without sending it directly over an unsecured channel.The algorithm relies on the computational complexity of calculating discrete logarithms to ensure the secrecy of the shared key.

Performance Evaluation and Simulation Setup

Simulation parameters play a crucial role in setting up and running simulations to study and analyze various aspects of a system or network. In the provided table, several simulation parameters are listed along with their corresponding values. The coverage area specifies the geographical extent of the simulated network. In this case, it is defined as 500 meters by 500 meters, indicating that the simulation focuses on a specific area of that size. The simulation time determines the duration of the simulated scenario. In this case, it is set to 350 milliseconds, indicating that the simulation captures events and activities within that time frame. The parameter has a minimum value of 50 and a maximum value of 100, indicating that the simulation considers scenarios with different node densities. In this case, it is set to "High" for traffic intensity and "Highway" for road type, indicating a scenario with high traffic load and highway-like road characteristics.

The transmission ranges are set as for On-Board Units (OBUs), the range is set to 50 meters, while for Roadside Units (RSUs), it is set to 100 meters. These values determine the communication coverage and connectivity between nodes. The packet size is set to 512 bytes, and a total of 100 packets are considered. These values affect the data transfer and transmission efficiency within the simulated network moreover the nodes can move at speeds up to 70 kilometers per hour. The proposed routing protocol ETBSRP parameter specifies the algorithm or protocol used for routing decisions in the simulated network. The chosen mobility model, manhattan suggests that the nodes move in a random manner along the street network. The MAC (Media Access Control) layer parameter specifies the protocol used for controlling access to the shared communication medium. The algorithms Dijkstra is used for identifying the shortest path in the network, next Trust Routing Protocol is used to detect trust nodes in the routing process and finally Diffie-Hellman used for secure message transmission by establishing shared secret keys.

Figure 8. Diffie Hellman key exchange protocol working

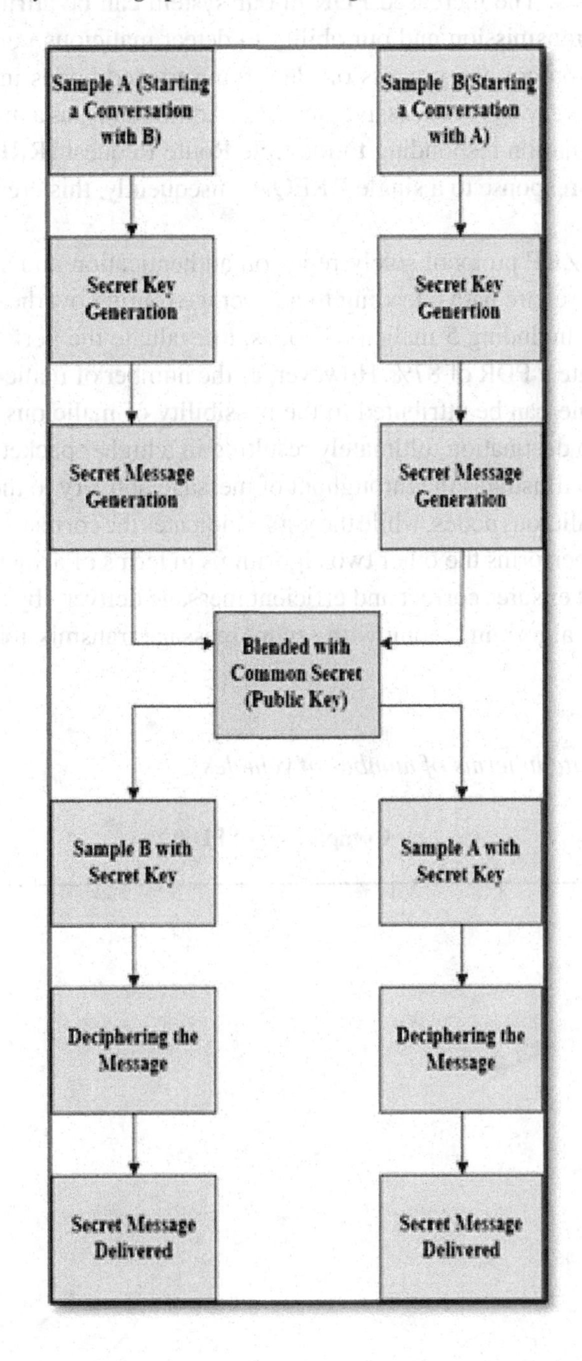

Performance Analysis

Packet delivery Rate: Figure 9 displays a graph depicting the number of nodes on the X-axis and the Packet Delivery Ratio (PDR) on the Y-axis. The graph demonstrates that our proposed protocol, the Emphasized Trust-Based Secure Routing Protocol (ETBSRP), achieves a higher PDR compared to the

AOMDV and ZRP protocols. The increased PDR in our system can be attributed to our selection of secure paths for message transmission and our ability to detect malicious nodes before sharing information. In the AOMDV protocol, the process of identifying trusted nodes incurs additional message overhead during route discovery due to excessive flooding. Additionally, as a multipath routing protocol, AOMDV leads to the destination responding to multiple Route Request (RREQ) packets, resulting in longer overhead packets in response to a single RREQ. Consequently, this creates a significant amount of control overhead.

On the other hand, the ZRP protocol solely relies on authentication and signatures to share information without employing secure paths, leading to an average routing overhead. In this study, focused a network with 100 nodes, including 5 malicious nodes, to evaluate the performance of our proposed protocol. The results indicate a PDR of 87%. However, as the number of malicious nodes increases, the PDR decreases. This decline can be attributed to the possibility of malicious nodes causing delays in forwarding messages to the destination, ultimately resulting in a higher packet loss ratio.

Throughput: Figure 10 illustrates the throughput of message delivery to the destination. The x-axis represents the number of malicious nodes, while the y-axis indicates the corresponding throughput values. The proposed protocol outperforms the other two algorithms in terms of achieving a higher throughput rate. The proposed protocol ensures correct and efficient message delivery by employing a shortest and most efficient path-finding algorithm, along with secure message transmission that minimizes routing

Figure 9. Packet delivery rate in terms of number of vehicles

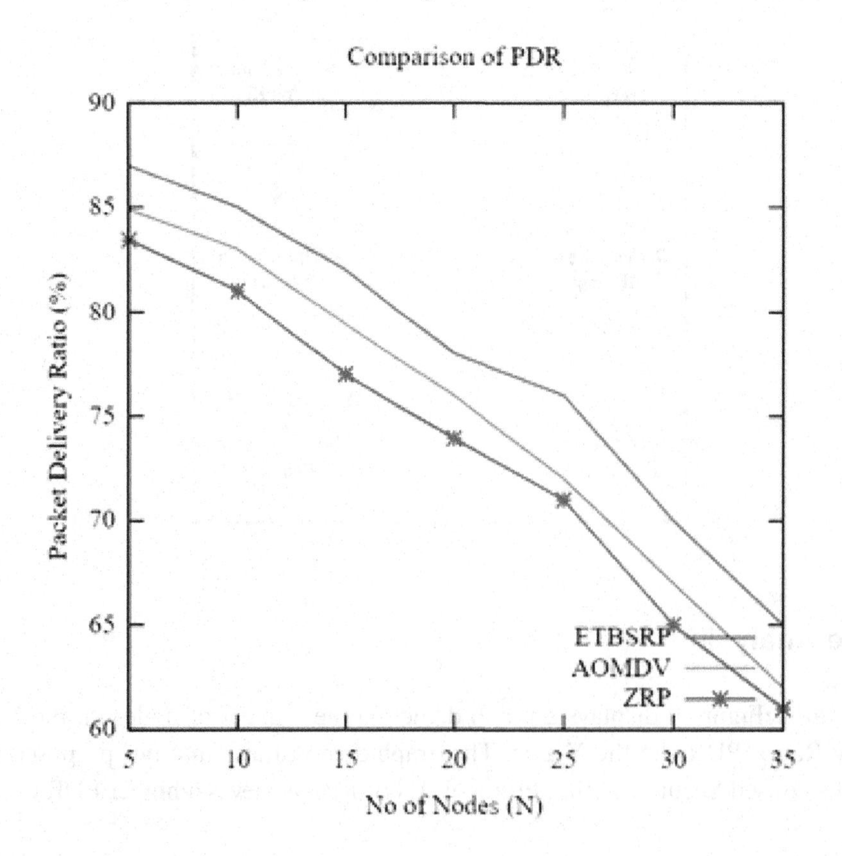

Figure 10. Throughput in terms of number of malicious vehicles

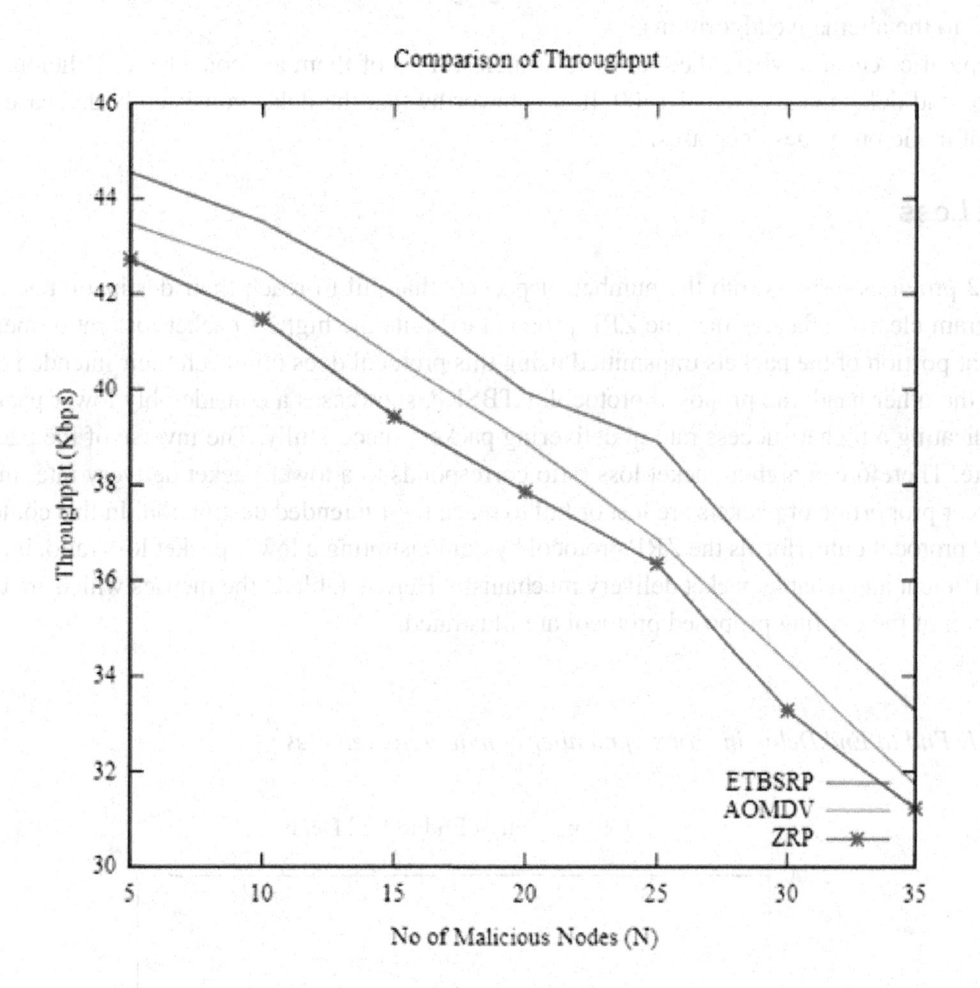

overhead compared to the other algorithms. This efficient routing strategy contributes to the improved throughput observed in the proposed protocol.

In the AOMDV algorithm, slight changes in throughput are observed compared to the proposed protocol. When a link split occurs in the present network topology, AOMDV attempts to discover alternative routes from available backup routes between the end to end node pairs. This process adds to the packet delivery time, resulting in a slightly reduced throughput. On the other hand, the ZRP algorithm prioritizes message security through encryption with signatures. While this enhances the security of message delivery, it sacrifices throughput as it does not prioritize finding the most efficient path. In this evaluation, the presence of malicious nodes was limited to a range of 35, with a difference of 5 between each range. In this scenario, the throughput achieved was 44.5. However, as the number of malicious nodes increases beyond this range, the throughput decreases accordingly.

End to End Delay : In this scenario, we examine different propagation models along with fixed routing protocols and changing node counts. Among the models, the Emphasized Trust-Based Secure Routing Protocol (ETBSRP) demonstrates the shortest end-to-end latency, as depicted in Figure 11. During comparison process of the proposed protocol, it exhibits superior performance than the other two protocols. The proposed protocol achieves slightly faster message delivery with reduced delays compared to the alternative algorithms.

In a specific scenario where there are 100 vehicles, if 35 of them are considered malicious nodes, the end-to-end delay ranges around 1400. It is noteworthy that the delay consistently decreases as the number of malicious nodes decreases.

Packet Loss

Figure 12 provides insights into the number of packets that fail to reach their destination accurately. The diagram clearly indicates that the ZRP protocol exhibits the highest packet loss ratio, meaning a significant portion of the packets transmitted using this protocol does not reach their intended destination. On the other hand, the proposed protocol, ETBSRP, showcases a considerably lower packet loss ratio, indicating a higher success rate in delivering packets successfully. The inverse of the packet delivery rate. Therefore, a higher packet loss ratio corresponds to a lower packet delivery rate, implying that a larger proportion of packets are lost or fail to reach their intended destination. In this context, the ETBSRP protocol outperforms the ZRP protocol by demonstrating a lower packet loss ratio, indicating a more efficient and reliable packet delivery mechanism. Here in table 2, the metrics which are used for comparison of the existing proposed protocol are illustrated.

Figure 11. End to End Delay in terms of number of malicious vehicles

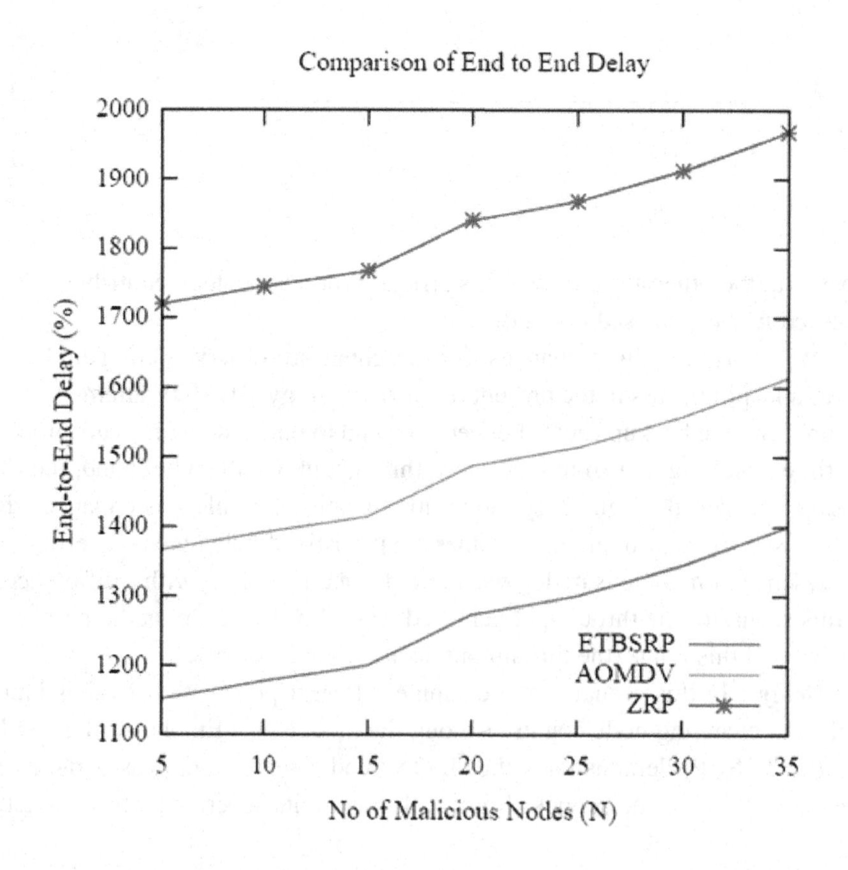

Figure 12. Packet Loss in terms of number of malicious vehicles

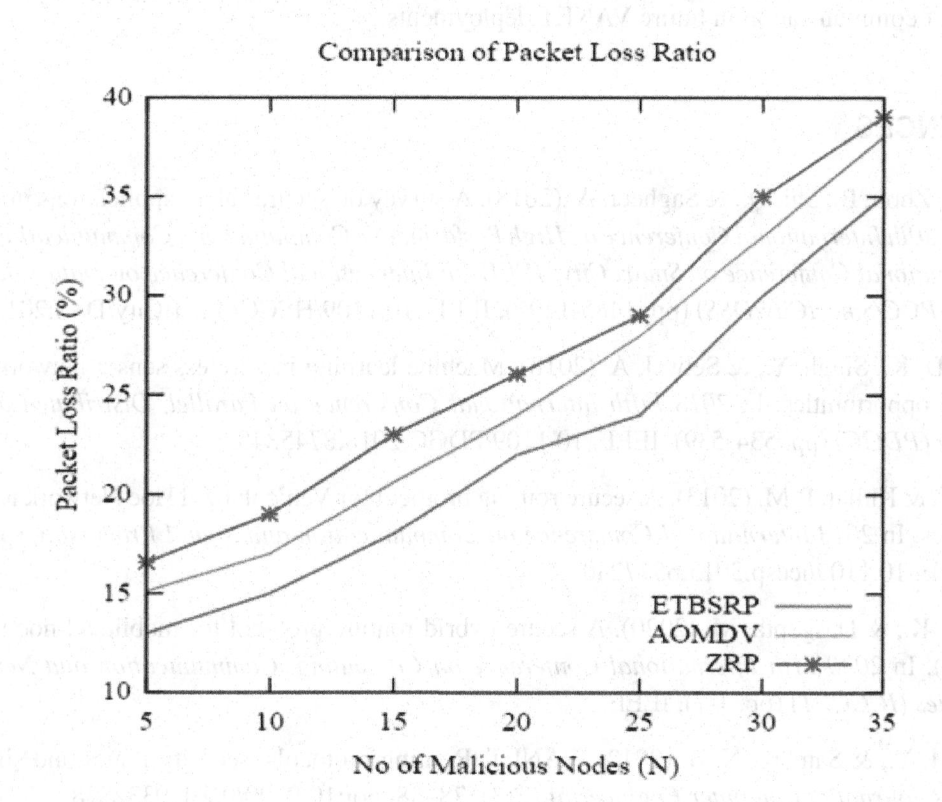

CONCLUSION

This paper introduced a secure and efficient routing protocol for message delivery in Vehicular Ad-Hoc Networks (VANETs). The proposed protocol, called Emphasized Trust-Based Secure Routing Protocol (ETBSRP), employs the Dijkstra algorithm to establish routes by creating a formulated graph to determine the most optimal path. To verify trusted nodes, we utilize RREQ and RREP requests, updating the routing table with the final values. Additionally, we ensure secure message transmission by employing the Diffie-Hellman key exchange algorithm for message encryption. Through this research, ETBSRP has been used as a routing protocol to enhance the quality of service (QoS) in VANETs. This protocol is well-suited for larger networks and considers various parameters such as configuration, node trust levels, and node battery life throughout the routing process. As a result, it improves speed and reduces overhead. Moreover, the proposed protocol incorporates multipath routing, which reduces unnecessary control messages during route formation in scenarios involving congestion or node failure. By identifying hostile nodes, ETBSRP ensures secure connections within the network. Through simulation experiments, we have demonstrated that the ETBSRP protocol outperforms existing routing strategies in terms of packet delivery ratio (PDR), packet loss ratio (PLR), average end-to-end time, and throughput. The results highlight the effectiveness of the proposed ETBSRP in enhancing the QoS of VANETs while ensuring secure communication. Looking ahead, we emphasize the implementation of advanced security algorithms, including encryption, decryption, and blockchain approaches, to provide even higher levels of security to VANETs. By incorporating these measures, we aim to further enhance the protection and reliability of communication in future VANET deployments.

REFERENCES

Al-Ani, R., Zhou, B., Shi, Q., & Sagheer, A. (2018). A survey on secure safety applications in vanet. In *2018 IEEE 20th International Conference on High Performance Computing and Communications; IEEE 16th International Conference on Smart City; IEEE 4th International Conference on Data Science and Systems (HPCC/SmartCity/DSS)* (pp. 1485-1490). IEEE. 10.1109/HPCC/SmartCity/DSS.2018.00245

Bangotra, D. K., Singh, Y., & Selwal, A. (2018). Machine learning in wireless sensor networks: Challenges and opportunities. In *2018 Fifth International Conference on Parallel, Distributed and Grid Computing (PDGC)* (pp. 534-539). IEEE. 10.1109/PDGC.2018.8745845

Bhoi, S. K., & Khilar, P. M. (2013). A secure routing protocol for Vehicular Ad Hoc Network to provide ITS services. In *2013 International Conference on Communication and Signal Processing* (pp. 1170-1174). IEEE. 10.1109/iccsp.2013.6577240

Biswas, A. K., & Dasgupta, M. (2020). A secure hybrid routing protocol for mobile ad-hoc networks (MANETs). In *2020 11th International Conference on Computing, Communication and Networking Technologies (ICCCNT)* (pp. 1-7). IEEE.

Gadkari, M. Y., & Sambre, N. B. (2012). VANET: Routing protocols, security issues and simulation tools. *IOSR Journal of Computer Engineering*, *3*(3), 28–38. doi:10.9790/0661-0332838

Ghosh, U., & Datta, R. (2014). SDRP: Secure and dynamic routing protocol for mobile ad-hoc networks. *IET Networks*, *3*(3), 235–243. doi:10.1049/iet-net.2013.0056

Hamdi, M. M., Audah, L., Rashid, S. A., & Alani, S. (2021). VANET-based traffic monitoring and incident detection system: A review. *International Journal of Electrical & Computer Engineering (2088-8708), 11*(4).

Hamdi, M. M., Audah, L., Rashid, S. A., Mohammed, A. H., Alani, S., & Mustafa, A. S. (2020). A review of applications, characteristics and challenges in vehicular ad hoc networks (VANETs). In 2020 international congress on human-computer interaction, optimization and robotic applications (HORA) (pp. 1-7). IEEE.

Hu, F., Chen, B., Shi, D., Zhang, X., & Pan, H. Z. (2020). Secure Routing Protocol in Wireless Ad Hoc Networks via Deep Learning. In 2020 IEEE Wireless Communications and Networking Conference (WCNC) (pp. 1-6). IEEE. doi:10.1109/WCNC45663.2020.9120545

Inam, M., Li, Z., Ali, A., & Zahoor, A. (2019). A novel protocol for vehicle cluster formation and vehicle head selection in vehicular ad-hoc networks. *International Journal of Electronics and Information Engineering*, *10*(2), 103–119.

Kamboj, S., & Chawla, S. (2014). Geocast routing in vehicular Ad Hoc networks: A survey. [IJCSIT]. *International Journal of Computer Science and Information Technologies*, *5*(4), 5365.

Kumar, K. R. (2010). VANET parameters and applications: A review. *Global Journal of Computer Science and Technology, 10*(7), 72–77.

Kumar, V., Mishra, S., & Chand, N. (2013). Applications of VANETs: present & future. *communications and network, 5*(01), 12-15.

Lai, W. K., Lin, M. T., & Yang, Y. H. (2015). A machine learning system for routing decision-making in urban vehicular ad hoc networks. *International Journal of Distributed Sensor Networks, 11*(3), 374391. doi:10.1155/2015/374391

Lazrag, H., Chaibi, H., Saadane, R., & Rahmani, M. D. (2018). An optimal and secure routing protocol for wireless sensor networks. In *2018 6th International Conference on Multimedia Computing and Systems (ICMCS)* (pp. 1-5). IEEE. 10.1109/ICMCS.2018.8525911

Liang, W., Li, Z., Zhang, H., Wang, S., & Bie, R. (2015). Vehicular ad hoc networks: Architectures, research issues, methodologies, challenges, and trends. *International Journal of Distributed Sensor Networks, 11*(8), 745303. doi:10.1155/2015/745303

Mejri, M. N., & Ben-Othman, J. (2016). GDVAN: A new greedy behavior attack detection algorithm for VANETs. *IEEE Transactions on Mobile Computing, 16*(3), 759–771. doi:10.1109/TMC.2016.2577035

Pandit, C. M., & Ladhe, S. A. (2014). Secure routing protocol in MANET using TAC. In *2014 First International Conference on Networks & Soft Computing (ICNSC2014)* (pp. 107-112). IEEE. 10.1109/CNSC.2014.6906693

Paramasivan, B., Prakash, M. J. V., & Kaliappan, M. (2015). Development of a secure routing protocol using game theory model in mobile ad hoc networks. *Journal of Communications and Networks (Seoul), 17*(1), 75–83. doi:10.1109/JCN.2015.000012

Patil, C., & Nataraj, K. R. (2014). A Secure Routing Protocol for VANET. *International Journal of Engine Research, 3*(5).

RadhaKrishna Karne, D. T. (2021). Review on vanet architecture and applications. [TURCOMAT]. *Turkish Journal of Computer and Mathematics Education, 12*(4), 1745–1749.

Rajkumar, M. N., Nithya, M., & HemaLatha, P. (2016). Overview of VANETs with its features and security attacks. *International Research Journal of Engineering and Technology, 3*(1).

Rehman, S. U., Khan, M., Zia, T., & Zheng, L. (2013). Vehicular ad-hoc networks (VANETs): an overview and challenges. *Journal of Wireless Networking and communications, 3*(3), 29-38.

Saggi, M. K., & Sandhu, R. K. (2014). A survey of vehicular ad hoc network on attacks and security threats in VANETs. In *International Conference on Research and Innovations in Engineering and Technology (ICRIET 2014)* (pp. 19-20). IEEE.

Saini, M., & Singh, H. (2016). VANET its characteristics attacks and routing techniques: A survey. *International Journal of Scientific Research, 5*(5), 1595–1599.

Sampoornam, K. P., Saranya, S., Vigneshwaran, S., Sofiarani, P., Sarmitha, S., & Sarumathi, N. (2020). A comparative study on reactive routing protocols in VANET. In *2020 4th International Conference on Electronics, Communication and Aerospace Technology (ICECA)* (pp. 726-731). IEEE. 10.1109/ICECA49313.2020.9297550

Sarma, A. H. K. D., Kar, B. A., & Mall, C. R. (2011). Secure routing protocol for mobile wireless sensor network. In *2011 IEEE Sensors Applications Symposium* (pp. 93-99). IEEE. 10.1109/SAS.2011.5739778

Shaktawat, R. S., Singh, D., & Choudhary, N. (2014). An efficient secure routing protocol in MANET Security-Enhanced AODV (SE-AODV). *International Journal of Computer Applications*, *97*(8). Advance online publication. doi:10.5120/17030-7329

Singh, G., Rohil, H., Rishi, R., & Ranga, V. (2019). LETSRP: A secure routing protocol for MANETs. *Int J Eng Adv Technol (IJEAT). ISSN*, *9*(1), 2249–8958.

Singh, R., Kathuria, K., & Sagar, A. K. (2018). Secure routing protocols for wireless sensor networks. In *2018 4th international conference on computing communication and automation (ICCCA)* (pp. 1-5). IEEE. 10.1109/CCAA.2018.8777557

Sonker, A., & Gupta, R. K. (2020). A new combination of machine learning algorithms using stacking approach for misbehavior detection in VANETs. *International Journal of Computer Science and Network Security*, *20*(10), 94–100.

Yadav, N., & Chug, U. (2019). Secure Routing in MANET: A Review. In *2019 International Conference on Machine Learning, Big Data, Cloud and Parallel Computing (COMITCon)* (pp. 375-379). IEEE.

Yadav, S., Rajput, N. K., Sagar, A. K., & Maheshwari, D. (2018). Secure and reliable routing protocols for VANETs. In *2018 4th International Conference on Computing Communication and Automation (ICCCA)* (pp. 1-5). IEEE. 10.1109/CCAA.2018.8777690

Yang, T., Xiangyang, X., Peng, L., Tonghui, L., & Leina, P. (2018). A secure routing of wireless sensor networks based on trust evaluation model. *Procedia Computer Science*, *131*, 1156–1163. doi:10.1016/j.procs.2018.04.289

Zhao, L., Li, Y., Meng, C., Gong, C., & Tang, X. (2016). A SVM based routing scheme in VANETs. In *2016 16th International Symposium on Communications and Information Technologies (ISCIT)* (pp. 380-383). IEEE. 10.1109/ISCIT.2016.7751655

Chapter 8
Autonomous Systems Revolutionizing Health Insurance Industry:
Achieving Operational Excellence in Services

Anupa Stanly
https://orcid.org/0009-0003-7374-8618
Karunya Institute of Technology and Sciences, India

K. Aruna
Karunya Institute of Technology and Sciences, India

ABSTRACT

Financial services, particularly the insurance service sector, are increasingly embracing technology. Autonomous systems, which include artificial intelligence (AI), machine learning, and automation, are driving a striking revolution in the health insurance industry. This research provides a comprehensive analysis of various facets, such as the use of autonomous systems, expediting claim processing, identifying, and preventing fraud, improving the customer experience, data-driven decision-making, and adherence to healthcare legislation. The study concludes by highlighting the sector's profound impact from autonomous systems and pointing to a promising future for health insurance that will be characterised by operational effectiveness and customer-centricity. This study sets out a thorough exploration of the dynamic world where autonomous systems are changing the laws governing health insurance. This investigation's main objectives are to achieve operational excellence and provide services that are utterly customer-centric.

INTRODUCTION

The integration of autonomous systems is driving a seismic upheaval in the health insurance sector. Artificial intelligence and machine learning are powering these cutting-edge technologies, which are

DOI: 10.4018/979-8-3693-1962-8.ch008

completely changing how insurers handle claims, control risks, and communicate with policyholders. Insurers are under tremendous pressure to improve efficiency, accuracy, and customer experience as healthcare costs continue to rise and regulatory requirements get more intricate. In this context, autonomous systems show promise as never before seen, providing previously unheard-of chances to boost productivity, enhance judgement, and spur industry innovation.

The health insurance industry is entering a new era of efficiency and efficacy with the integration of autonomous systems. Insurance companies may speed workflows, minimise administrative hassles, and minimise risks by using these technologies to automate regular tasks like fraud detection and claims processing. Furthermore, insurers can use massive volumes of data to customise services, allocate resources more efficiently, and make data-driven decisions that spur strategic expansion and innovation thanks to autonomous systems. Insurance companies are well-positioned to provide more value to policyholders, improve operational resilience, and succeed in a market that is becoming more and more competitive as they adopt these game-changing technologies.

LITERATURE REVIEW

The relentless pace of technological advancement is pushing the health insurance sector towards a significant shift. Autonomous systems, a fusion of artificial intelligence (AI), machine learning, and automation, are at the centre of this metamorphosis and are poised to revolutionise how health insurance is created, provided, and experienced (*The Impact of Big Data and Artificial Intelligence (AI) in the Insurance Sector*, 2020). The development of autonomous systems shines as a beacon of hope in a field frequently plagued by bureaucratic delays, rising prices, and the persistent search for operational efficiency. Increased enrollment in insurance, particularly among low-income households, can be facilitated by digital health insurance management systems and an increase in new digital mediators in the insurance sector is indicative of technology's immediate impact (Stoeckli et al., 2018). The expert interviews have unambiguously shown that organizational culture, technology policy, and strategic philosophy do affect how widely health insurance companies are utilizing autonomous systems (Akter et al., 2022; *Technology and Innovation in the Insurance Sector*, n.d.). Health insurers should concentrate on raising their service standards and come up with ways to preserve the tacit base of knowledge generated by wearable technology for long-term competitive advantage (Nayak et al., 2019).

The history of health insurance is replete with difficulties, from the difficulty of claims processing to the consistently rising price of healthcare. These ongoing problems have long hindered the industry's ability to deliver services that are easily accessible, effective, and client-focused. Health insurers, however, are now in a position to overcome these obstacles and usher in a new era of operational excellence due to the integration of autonomous systems. The insurance firms, not investing in new technology may not seem like a realistic option and should let the future digital business be performed as new or separate divisions that are evolved from concept to completion without legacy applications if it desires to survive into the next few decades (Eling & Lehmann, 2018). By mandating quick reaction times, these new rivals encourage innovation and hasten industry transition. They go beyond simply digitizing the business model to anticipate consumers' demands and offer intelligent solutions and services rather than traditional products and services (Cappiello, 2018). Incumbents work with Insurtechs and digital platforms to start making the best use of the opportunity, as these alliances have the potential to enable more customized online distribution, predictive underwriting, and much more effective management of insurance claims and significantly boost the insurance penetration in India (Alpesh Shah et al., 2021). Regulators need to realize how technologies function to facilitate innovations that the policyholder stands to benefit. Technologies that enable easier business for both the insured and the insurer enhances more scientific pricing and helps both parties should be encouraged (Report on InsurTech in the Context of Risk Assessment, Product Design and Pricing, n.d.).

THE CURRENT LANDSCAPE OF THE HEALTH INSURANCE INDUSTRY

Many important reasons have influenced the dynamic landscape that the health insurance sector is navigating today. More and more people are facing financial hardship as a result of rising healthcare expenditures that are outpacing inflation. Insurance companies have to deal with increased administrative costs in addition to rising medical costs, which are partly caused by the complexity of processing claims and the need for regulatory compliance. Insurance companies now face an even more challenging operating environment as a result of the introduction of new rules and mandates intended to increase access to treatment and improve quality, such as the Affordable treatment Act in the United States.

Further, policyholders' expectations are changing, and they now want their insurers to provide them with more ease, transparency, and personalised service. Digitalization and technological advancements have given customers access to resources and information, empowering them to compare insurance plans, look for alternative care, and take a more active role in managing their health and insurance coverage. Insurance companies face pressure to innovate and set themselves apart in this more competitive market by offering value-added services, boosting customer happiness, and increasing consumer engagement.

The COVID-19 pandemic has highlighted the value of adaptability, resiliency, and creativity in the health insurance sector amidst these difficulties. The pandemic has changed how healthcare services are provided and used, hastening the use of telehealth and virtual care solutions. In order to satisfy the evolving needs of policyholders, insurers are adjusting their products and business practices. At the same time, they are managing the financial effects of the pandemic, which include rising claim volumes and unstable healthcare utilisation trends.

Insurance companies that want to increase productivity, accuracy, and customer satisfaction in this situation must strategically integrate autonomous technologies into their claims processing operations. Insurance companies may improve their decision-making powers, streamline processes, and lower administrative costs by utilising cutting-edge technology like artificial intelligence, machine learning, and data analytics.

Autonomous Systems in the Health Insurance Industry

The health insurance sector uses a variety of autonomous technologies to improve operational efficiency overall, expedite operations, and improve consumer experiences. These autonomous systems make use of data analytics, automation, machine learning, and artificial intelligence (AI). The major autonomous systems frequently used in health insurance are:

- **AI-Powered Chatbots:**

The application of AI-powered chatbots, which improve customer relationships, expedite processes, and raise overall efficiency, is highly advantageous to the health insurance industry. Artificial intelligence (AI)-driven chatbots have become indispensable resources in the health insurance industry, dramatically altering both client relations and operational effectiveness (Eling et al., 2022). These sophisticated chatbots offer several advantages that improve user experience and expedite certain business operations.

The capacity of AI-powered chatbots to provide immediate and reachable 24/7 customer service is one of their main advantages in the health insurance market. These chatbots, which are always on and available to policyholders, guarantee that they can always get support and information, making the experience more responsive and user-friendly. This is especially important in the healthcare industry, as prompt assistance and information might be vital.

When it comes to promptly answering commonly requested questions concerning policy details, managing claims, paying premiums, and other common difficulties, these chatbots thrive. Because chatbots answer frequently asked questions automatically, they reduce the need for customers to wait for assistance or negotiate complex phone interfaces, which improves both the efficacy and pleasure of customer care.

In the claims processing sector, chatbots equipped with artificial intelligence (AI) functionalities are indispensable for aiding users. They provide claim status updates, simplify the required paperwork, and generally improve the efficiency of the claims process. Customer satisfaction is increased and the claims process is sped up by keeping policyholders informed and participating.

Additionally, policyholders' general health literacy is increased by employing these chatbots. They can communicate complicated insurance plans, coverage details, and exclusions in an easy-to-understand manner. Chatbots assist customers in making informed decisions about their coverage by translating complex insurance jargon.

Artificial intelligence (AI)-powered chatbots are crucial for automatically reminding and informing policyholders (Akter et al., 2022). These reminders may include dates that are significant for future renewals, premium payment deadlines, and other events. This proactive approach guarantees that insurance is up to date and that policyholders are advised of important actions they need to take.

Chatbots that use AI can provide personalized wellness and health recommendations in addition to answering queries. These chatbots can analyze individual health data to provide tailored advice, encourage healthy living choices, and offer information on wellness programs supported by insurance companies. This fosters care for the well-being of policyholders and enhances long-term health outcomes.

In short, Artificial intelligence (AI)-driven chatbots are revolutionizing the health insurance sector by optimizing customer support, streamlining the claims process, and supporting wellness and health-related programs. These chatbots have a significant potential to improve customer experiences and boost the effectiveness of health insurance operations as technology develops.

- **Predictive Analytics and Machine Learning Models:**

In the health insurance sector, predictive analytics and machine learning models have emerged as indispensable instruments, providing features that improve risk assessment, decision-making, and overall efficiency in operation. These technologies' capacity to evaluate and forecast health concerns is one of its main advantages. Insurance companies can predict the possibility of certain health problems in their policyholders by using predictive analytics, which can evaluate large datasets to find patterns and trends linked to particular medical disorders. These models use past data to help create more precise risk profiles for individuals, which enables insurers to adjust rates and policies accordingly.

A subset of artificial intelligence called machine learning models plays a major role in the underwriting procedures used in the health insurance industry. These models may independently assess a wide range of variables, including medical history, lifestyle decisions, and demographic data, to ascertain if a prospective policyholder is insurable. This reduces the possibility of human bias and speeds up the underwriting process while ensuring a more complex and data-driven assessment of risk. Predictive analytics is also essential for spotting fraud in the health insurance market. These models can detect potentially fraudulent activity by analysing claim trends and spotting anomalies, which enables insurers to take preventative action (Yaneva, n.d.). This preserves the financial interests of insurance companies and upholds the integrity of the insurance system as a whole.

Machine learning models simplify the examination of insurance claims in the field of claims processing. By automatically comparing claims to pre-established standards, these models can streamline the procedure and reduce the chance of inaccuracy. Furthermore, they aid in the identification of anomalies or discrepancies, enhancing the precision of choices made about the approval or rejection of claims.

Initiatives for personalised health and wellbeing are also revolutionised by the application of machine learning and predictive analytics. With the use of these technologies, insurers may evaluate policyholders' specific health data and deliver them personalised suggestions. This proactive strategy can result in fewer claims and lower expenses for the insurance company, in addition to cultivating a healthier clientele. Additionally, these technologies support health insurance firms' overall financial viability. Insurers can more effectively plan and distribute resources by precisely forecasting patterns of healthcare utilisation and related costs. They may then optimise premium pricing, properly manage risk, and guarantee the long-term profitability of their offers as a result.

- **Automated claims processing:**

In the health insurance sector, automated claims processing is a crucial development that has completely changed the way insurance claims are managed and greatly increased productivity. This creative method uses automated technologies and processes to expedite the whole claims process, from filing to settlement. Accelerated claim adjudication is one of the main advantages of automated claims processing. Automated systems evaluate claims based on predetermined standards using algorithms, allowing for quick and precise decision-making (Guzmán-Ortiz et al., 2020). This guarantees that valid claims are handled quickly and speeds up the policyholders' compensation procedure.

Moreover, the danger of errors resulting from manual data entry and evaluation is reduced by automated claims processing. These systems can identify discrepancies, errors, or perhaps fraudulent activity in claims submissions by utilising machine learning and data analytics. This helps to maintain the claims adjudication process's general integrity and correctness.

Policyholders benefit from a more smooth and open experience as a result of the automation of claims processing. Electronic documentary submission is available to claimants, who can also get automated updates on the status of their submissions and monitor the status of their claims in real-time. Policyholders value the convenience and openness provided by automated systems, and this increased visibility helps to build their trust and transparency provided by automated frameworks (Eckert et al., 2022). Additionally, automated techniques for processing claims help health insurance providers save money.

Insurers can maximise their operating costs by eliminating the need for a great deal of physical labour and minimising the likelihood of errors. This cost-effectiveness is especially important in an industry where total financial performance can be greatly impacted by administrative costs.

The adoption of automated claims processing not only saves money but also fits in with the larger industry trend towards digital transformation. Health insurance firms can maintain their competitiveness by adopting technology innovations that enhance their operational efficiency and establish them as modern, user-friendly service providers. For sensitive health data to be protected, strong cybersecurity safeguards must be in place when implementing automated claims processing, even with these benefits. Safeguarding patient privacy and security during the claims process necessitates strict adherence to healthcare standards, including the Health Insurance Portability and Accountability Act (HIPAA). In short, automated claims processing is a revolutionary development in the health insurance industry that brings advantages including faster processing, lower error rates, more cost-effectiveness, and higher customer satisfaction. The incorporation of automated technologies is probably going to be a major factor in determining how claims administration in the health insurance sector develops in the future.

- **Fraud Detection Algorithms:**

Through the identification and prevention of fraudulent activity, fraud detection algorithms are essential in preserving the financial stability of health insurance systems. Within the realm of health insurance, fraud can take many different forms, such as fraudulent claims, identity theft, and policyholder-provider cooperation. In addition to addressing these issues, the use of sophisticated fraud detection algorithms has other important advantages. Pattern and anomaly analysis within big datasets is one of the main purposes of fraud detection systems. These algorithms detect possibly fraudulent actions by looking at past claim data and highlighting anomalies. These algorithms can adjust and change over time, picking up new information and increasing in accuracy thanks to machine learning techniques. These algorithms support a proactive strategy for reducing the likelihood of fraud. Detecting fraud in real-time is possible because to fraud detection algorithms, which highlight questionable behaviours as they happen. This is preferable to depending just on post-claim audits, which can be laborious and resource-intensive. This prompt action safeguards policyholders and insurers by preventing money from being paid out for fraudulent claims. Furthermore, fraud detection algorithms support health insurance companies' broader efforts to control costs. Insurance companies can lessen the financial losses brought on by paying out on inflated or fraudulent claims by spotting and stopping fraudulent claims. As a result, insurance programmes continue to be financially viable, which may result in more consistent premium prices for policyholders.

The advanced algorithms used for fraud detection enable them to take into account a wide range of parameters. These could include patient demographics, provider behaviour, geographic trends, and claim history. These algorithms can improve their ability to detect fraud by developing a sophisticated knowledge of what defines normal or suspicious behaviour by looking at a large number of data points. Successful fraud detection requires industry-wide collaboration. Information concerning fraudulent activity and anonymised data are frequently shared by insurance firms. The ability of the industry to fight fraud on a larger scale is improved by this collaborative approach, which makes it possible to develop more reliable fraud detection algorithms that can spot patterns that may affect several insurers. Striking a balance between thorough fraud detection and quick processing of valid claims is crucial, though. Inadvertently flagging legitimate claims by overly aggressive algorithms might cause delays and problems for policyholders (Gellweiler & Krishnamurthi, 2020). To maintain accuracy while reducing false positives, fraud detection algorithms must be continuously improved and adjusted. Algorithms for detecting fraud are essential resources for the health insurance sector, as they aid in the detection and prevention of fraudulent activity. Their flexibility, suppleness, and capacity to take into account a wide range of factors render them indispensable for preserving the fiscal soundness of health insurance schemes and cultivating policyholder confidence. These algorithms will need to be continuously improved in order to keep up with changing fraud techniques as technology develops.

- **Data Analytics Platforms:**

The health insurance industry has evolved to rely extensively on data analytics tools, which enable providers to glean insights from large and intricate datasets that may be put to use. These platforms offer several benefits in terms of decision-making, risk management, and operational efficiency. They do this by utilising cutting-edge analytics tools and technologies to analyse, interpret, and visualise data. Finding relevant patterns and trends in a variety of datasets is one of the main functions of data analytics platforms in the health insurance industry. These insights can be used by insurers to improve pricing tactics, customise policy offerings, and adjust to shifting market conditions. These systems facilitate a data-driven approach to business decisions by utilising real-time and historical data, which eventually improves the competitiveness and sustainability of health insurance products.

Another crucial area where data analytics tools come in handy is risk management. Through the analysis of vast amounts of policyholder, healthcare provider, and healthcare trend data, these platforms assist insurers in recognising possible hazards and forecasting upcoming difficulties. These platforms' predictive analytics make it possible to identify high-risk individuals or groups, allowing insurers to proactively carry out focused interventions and preventive actions (Ostrowska, 2021). Data analytics tools greatly simplify the process of managing claims. These platforms can examine past claim data to spot trends, evaluate the veracity of claims, and spot possible fraud. By lowering reimbursements linked to inflated or fraudulent claims, this helps with cost control efforts in addition to speeding up the claims processing workflow. Additionally, in the health insurance sector, data analytics systems are crucial to improving customer experience. Insurers can obtain knowledge about preferences, behaviours, and expectations by examining client data and interactions. In order to increase customer satisfaction and retention rates, this data can be used to improve service offerings, personalise communications, and optimise customer interaction methods.

A new field where data analytics tools are having a big influence is population health management. Large-scale data aggregation and analysis capabilities of these systems enable insurers to see patterns, use resources effectively, and create focused interventions aimed at enhancing population health as a whole. In addition to helping policyholders, this proactive strategy also helps ensure the long-term viability of health insurance programmes (Flückiger & Duygun, 2022). Adopting data analytics platforms in the health insurance industry is closely linked to privacy and data security concerns. To protect the confidentiality and integrity of patient data, insurers must prioritise strong cybersecurity measures and adherence to laws like the Health Insurance Portability and Accountability Act (HIPAA). This is because health-related information is sensitive. By utilising data to support decision-making, improve risk management, expedite claims processing, enhance customer experience, and support population health management, data analytics systems play a critical role in transforming the health insurance market.

The use of data analytics in health insurance is anticipated to grow as technology develops, spurring additional advancements and enhancements in the sector.

- **Blockchain for Claims Management:**

Blockchain technology is revolutionising the health insurance industry by providing a safe and decentralised method of managing claims. In this case, the entire claims processing procedure is made more transparent, efficient, and secure by using blockchain technology. Dispersed ledger technology, which is the foundation of blockchain, allows transactions to be recorded securely and transparently over a network of computers. Blockchain technology is used in health insurance to produce an unchangeable, tamper-proof record of each stage in the claims handling procedure. This will allow all relevant parties, including policyholders, healthcare providers, and insurers, to have access to a single version of the truth.

The increased transparency that blockchain offers is among the main advantages of using technology for health insurance claims processing. Every claim-related transaction or modification is documented in a block, which is then connected chronologically. Because there is a shared source of truth and all authorised parties have real-time access to the same information, this transparency lowers the likelihood of disagreements. When handling sensitive health data, security is of utmost importance. Blockchain uses sophisticated encryption algorithms to address this issue. Claims information and patient data are kept in encrypted blocks, and consensus processes make sure that any modifications to the data need network approval (Marano, 2019). The possibility of illegal access and data breaches is greatly decreased with this degree of protection.

Smart contracts are essential to blockchain-based claims management because they are self-executing contracts with explicit terms encoded into the code. Smart contracts have the potential to automate several claims process steps in the health insurance industry, including fund disbursement, eligibility verification, and claims validation. This automation lowers the possibility of errors and fraud while also speeding up the claims processing timeframe. Blockchain provides a strong answer to the enduring problem of fraud detection in the insurance sector. The blockchain ledger's transparency and immutability facilitate the process of tracking down the source of a claim, spotting anomalies, and spotting possible fraudulent activity. Insurance companies are better equipped to detect and stop fraudulent claims because of this increased visibility. Increased operational efficiency results from blockchain's decentralised structure, which lessens reliance on a central authority. The claims management process can be made more efficient by minimising the need for intermediaries and third-party administrators. This will also lead to faster decision-making and lower administrative expenses. Although the use of blockchain in health insurance is uplifting, issues including regulatory compliance, industry standards, and interoperability with current systems need to be resolved. Blockchain has the potential to completely transform health insurance claims management, offering not only a more secure and efficient system but also one that is more transparent and accountable as stakeholders work together to overcome these obstacles and the technology advances.

- **Telemedicine platforms:**

By revolutionising the way healthcare is delivered and enhancing policyholder accessible, telemedicine platforms have emerged as essential parts of health insurance. With a number of advantages that improve healthcare as a whole, these platforms use technology to enable remote medical consultations. The wider accessibility of healthcare services is one of the main benefits of telemedicine platforms when it comes to health insurance. Through the use of these platforms, policyholders can communicate virtually with medical experts, removing geographical limitations and facilitating prompt medical advice without requiring in-person visits to healthcare institutions. This is especially helpful for policyholders who may have trouble getting to healthcare professionals since they live in rural or underserved locations.

Platforms for telemedicine reduce costs for policyholders as well as insurers. Enabling virtual consultations can save healthcare costs by preventing needless ER visits and urgent care sessions. Policyholders save money by paying less out-of-pocket for in-person visits, and insurers profit from decreased claims costs. One major aspect propelling telemedicine systems' adoption in health insurance is the ease they provide. Policyholders can receive medical consultations without having to travel or wait in queue by visiting the convenience of their homes or places of employment. For minor health conditions, routine check-ups, or follow-up visits, this convenience is especially helpful as it enables people to receive care on time without interfering with their everyday life.

Early intervention and preventive care are greatly aided by telemedicine. Healthcare providers can regularly interact with policyholders via virtual consultations and remote monitoring to monitor chronic conditions, evaluate the policyholders' state of health, and provide timely interventions. Better health outcomes and lower long-term healthcare expenditures for policyholders and insurers can result from this proactive strategy. Telemedicine systems are beneficial for health insurance since they increase client satisfaction. Insurers who incorporate telemedicine into their products frequently discover that it improves their overall value proposition, and policyholders like the flexibility and accessibility of virtual healthcare services. Increased client retention and loyalty can result from positive telemedicine encounters. It's crucial to remember that, in order to secure sensitive health information, privacy and security precautions must be taken into account when integrating telemedicine into health insurance. In virtual healthcare exchanges, maintaining the confidentiality and integrity of patient data requires adherence to standards such as the Health Insurance Portability and Accountability Act (HIPAA).

- **Wearables and health apps for policyholders:**

In the world of health insurance, wearables and health apps have grown to be indispensable tools that give policyholders greater control over their health and give insurers useful information for individualised treatment plans. Health apps on smartphones and wearable technology like smartwatches and fitness trackers offer a host of advantages for insurance companies as well as customers. For policyholders, the ability to track and monitor multiple aspects of their health in real-time is a major benefit of wearables and health applications. These gadgets have the ability to record information on heart rate, sleep patterns, physical activity, and other health parameters (Nayak et al., 2019). Policyholders can use this information to better understand their lifestyle and make changes that will enhance their general health and well-being.

Insurers can gain a more precise and up-to-date picture of policyholders' health thanks to the data produced by wearables and health applications. With the use of this data, insurers can use personalised risk assessment to customise coverage and rates according to the health habits and results of each policyholder. The move to customised insurance plans encourages a more fair sharing of risk and could encourage subscribers to lead healthier lives. Wearables and health applications greatly aid preventive care. These gadgets can offer timely alerts or warnings for possible health problems by constantly tracking health parameters. With this information, insurers may better serve their policyholders' long-term health by lowering the likelihood of costly medical procedures and by providing proactive health interventions like wellness programmes and specialised health coaching. Policyholders' participation and accountability are increased when wearables and health apps are incorporated into health insurance policies. As a result, insurers gain from a healthier and possibly lower-risk client base, while policyholders are encouraged to adopt healthy habits. The usage of wearables and health applications in the context of health insurance requires careful consideration of privacy and data security. Insurers must guarantee that stringent protocols are implemented to safeguard the privacy and accuracy of health information gathered from policyholders. Establishing and preserving trust with policyholders requires adherence to pertinent legislation, such as the General Data Protection Regulation (GDPR), as well as regulations unique to the healthcare industry, such as the Health Insurance Portability and Accountability Act (HIPAA). By enabling policyholders to take an active role in their health management and enabling insurers to transition to more individualised and preventive healthcare models, wearables and health apps have completely changed the landscape of health insurance. These technologies are expected to become more deeply integrated into health insurance policies as they develop, opening up new avenues for better health outcomes and a more customer-focused and sustainable insurance sector.

- **Automated processes for customer onboarding:**

The health insurance sector has evolved to rely heavily on automated customer onboarding procedures, which expedite the accurate and efficient enrollment of new subscribers. Using cutting-edge systems and technology, automated onboarding procedures provide several benefits regarding efficiency, precision, and general client satisfaction. Enrollment process acceleration is one of the main advantages of automated customer onboarding in health insurance (Malika Shuxratovna & Nuriddin Rustam Ugli, 2021). By using electronic signatures, digital forms, and automated document verification, new policyholders can expedite the onboarding process and eliminate the need for copious paperwork. This not only shortens the enrollment process but also improves the client experience by offering a smooth and intuitive onboarding process. Accuracy and consistency in the data gathered from policyholders are guaranteed via automation in customer onboarding. The possibility of errors resulting from manual data entry is reduced when data inputs are validated in real-time by automated systems. This precision is crucial when it comes to health insurance, as it is required for the construction of policies, the computation of premiums, and the precise assessment of coverage.

Automated onboarding processes aid compliance with regulatory regulations. Adherence to data protection standards, such as the Health Insurance Portability and Accountability Act (HIPAA), is crucial since health insurance entails the acquisition of sensitive personal and medical information. In order to lower the risk of regulatory problems and guarantee the security and confidentiality of client data, automated systems can be built with the ability to enforce and audit compliance procedures. Automated workflows can guide new policyholders through the process, giving them the appropriate documentation and clear instructions. This enhances the onboarding process and lessens the possibility of incomplete or inaccurate submissions, which lowers the need for additional follow-ups and adjustments. Health insurance companies' scalability is improved via automated onboarding procedures. Automated systems can handle higher enrollment numbers as the client base expands without causing a corresponding rise in administrative workload. For insurance companies looking to effectively handle a sizable and varied customer, this scalability is essential. A human-centric approach and automation must coexist in harmony. Although common activities can be handled by automated procedures, complicated inquiries, individualised support, and making sure that clients feel supported throughout the onboarding process still requires human intervention.

- **Automating KYC/AML Compliance:**

In the health insurance industry, automating Know Your Customer (KYC) and Anti-Money Laundering (AML) compliance procedures has become a crucial endeavour, as it improves customer onboarding speed and conforms to regulatory standards. Automated KYC/AML compliance in health insurance enables a comprehensive and standardised approach to customer identity verification and potential financial crime detection by utilising technology and sophisticated algorithms. The speed and accuracy of customer identity verification is one of the main benefits of automating KYC/AML compliance in health insurance. Automated solutions could evaluate and cross-reference vast amounts of client data quickly across many databases, which guarantees an accurate and timely identification assessment. This reduces the possibility of human error connected with conventional verification techniques while simultaneously speeding up the onboarding process. Overall fraud detection efficacy is improved by automation in KYC/AML compliance. A more advanced and proactive method of spotting possible money laundering or fraudulent activity is to use machine learning algorithms to examine trends, abnormalities, and warning signs in client data. In industries like healthcare and finance where preventing financial crimes and adhering to regulations are critical, this degree of monitoring is crucial.

Health insurance companies can reduce costs by implementing automated KYC/AML compliance procedures. Insurers can minimise the operating costs related to compliance by decreasing the amount of documentation required and the need on manual labour. Because of its affordability, insurers can spend resources more effectively while maintaining constant compliance with regulatory standards. Automation guarantees that KYC/AML compliance is handled uniformly and consistently for all of the clientele. Insurers can lower the risk of errors or overlook by establishing a consistent standard of due diligence by doing the same thorough checks on every customer. This strengthens the compliance process's overall integrity and assists in meeting regulatory requirements. Since health insurance data is sensitive and strong privacy protections are required, the industry needs to integrate automation into KYC/AML compliance. To guarantee that client data is handled securely and privately, automated systems can be made to abide by healthcare laws. Even though automation makes routine compliance operations more efficient, human oversight is still necessary. Human knowledge should be added to automated systems, particularly when managing difficult cases, dealing with unusual situations, and reaching well-informed choices that could call for delicate judgment. A complete and comprehensive compliance framework is ensured by a balanced approach that blends automation and human intuition. To comply with regulations, improve operational effectiveness, and enhance fraud prevention strategies, health insurance companies must automate KYC/AML compliance. The landscape of health insurance will become safer and more compliant as technology develops and intelligent automation in compliance processes becomes more sophisticatedly integrated.

- **Tools for Data Security and Privacy:**

In the health insurance sector, where sensitive personal and medical data is handled, maintaining strong data security and privacy is crucial. To protect information, adhere to legal requirements, and foster confidence with policyholders, a range of instruments and technologies are used. One essential tool for safeguarding data while it's in transit and at rest is encryption. Information including patient records, policy details, and financial transactions can be safely encoded using sophisticated encryption methods, limiting illegal access even if the data is intercepted. The management and restriction of user access to sensitive health data requires the use of access controls and authentication procedures. The use of multi-factor authentication, biometric verification, and role-based access restrictions can effectively mitigate the risk of data breaches by limiting access to specific information to only authorised persons. These instruments are essential for preserving patient record integrity and confidentiality in health insurance systems.

The prevention of sensitive data from being transmitted without authorization is greatly aided by data loss prevention (DLP) solutions (Shevchuk et al., 2020). These instruments keep an eye on and regulate information flow, preventing unintentional or deliberate data leaks. DLP tools reduce risks associated with human mistake or malicious intent by preventing the unintentional disclosure of private information, which is crucial in the health insurance industry where patient privacy is of utmost importance. In the health insurance industry, privacy-enhancing technologies (PETs) like homomorphic encryption and differential privacy are becoming more and more popular. These technologies enable the study of encrypted data without the need to decrypt it, protecting personal information while still yielding insightful results. This is especially important when exchanging data is required for analysis or study without jeopardising the privacy of the individual. Programmes for employee awareness and training are essential for encouraging a data security and privacy-conscious culture in health insurance companies. The first line of defence against possible security threats is knowledgeable people, and training initiatives make sure that workers are prepared to identify and address security issues (Gorchakova, 2019). Strong data security and privacy in the health insurance sector require a multipronged strategy that includes encryption, access controls, DLP, blockchain, privacy-enhancing technology, compliance management, and employee training. To keep ahead of potential dangers and safeguard the private data entrusted to health insurance carriers, it is imperative to continuously integrate cutting-edge tools and technology.

- **Mobile Apps and Portals:**

Mobile applications and portals, which enable policyholders to manage their healthcare needs simply and effectively, have become vital components of the health insurance landscape. These digital platforms offer several features that increase overall customer happiness, speed up administrative processes, and give customers the ability to take control of their health and insurance coverage. The convenience that mobile apps and portals provide to policyholders is one of the main benefits of health insurance. From the palm of their hands, users may access these platforms' insurance information, policy data, and pertinent health resources. Mobile apps offer real-time information for verifying coverage, monitoring claims status, and accessing digital insurance cards. This eliminates the need for policyholders to negotiate convoluted phone systems or wait for help. Mobile apps for health insurance often provide features that make processing claims easier. Through the app, policyholders can easily file claims, add supporting documents, and monitor the progress of their claims. This improves transparency by updating users on the status of their submissions and speeds up the claims process.

To improve communication between insurers and policyholders, mobile apps and portals are essential. Users are alerted about impending renewals, critical updates, and payment reminders using push notifications and alerts. Furthermore, by giving insurers a direct line of communication for pertinent health and wellness information, these platforms encourage policyholders to lead healthy lifestyles and receive preventive treatment. Policyholders can obtain virtual healthcare consultations thanks to the growing integration of telemedicine technologies into health insurance smartphone apps. This feature offers policyholders a prompt and simple substitute for in-person visits in the event of non-emergency medical difficulties. This helps policyholders and insurers save money while also improving accessibility to healthcare services. Mobile apps and portals greatly aid the promotion of health and wellness programmes. Features that measure exercise, track fitness objectives and offer individualised health advice are available from many health insurance companies. These resources motivate policyholders to pursue healthier lifestyles, which may have long-term advantages including lower medical expenses and enhanced general health. The way that policyholders interact with their health insurance plans has evolved through the advent of mobile apps and portals. These online resources provide an intuitive user interface for managing tasks linked to insurance, accessing healthcare services, and learning about health and wellness (Lin & Chen, 2020). As technology advances, it is projected that adding new features and functionality to health insurance apps would encourage a more proactive approach to healthcare management and enhance the user experience in general.

The Role of Autonomous Systems in Operational Excellence and Customer-Centric Services of the Health Insurance Industry

The use of autonomous systems in the health insurance sector is a paradigm-shifting development that will fundamentally alter how healthcare is handled, delivered, and experienced. These autonomous systems, which use a variety of technologies including automation, machine learning, and artificial intelligence (AI), have played a crucial role in tackling persistent issues and advancing the sector's focus on operational excellence and customer-centric services.

- Claims Processing Streamlining:

The integration of diverse technologies, such as automation, machine learning, and artificial intelligence, is causing a disruptive shift in the health insurance sector by streamlining many parts of operations. Claims processing is one area where these technologies have had a big influence. Specifically, automation has sped up the appraisal and settlement of claims by decreasing manual intervention and streamlining the claims procedure. As a result, compared to previous, manual processing methods, there has been a decrease in errors and an increase in operational efficiency. An innovative factor in the healthcare sector is the ability of autonomous systems to expedite the processing of health insurance claims (Kumar, 2017). These solutions bring in a new era of efficacy and openness by utilising artificial intelligence to deliver previously unheard-of speed, precision, and efficiency. A more responsive and user-friendly ecosystem is anticipated for the health insurance market in the future thanks to the convergence of human expertise and autonomous capabilities brought about by technological advancements.

The conventional method of processing health insurance claims has long been linked to complications, hold-ups, and the possibility of mistakes (Mueller, n.d.). Due to the increased volume of healthcare data and the complex nature of medical records, manual processing frequently results in inefficiencies, which aggravate stakeholders and cause reimbursements to be delayed. Acknowledging these difficulties, the sector has resorted to self-governing platforms to transform the complete claims-handling process.

The Insurtech startup Lemonade, situated in New York, put autonomous technologies in place to improve customer satisfaction and expedite the processing of claims. Their AI-powered technology automates certain parts of managing claims by utilising machine learning algorithms and natural language processing. For example, in a matter of seconds, the AI system evaluates a policyholder's claim submitted via Lemonade's app, confirms the policy's coverage, and determines the payout amounts. Lemonade dramatically speeds up the claims process by removing the need for human interaction in common chores; many claims are handled quickly. Lemonade has received recognition for this novel strategy's effectiveness and openness, which has raised the bar for the sector.

One of the biggest insurers in China, Ping An Insurance Group, uses autonomous systems to reduce insurance fraud and improve operational effectiveness. To find suspicious trends and anomalies suggestive of fraud, Ping An's AI-driven fraud detection system examines enormous volumes of data, including claim histories, medical records, and social media activity. Ping An has been able to save billions of dollars a year by drastically reducing the amount of money paid out for false claims by automating the detection process. Furthermore, Ping An can keep ahead of new fraud schemes and successfully defend the interests of its policyholders since the AI system is always learning from fresh data.

These case studies show how autonomous systems are revolutionising the health insurance market by increasing consumer satisfaction, speeding up the processing of claims, and boosting accuracy. The insurance business has enormous room for innovation and disruption as long as insurers keep investing in AI and machine learning technology. This bodes well for a day when insurance services will be more streamlined, easily accessed, and individually tailored than ever before.

- **Fraud Detection and Prevention:**

Artificial intelligence and machine learning have been incorporated into health insurance, resulting in more advanced fraud detection and prevention. In order to spot trends, abnormalities, and warning signs that can point to possible fraud, these systems examine enormous datasets. Insurance companies can protect the integrity of the insurance market and safeguard their financial interests by using predictive analytics to proactively identify and stop fraudulent claims (Prakhar Harit, 2021). Machine learning models are able to adjust to changing fraud strategies due to their continual learning capabilities, which makes fraud detection methods more effective overall.

The health insurance sector has seen a major transformation in fraud detection because to artificial intelligence (AI), which has significantly increased efficiency and accuracy. AI-powered fraud detection systems can analyse enormous volumes of data with unmatched speed and precision by utilising cutting-edge algorithms and machine learning techniques. This allows insurers to identify and stop fraudulent activity more successfully than in the past. These real-time systems can adjust and learn from fresh data over time to increase their accuracy. They also continuously monitor incoming claims and transactions for suspicious activity.

- **Enhancing Customer Experience:**

One of the top priorities for health insurance companies is improving the client experience, and technology is essential to reaching this objective. The relationships between insurers and policyholders are streamlined via chatbots, automated communication systems, and self-service portals driven by artificial intelligence (Joshi et al., 2020). In addition to facilitating quicker inquiry resolution and instantaneous information access, these technologies also improve customer responsiveness and personalisation. Policyholders gain from this increased sense of participation and satisfaction with their insurance companies.

With their ability to provide policyholders with individualised services, expedited response times, and innovative features, autonomous systems are essential to improving the customer experience in the health insurance sector. These systems provide customised healthcare plans or wellness programmes based on the medical histories and lifestyle characteristics of individual policyholders, via the use of artificial intelligence and powerful data analytics. Additionally, policyholder concerns can be instantly answered by AI-powered chatbots and virtual assistants, which speeds up response times and simplifies the claims procedure. Cutting-edge elements that enhance the policyholder experience further include personalised health insights and incentives for continuing healthy behaviours, such as wearable technology and health-tracking apps coupled with autonomous systems. All things considered, policyholders benefit from a smooth and frictionless experience provided by autonomous systems, which strengthens their bonds with insurers and eventually raises customer satisfaction and loyalty.

- **Data-Driven Decision Making:**

In the health insurance sector, data-driven decision-making is becoming essential to efficient management practices. This paradigm is supported by automation, machine learning, and artificial intelligence, which offer useful insights from enormous datasets. With the use of these technologies, insurers may better understand individual and population health data, evaluate risk, adjust pricing, and customise insurance plans (Cappiello, 2018). Making well-informed decisions helps to improve risk management, allocate resources more effectively, and provide insurance products that are focused on the needs of the consumer.

Data-driven decision-making in the health insurance industry has significantly evolved with the integration of autonomous systems, allowing insurers to use data in previously unheard-of ways. These cutting-edge technologies examine enormous volumes of organised and unstructured data, from claim histories and medical records to demographic data and socioeconomic determinants of health, by utilising artificial intelligence, machine learning, and data analytics.

- Autonomous systems give insurers important insights into trends, patterns, and risk factors by processing and interpreting this data in real-time. This allows insurers to make well-informed decisions that spur strategic growth and innovation. Predictive analytics, for instance, can be used by insurers to anticipate patterns in healthcare utilisation, allocate resources optimally, and provide focused treatments that enhance health outcomes and reduce risks. Additionally, based on the unique needs and preferences of each policyholder, autonomous systems allow insurers to improve customer experience, customise offerings, and personalise services. The health insurance industry has seen a revolution in data-driven decision-making due to the integration of autonomous systems. This has allowed insurers to stay ahead of the curve, adjust to shifting market dynamics, and provide better value to stakeholders and policyholders alike.

Compliance and Regulation

In the health insurance business, adherence to laws like the Health Insurance Portability and Accountability Act (HIPAA) is a requirement that cannot be compromised. By automating regulatory inspections, keeping an eye on compliance with data protection regulations, and enabling audit trails, automation and artificial intelligence help to ensure compliance. By informing policyholders that their sensitive health data is managed in compliance with strict regulatory criteria, this not only helps insurers avoid legal ramifications but also builds trust.

Several facets of the health insurance sector are undergoing radical change as a result of the industry's adoption of automation, machine learning, and artificial intelligence. These technologies, which range from claims processing to fraud detection, customer experience enhancement, data-driven decision-making, and compliance management, all help to make the health insurance market more effective, safe, and focused on the needs of the consumer. Future developments in these fields are probably going to influence how health insurance is run as technology progresses.

Regulatory Framework for Autonomous Systems in Health Insurance Industry

The regulation of autonomous systems that are redefining health insurance is essential to ensuring that these innovations uphold moral and legal obligations, safeguard patient privacy and security, and keep policyholders' trust. Providing customer-centric services and achieving operational excellence must take place within a clear regulatory framework. Here are some key regulations and considerations for autonomous systems in health insurance:

- **Data Privacy and Security Regulations:**

Health insurance autonomous systems need to abide by data privacy laws, like the Health Insurance Portability and Accountability Act (HIPAA) in the US. Strict requirements are established by HIPAA to secure sensitive health information. These criteria require the use of security measures, risk assessments, and safeguards to guarantee the confidentiality, availability, and integrity of patient data.

- **Consent Mechanisms:**

Before setting up autonomous systems that process personal health data, obtaining informed consent from policyholders is imperative. Consent procedures that are transparent and unambiguous must be in place, outlining how self-governing systems will handle and interpret patient data. Adherence to laws such as the General Data Protection Regulation (GDPR) in the European Union is very important in this context.

- **Healthcare Provider Compliance:**

Autonomous systems must comply with rules regulating communications with healthcare providers. To guarantee a smooth integration with the current healthcare infrastructure, these regulations cover data sharing, interoperability, and adherence to Electronic Health Record (EHR) standards.

- **Standards and Best Practices:**

The deployment of autonomous systems ethically and efficiently depends on adherence to industry standards and best practices. A standardised and responsible approach is ensured by adhering to recommendations from organisations such as the International Organisation for Standardisation (ISO) and industry-specific agencies.

- **Monitoring Regulatory Compliance:**

It's critical to keep an eye on any updates or modifications to regulations. Regulations change over time, and autonomous systems need to be flexible enough to modify accordingly. Compliance monitoring makes sure that changes are done on time to keep adherence to the law.

- **Regulatory Reporting:**

Reporting protocols need to be in place to adhere to the transparency requirements outlined by regulations. This includes reporting any incidents, like as data breaches and system failures, that might have an impact on the security or privacy of policyholders' health information.

- **Anti-Fraud Regulations:**

Autonomous systems ought to abide by anti-fraud guidelines in the context of health insurance. To identify and stop fraudulent activity, these systems should use machine learning and advanced analytics, in accordance with industry-specific anti-fraud regulations.

- **International Regulations:**

When operating internationally, health insurers have to pay attention to international legislation. Comprehending and adhering to the diverse nations' health data protection regulations is imperative to guarantee the legitimate handling and international transfer of medical records.

- **Fraud Detection and Reporting:**

Autonomous systems should adhere to anti-fraud legislation as well as have procedures in place for identifying and reporting possible fraudulent activity. This involves collaborating with industry networks to exchange information about emerging fraud patterns, real-time surveillance, and pattern recognition.

- **Ethical Considerations:**

Ethical considerations, beyond legal requirements, are paramount. The utilisation of autonomous systems requires accountability, justice, and transparency. Establishing and upholding ethical standards requires cooperation between policymakers, insurers, and technologists.

- **Consumer Protection:**

A primary concern is making sure that customers are protected. Policyholders' well-being should be improved by autonomous systems, not jeopardized. The provision of channels for dispute resolution and decision-making transparency are essential elements of consumer protection.

- **Transparency and Explainability:**

AI systems for health insurance must be transparent and understandable. The right to know how decisions made by autonomous systems affect policyholders' coverage or care is guaranteed. People should be able to understand the reasoning behind algorithmic judgments through the integration of comprehension methods.

- **Access Control and Encryption:**

Strong encryption techniques and access control systems are essential for protecting health information. Strong access controls guarantee that sensitive information can only be accessed by those who are permitted, and encryption protects data both during communication and storage.

The implementation of autonomous systems in the health insurance industry necessitates a thorough comprehension of and adherence to a wide range of rules and factors. This entails abiding by the law as well as adhering to moral standards, openness, and the defence of consumer rights. Continued attention to detail and flexibility are crucial to the responsible application of autonomous systems in the health insurance industry as the regulatory environment changes. Insurance companies are obligated to make sure that their autonomous systems not only increase efficiency but also comply with the intricate web of healthcare and insurance rules to achieve operational excellence and deliver customer-centric services. Respecting these rules promotes trust, safeguards policyholders' rights, and upholds the integrity of the health insurance sector.

Key Challenges of Increased Automation in the Health Insurance Sector

The increase in automation in the health insurance industry offers both great opportunities and serious moral conundrums. The most significant of these difficulties is the possibility of biassed decision-making in automated systems, which could result in unequal access to healthcare services and coverage. Existing biases in healthcare data, such as inequities based on race, ethnicity, or socioeconomic position, run the risk of being perpetuated as algorithms analyse massive volumes of data to decide insurance rates, coverage eligibility, and claims processing. It is critical to maintain social justice ideals and prevent escalating already-existing healthcare disparities by ensuring fairness and equity in automated decision-making systems.

Concerns regarding data security and privacy are also raised by the automation of health insurance procedures. The growing digitization of medical records and personal health data makes it morally important to protect sensitive information against abuse, breaches, and unauthorised access. Patients need to have confidence that automated systems respect strict privacy standards and laws, like the Health Insurance Portability and Accountability Act (HIPAA) in the US, and that their private health information is handled safely. Retaining confidence in the healthcare system and assuring ethical behaviour in the health insurance industry requires striking a balance between the advantages of automation in expediting administrative procedures and enhancing productivity and the requirement to safeguard patient privacy and data security.

Findings of the Study

After extensive research, several important results come to light regarding how autonomous systems affect the health insurance industry. First of all, the claims processing sector has seen a radical transformation thanks to the quick response times and streamlined operations provided by these state-of-the-art technologies. Routine chores like data input and document processing can be handled by automated systems with previously unheard-of speed and accuracy, which benefits policyholder happiness and speeds up payments for healthcare providers. Second, by using AI algorithms to analyse enormous volumes of data and find patterns suggestive of fraudulent conduct, autonomous systems have revolutionised the field of fraud detection. By taking a proactive stance, insurers can protect the integrity of the insurance market and lessen the financial losses caused by fraud. Insurance companies are now able to improve client experience by offering personalised services, faster response times, and cutting-edge features thanks to the integration of autonomous systems. Using AI-driven insights, insurers can better make decisions, customise products to meet the demands of specific clients, and build closer bonds with policyholders.

Autonomous systems have a wide-ranging and significant effect on health insurance overall. These technologies have enabled insurers to provide policyholders with a more individualised and responsive level of service, in addition to increasing efficiency and accuracy in claims processing and fraud detection. Insurance companies are now able to maintain their competitiveness in a market that is changing quickly thanks to the industry's increased use of autonomous systems, which has also encouraged innovation and distinction. Therefore, autonomous systems have a transformative overall effect on health insurance, resulting in improvements to consumer experience, operational efficiency, and, eventually, healthcare service-delivery.

FUTURE RESEARCH DIRECTIONS

Future research will have a wide range of areas to explore, including applications of autonomous systems in the health insurance industry. Promising research areas include the investigation of robotic process automation (RPA) in claims processing to improve efficiency and accuracy, AI-driven disease prediction and prevention based on health data of policyholders, and the adoption of blockchain for safe and transparent health data sharing. Further, the integration of IoT devices for real-time health monitoring and the effect of telemedicine adoption on healthcare access and policyholder satisfaction present exciting research directions. The notion of investigating standardised data interoperability protocols, utilising AI for fraud investigation, and creating predictive models for pandemic preparedness is quite promising. Finally, insurers using autonomous systems can monitor and improve customer happiness and loyalty by developing service measures that are centred on the needs of the consumer. Together, these study areas help to advance the application of autonomous systems in health insurance by promoting improved effectiveness, client-centeredness, and data-driven services.

CONCLUSION

In essence, the use of autonomous systems in the health insurance industry represents a major change. It is a journey towards improved operational effectiveness, client-focused services, and a reaffirmed commitment to serving policyholder demands. In order to support a new era of healthcare coverage that puts the patient at its centre, the health insurance sector is poised to become more effective, responsive, and technologically sophisticated as autonomous systems continue to develop and broaden their capabilities. The study reveals the vital role of autonomous systems in expediting fraud detection, optimising claims processing, improving customer satisfaction, enabling data-driven decision-making, and assuring compliance with the complex web of healthcare regulations. We will shed light on the revolutionary power that autonomous systems represent by investigating these aspects, not just as instruments for automation but also as the designers of a paradigm shift in the health insurance ecosystem.

REFERENCES

Akter, S., Michael, K., Uddin, M. R., McCarthy, G., & Rahman, M. (2022). Transforming business using digital innovations: The application of AI, blockchain, cloud and data analytics. *Annals of Operations Research, 308*(1–2), 7–39. doi:10.1007/s10479-020-03620-w

Cappiello, A. (2018). Digital Disruption and InsurTech Start-ups: Risks and Challenges. In Technology and the Insurance Industry (pp. 29–50). Springer International Publishing. doi:10.1007/978-3-319-74712-5_3

Eckert, C., Neunsinger, C., & Osterrieder, K. (2022). (Repeat J)Managing customer satisfaction: Digital applications for insurance companies. *The Geneva Papers on Risk and Insurance. Issues and Practice, 47*(3), 569–602. doi:10.1057/s41288-021-00257-z

Eling, M., Nuessle, D., & Staubli, J. (2022). The impact of artificial intelligence along the insurance value chain and on the insurability of risks. *The Geneva Papers on Risk and Insurance. Issues and Practice, 47*(2), 205–241. doi:10.1057/s41288-020-00201-7

Flückiger, I., & Duygun, M. (2022). (LR 111)New technologies and data in insurance. In Geneva Papers on Risk and Insurance: Issues and Practice, 47(3). Palgrave Macmillan. doi:10.1057/s41288-022-00274-6

Gellweiler, C., & Krishnamurthi, L. (2020). Editorial: How digital innovators achieve customer value. In Journal of Theoretical and Applied Electronic Commerce Research, 15(1). Universidad de Talca. doi:10.4067/S0718-18762020000100101

Gorchakova, E. R. (2019). *The impact of digitalization on the health insurance system.* doi:10.18411/lj-03-2019-48

Guzmán-Ortiz, C. V., Navarro-Acosta, N. G., Florez-Garcia, W., & Vicente-Ramos, W. (2020). (LR 122)Impact of digital transformation on the individual job performance of insurance companies in peru. *International Journal of Data and Network Science, 4*(4), 337–346. doi:10.5267/j.ijdns.2020.9.005

Harit, P. (2021). (LR 83)The Rise of Insurtech: The Ups and Downs of New Trend. SSRN *Electronic Journal*. doi:10.2139/ssrn.3799576

Joshi, V., Joshi, V. V., Pawar, N., & Acharya, S. (2020). *Digitalisation of Health Care in India: Initiatives and Challenges. 40*. https://www.researchgate.net/publication/354586176

Kumar, A. (2017). (LR 117) Impact of Digitalisation on Service Sector in India. *Reviewed Refereed Research Journal*, *8*(15). https://www.researchgate.net/publication/360782465

Lin, L., & Chen, C. (2020). (LR 82)The promise and perils of insurtech. *SSRN*, *2020*, 115–142. doi:10.2139/ssrn.3463533

Malika Shuxratovna, U., & Nuriddin Rustam Ugli, A. (2021). (LR 114)DIGITALIZATION OF THE INSURANCE MARKET. In Multidisciplinary Peer Reviewed Journal ISSN, 7(2).

Marano, P. (2019). (LR 112)Navigating insurtech: The digital intermediaries of insurance products and customer protection in the EU. *Maastricht Journal of European and Comparative Law*, *26*(2), 294–315. doi:10.1177/1023263X19830345

Mueller, J. (n.d.). *InsurTech Rising: A Profile of the InsurTech Landscape DECEMBER 2018.*

Nayak, B., Bhattacharyya, S. S., & Krishnamoorthy, B. (2019). LR 77: Integrating wearable technology products and big data analytics in business strategy: A study of health insurance firms. *Journal of Systems and Information Technology*, *21*(2), 255–275. doi:10.1108/JSIT-08-2018-0109

Ostrowska, M. (2021). (LR 110)Does new technology put an end to policyholder risk declaration? The impact of digitalisation on insurance relationships. *The Geneva Papers on Risk and Insurance. Issues and Practice*, *46*(4), 573–592. doi:10.1057/s41288-020-00191-6

Sarkar, S. (2021). LR 81: The Evolving Role of Insurtech in India: Trends, Challenges and The Road Ahead. *The Management Accountant Journal*, *56*(12), 30–37. doi:10.33516/maj.v56i12.30-37p

Shah, A., Mehrotra, P., Sinha, S., & Sha, J. (2021). Bcg Insurtech Report India Insurtech Landscape And Trends. In Boston Consulting Group.

Shevchuk, O., Kondrat, I., & Stanienda, J. (2020). (LR 115)Pandemic as an accelerator of digital transformation in the insurance industry: Evidence from Ukraine. *Insurance Markets and Companies*, *11*(1), 30–41. doi:10.21511/ins.11(1).2020.04

Stoeckli, E., Dremel, C., & Uebernickel, F. (2018). LR 63; Exploring characteristics and transformational capabilities of InsurTech innovations to understand insurance value creation in a digital world. *Electronic Markets*, *28*(3), 287–305. Advance online publication. doi:10.1007/s12525-018-0304-7

Yaneva, T. (n.d.). *Digital Transformation of Insurance Sector*.

Chapter 11
Artificial Intelligence in Robotics

Preethiya T.
https://orcid.org/0000-0003-3504-1884
SRM Institute of Science and Technology, India

Priyanga Subbiah
https://orcid.org/0000-0002-2395-7492
SRM Institute of Science and Technology, India

Pandiarajan T.
https://orcid.org/0009-0007-7808-8961
Rajalakshmi Institute of Technology, India

Stephen Ojo
College of Engineering, Anderson University, USA

Vijayalakshmi S.
SNS College of Engineering, India

ABSTRACT

The advent of artificial intelligence (AI) has had a profound impact on the realm of robotics, fundamentally altering the capabilities of self-governing devices. This abstract examines the significant influence of artificial intelligence (AI) on the field of robotics, emphasizing notable progress and practical implementations. Artificial intelligence (AI)-powered robots demonstrate improved capabilities in perception, decision-making, and adaptability, which allows them to thrive in a wide range of jobs across several domains such as industry, healthcare, space exploration, and autonomous vehicles. Machine learning methodologies, such as deep learning and reinforcement learning, enable robots to acquire aptitudes, enhance their performance, and engage in intelligent interactions with their surroundings. The ethical considerations, safety measures, and societal repercussions pertaining to AI-driven robots are also examined and analyzed.

DOI: 10.4018/979-8-3693-1962-8.ch011

INTRODUCTION

The incorporation of AI has resulted in significant shifts in the ways in which we interact with and make use of robotic systems as a result of its profound impact on the area of robotics. Robotics, which examines and advances intelligent systems that are capable of mimicking human activities, is currently in a leading position in the field of AI research and practical application (Nehal, 2023). This is due to the fact that robotics comprises the examination and improvement of such systems. The field of robotics, which is well-known for having qualities that are both dynamic and transdisciplinary, has been witness to considerable breakthroughs in recent years, which has led to the production of increasingly sophisticated robots with expanded capabilities. A new era that is characterised by the adoption of intelligent automation has begun with the introduction of AI into the field of robotics. This new age will be known as the Fourth Industrial Revolution. Robotic creatures have experienced a major transformation since their original purpose as simply mechanical tools. They have transitioned into cognitive beings capable of sensing, learning knowledge, and modifying their behaviour in reaction to the environment in which they find themselves. The shift that was described earlier has had a significant influence on a variety of spheres of activity, including manufacturing, healthcare, the exploration of space, and everyday life.

The purpose of this statement is to offer a framework for investigating the intricate and multidimensional nature of the connection that exists between robotics and artificial intelligence (Gurjeet, 2022). The purpose of this research is to analyse the fundamental components and technological progressions that are driving the incorporation of artificial intelligence inside the domain of robotics. In addition to this, the objective is to investigate the real-world uses of this technology in a variety of contexts and to kick off a conversation about the potential benefits and difficulties that are associated with this rapidly developing area of research. The combination of AI and Robotics is more than simply a technological convergence; it is a profound and revolutionary shift that is redefining a wide range of industries as well as significantly altering our way of life and the professional techniques that we use.

THE ROLE OF NLP IN HUMAN-ROBOT COMMUNICATION

The field of human-robot interaction has witnessed the emergence of Natural Language Processing (NLP) as a crucial technology. NLP facilitates the ability of computers to understand, produce, and react to human language in a manner that is more intuitive and akin to human behaviour. The advent of this transformative invention has inaugurated a novel epoch of communication between human beings and robots, which exhibits considerable potential for many applications across several industries and in our everyday existence.

This investigation examines the significance of natural language processing (NLP) in facilitating communication between people and robots, highlighting its tremendous influence on bridging this gap (Jonathan, 2023). This study aims to investigate the fundamental technologies, strategies, and algorithms that enable robots to comprehend and produce human language. Natural Language Processing (NLP) is of utmost importance in facilitating smooth communication between humans and various machines, including chatbots, virtual assistants, service robots, and autonomous cars.

Moreover, this discourse underscores the practical implementations of Natural Language Processing (NLP) within the field of robotics, placing particular emphasis on its ability to augment human-robot cooperation, improve customer service, better healthcare practises, and facilitate educational endeavours, among various other domains. This study also addresses the emerging difficulties and ethical implications related to natural language processing (NLP) in the context of human-robot communication, as we explore the dynamic intersection of technology and human contact.

The incorporation of natural language processing (NLP) into the realm of human-robot communication is fundamentally transforming the manner in which we engage with and derive advantages from robotic systems. AI has the capacity to enhance our interactions with machines by making them more intuitive, efficient, and engaging (Nam et al., 2021). This has the potential to create a future in which robots possess a genuine understanding of human behaviour and can respond to us in a manner that is both natural and reminiscent of human-to-human communication.

CONTROL OF THE ROBOT AND DETAILED ROUTE PLANNING

The optimisation of robotic control and the meticulous design of routes are fundamental components of robotics that have a direct influence on their operational efficiency and practicality. Within the domain of robotics, the term "control" pertains to the assortment of techniques and algorithms employed for the purpose of regulating a robot's locomotion, behaviours, and engagements with its surrounding milieu (Nagadevi, 2021). In contrast, route planning is the meticulous determination of pathways and trajectories that a robotic system must adhere to in order to accomplish its designated goals. These elements play a crucial role in a wide range of applications, including as industrial automation, autonomous cars, and even space exploration.

The Management of the Robot

The process of robot control entails the integration of both hardware and software elements. The hardware generally encompasses sensory components for perception and actuation mechanisms for motion. In addition, the software component contains control algorithms, feedback systems, and decision-making processes. These various components collaborate to facilitate the locomotion, execution of tasks, and responsiveness of the robot to its environment. Control tactics encompass a spectrum of approaches, spanning from rudimentary teleoperation facilitated by a human operator to comprehensive autonomy, wherein the robot independently makes decisions.

Control systems can be classified into two main categories: open-loop systems and closed-loop systems. Open-loop control refers to a control system in which predetermined actions are conducted without receiving feedback from the surrounding environment. In contrast, closed-loop control refers to the utilisation of ongoing feedback from sensors to dynamically modify the robot's operations, hence augmenting precision and flexibility.

In-Depth Route Planning

Thorough route planning plays a pivotal role in facilitating the navigation of robots in intricate situations. The task encompasses the process of ascertaining the most appropriate routes, circumventing hindrances, and optimising for variables such as velocity, energy conservation, and security. The significance of route planning is particularly pronounced in the context of autonomous vehicles, unmanned aerial vehicles (UAVs), and industrial robotic systems.

The process of planning routes for robots frequently utilises sophisticated algorithms and methodologies derived from the fields of computational geometry, artificial intelligence, and machine learning. Prominent methodologies encompass the utilisation of A* and Dijkstra algorithms in the context of graph-based pathfinding, the employment of rapidly exploring random trees (RRT) in scenarios involving dynamic surroundings, and the application of artificial neural networks to acquire knowledge regarding ideal paths. In addition, the process of route planning involves the incorporation of the robot's environmental perception, which is facilitated by sensors like as cameras, lidar, or radar (Gaur et al., 2021). The aforementioned information is included into the decision-making process in order to effectively adjust to real-world circumstances.

The management of robotic systems and the meticulous design of navigation paths play a crucial role in the domain of robotics. Several factors influence the ability of robots to efficiently carry out tasks, successfully navigate complex situations, and effectively interact with their surroundings. The ongoing development of control systems and route planning algorithms has been crucial in the advancement of robots, leading to their enhanced versatility and increased capabilities across many applications.

THE PERCEPTION OF ROBOTS

The presence of robots, which was formerly restricted to the realm of fantastic fiction and entertainment, has now become profoundly interwoven in our day-to-day lives. A discernible shift is taking place in our understanding and interpretation of robots as a result of the growing prevalence of intelligent devices across all spheres of society. The current shifts in how robots are perceived, how they are involved in society, and what roles they play hold a tremendous amount of importance (Yinong, 2021). The shift in perspective that has been described above gives rise to questions and complications that are intellectually stimulating and are related to technology, ethics, and the dynamics of social interaction.

During the modern era, robots were most commonly associated with operations that took place within controlled environments, such as assembly lines located within manufacturing facilities. These entities were typically seen as rigid, impersonal constructions that possessed limited capabilities. The fields of artificial intelligence, machine learning, and robotics have made major strides in recent years, which has resulted in considerable advancements in the capabilities and possible uses of robotic systems. One such improvement is the ability to learn on their own. At this time, people are actively participating in a wide variety of fields, including healthcare, service industries, autonomous transportation, and taking on duties as carers for individuals of all ages, including the elderly and the young. The way in which people of different age groups and cultural backgrounds think about robots differs significantly from one another. There are some people who view robots as a way to improve ease of use and productivity, which would, in turn, optimise day-to-day activities and raise the overall level of living. Individuals are finding themselves confronted with ethical and existential questions as a result of concerns regarding the displacement of employment, issues linked to privacy, and the prospect of robots surpassing humans in certain fields. These anxieties and challenges have led to the emergence of these questions.

In this study, a variety of sociological perspectives on robots are investigated from different angles. The major purpose of this research is to analyse the impacts of human-robot contact, the psychological phenomena known as anthropomorphism, and the role that the media and popular culture play in the process of influencing our perspectives (Shaukat et al., 2020). In the modern setting of robotics, it is absolutely necessary to have an understanding of how robots are perceived. This is because the perception of robots is not only significant as an academic endeavour, but also as a crucial determinant in the development and application of these intelligent machines. The acquisition of a complete grasp of this idea is essential if one wants to be certain that the use of robots will result in positive outcomes rather than in outcomes that will have a negative impact on our society and the activities we engage in on a daily basis.

LEARNING AND ADAPTATION CAPABILITIES OF ROBOTS

The learning and adaption skills exhibited by robots signify a significant breakthrough within the realm of robotics. Historically, robots have been linked to pre-programmed and repetitive duties. However, the incorporation of AI has enabled them to acquire knowledge from their interactions and adjust their behaviour accordingly in novel circumstances. This significant advancement presents a multitude of potentialities and practical implementations. Robotic systems that are equipped with AI utilise machine learning methodologies in order to augment their functionalities (Mir et al., 2020). This encompasses computational methods such as deep learning and reinforcement learning. Deep learning enables robots to efficiently analyse large volumes of data, identify recurring patterns, and subsequently make informed decisions using this acquired knowledge. Reinforcement learning facilitates the acquisition of knowledge by robots through iterative experimentation, leading to progressive enhancement of their operational capabilities.

Robotic systems are endowed with the capability to detect their surroundings through the utilisation of sophisticated sensors, including cameras, lidar, and touch sensors. These entities possess the capability to discern and recognise various entities, including items, individuals, and barriers, hence facilitating their ability to traverse intricate environments and engage with objects in a cognitively advanced manner. Robots possessing the capacity to learn are capable of modifying their behaviour in accordance with dynamic circumstances. An instance of a robot developed for the purpose of environmental monitoring possesses the capability to adapt its data collecting methodologies in accordance with the distinct attributes of the environment within which it is functioning. Learning robots have the ability to partake in interactions with people that are more natural and intuitive in nature. These entities possess the ability to perceive and react to human movements, verbal communication, and even emotional cues. This is particularly advantageous in domains like as healthcare, wherein robots may offer aid to patients and offer companionship. Learning robots has the ability to autonomously make real-time decisions. Autonomous vehicles employ AI algorithms to effectively manoeuvre through intricate traffic scenarios and promptly execute judgements in order to uphold safety standards. Robotic systems have the capacity to acquire novel abilities and enhance their overall performance through iterative learning processes. This capability holds significant value in industrial contexts such as manufacturing, as it enables robots to effectively adjust to novel production procedures and engage in cooperative endeavours alongside human operators (Bartneck et al., 2021). Learning robots has the capability to identify anomalies or malfunctions within their operational frameworks and subsequently adjust their behaviour in order to address and minimise these concerns. Individuals have the capability to engage in self-diagnosis and deploy backup procedures, so augmenting their overall reliability. Learning robots continuously strive for improvement. Through repeated interactions and experiences, individuals enhance their abilities, gradually improving their proficiency in the tasks allocated to them.

Learning robots possess the capability to be tailored to certain activities and surroundings. The inherent adaptability of these entities enables them to thrive in a wide range of contexts, spanning from the exploration of outer space to the realm of agriculture. The potential of robots' learning and adaptability skills is vast, however, it is accompanied by various challenges. In order to enable responsible and safe deployment of learning robots, it is imperative to address ethical aspects, including decision-making openness, accountability, and safety precautions. The learning and adaptability skills exhibited by robots are significantly influencing the trajectory of automation, hence enhancing their versatility, responsiveness, and intelligence. With the continuous advancement of AI, the proficiency of robots in acquiring knowledge from their experiences is expected to increase, leading to a greater convergence of human and machine skills.

COLLABORATIVE EFFORTS BETWEEN HUMANS AND ROBOTS

The cutting edge of technological progress is currently experiencing a substantial concentration on joint undertakings between humans and robots, with particular attention placed on the fields of manufacturing, healthcare, and autonomous systems. This coordinated effort will involve the integration of human workers and robots in the workplace, with the goal of capitalising on each group's unique capabilities and advantages to improve operational efficacy, worker safety, and output.

The industrial industry has undergone a profound sea change as a direct result of the widespread adoption of collaborative efforts between humans and robots. Robots and human workers are able to work together in modern industrial settings thanks to the integration of complex sensors and artificial intelligence (Belk, 2021). This allows for the efficient completion of tasks that are both repetitive and taxing on the body. The use of automation not only reduces the amount of physical labour that workers are required to perform, but it also increases accuracy and uniformity, which ultimately leads to the creation of higher-quality items. In contrast, human labourers take on the responsibility of managing operations, participating in complex decision-making processes, and carrying out tasks that require innovative thinking and flexibility in order to complete successfully. The establishment of this collaboration has resulted in substantial changes to modern manufacturing practises, resulting in an increase in both the competitiveness of these practises and their flexibility in response to the ever-changing demands of the market.

Within the field of healthcare, robots are quickly becoming indispensable helpers for medical personnel, playing an increasingly significant part in the sector overall. Surgical robots, like the ones that were mentioned, make it possible to execute minimally invasive surgeries with a level of precision that has never been seen before. The use of these robots, which are operated by surgeons, enables them to do sophisticated surgical procedures, which in turn reduces patient trauma and shortens the amount of time needed for recovery. In the realm of eldercare, robots play an important role in assisting with the lifting of patients, distributing medication, and providing companionship to people who are in need. It has been demonstrated that the utilisation of collaborative efforts can improve the standard of care provided to patients while also reducing the amount of work required of healthcare professionals.

The field of autonomous systems, which includes things like self-driving automobiles and unmanned aerial aircraft, is a field in which people and robots work together on various projects. This is because autonomous systems comprise entities like these. Robots that are driven by AI do in-depth analyses on the most recent and relevant data that is acquired from sensors in order to make crucial decisions. In exceptional situations or in situations that are outside the scope of the capabilities of the robots' programming, the participation of humans is still absolutely necessary. The convergence of human and machine intelligence is propelling forward progress in a variety of industries, including transportation, agriculture, and logistics.

However, there are still challenges to overcome in this attempt involving multiple people. It is of the utmost importance that, wherever robots are present, the protection of people be given top priority. Especially in the fields of healthcare and autonomous systems, it is absolutely necessary to address ethical problems (Oksanen et al., 2020). In addition, it is projected that the ongoing development of AI and robotics will make it increasingly difficult to differentiate the skills of humans and machines, which could lead to a re-evaluation of job responsibilities and duties.

The collaboration of humans and robots is an intriguing field that holds the promise of bringing about profound changes in the ways in which we perform work, the amount of output we achieve, and the way we live our lives. These phenomena will continue to exist as a result of the continued advancement of technology, which will eventually lead to a change in our attitude towards a variety of enterprises and activities. Table1, a streamlined table that compares the amount of effort put forward by humans versus robots in a variety of contexts.

ETHICS OF ROBOTS AND THE PROTECTION OF AI

The topic of "Ethics of Robots and the Protection of AI" is strongly rooted in the significant influence that AI and robotics have on different elements of our society, the economy, and our day-to-day lives. This influence can be seen in the title of the course, which is "Ethics of Robots and the Protection of AI."

Table 1. Contrasting human versus robotic efforts in several domains

Aspect	Humans	Robots
Physical Strength	Variable, depending on individual strength	Consistent and can be tailored for tasks
Repetitive Tasks	Fatigue-prone over time	Efficient and consistent
Precision	Prone to errors, especially with fatigue	Extremely precise and accurate
Speed	Variable, may decrease with fatigue	Consistently fast
Learning	Learning curve, variable learning speed	Rapid learning through AI algorithms
Adaptability	Learn and adapt, but with limitations	Quick adaptation based on programming
24/7 Availability	Requires rest and sleep	Can operate continuously
Safety in Hazardous Environments	Limited without protective gear	Designed for safety in hazardous settings
Endurance in Repetitive Tasks	Fatigue sets in over time	Unaffected by repetitive tasks
Collaboration	Effective but may have interpersonal issues	Collaborate efficiently with other robots
Emotional Intelligence	Emotionally intelligent, understand context	Lacks emotional understanding
Creativity	Highly creative and innovative	Performs predefined tasks
Scalability	Limited by human workforce availability	Scalable by adding more robots

The relentless advancement of artificial intelligence technology provides us with important ethical and practical questions regarding the utilisation of AI systems, the regulation of AI systems, and the protection of AI systems. The history of and current circumstances surrounding this topic are summarised in the following paragraphs for your convenience.

AI Advancements: The fast advancements made in the fields of AI and robotics have led to the creation of systems that are both extremely advanced and completely autonomous. The application of AI in a wide variety of fields, including healthcare, finance, transportation, and others, calls for a thorough investigation into the ethical issues that are raised by this progress.

Autonomous Decision-Making: Robotics, self-driving cars, and unmanned aerial vehicles (UAVs) that are powered by AI have the potential to independently form opinions and make decisions. The aforementioned conditions give rise to ethical questions regarding the probable outcomes of these decisions and the people or organisations who are responsible for them.

Safety and Security: The provision of assurances on the safety and security of AI systems is of the utmost significance. Failures in the decision-making processes of AI have the potential to lead to a variety of undesirable consequences, such as accidents, data breaches, or unforeseen effects. It is of the utmost significance to protect AI systems from being compromised by cyberattacks.

Bias and Fairness: It is possible for AI algorithms to pick up biases from the data that they are trained on, which could lead to results that are unfair or discriminatory. In order to fulfil our ethical commitment, we need to address these shortcomings and ensure that the decision-making process that artificial intelligence uses is fair.

Human-AI Collaboration: The integration of human and artificial intelligence in working together is becoming increasingly common in today's society. Ethical questions are raised both by the assignment of responsibilities to AI and by the dynamics of the relationships that exist between humans and AI.

Long-Term Impact: When thinking about the future, it is essential to give serious consideration to the long-term effects that AI will have on the job market, the workings of society, and the possibility of a peaceful cohabitation between humans and AI.

Research and Development: The field of artificial intelligence research and development falls under the purview of ethical considerations as well. It is of the utmost importance to place a high priority on responsible innovation and to take preventative measures against the development of artificial intelligence systems that could be harmful.

Education and Awareness: In order to support informed decision-making, it is necessary for developers, users, and policymakers to create a complete grasp of AI ethics. This understanding will facilitate informed decision-making.

THE ROLE OF ROBOTS IN MEDICAL CARE

The utilisation of robots in the sphere of medical care has seen significant development over the course of time. This change has been spurred by advances in technology, an increasing demand for healthcare services, and the urge to improve patient outcomes. It is becoming increasingly common practise to employ robots in a variety of aspects of patient care, including diagnosis, treatment, surgery, rehabilitation, and support for patients. These robots are playing an increasingly important role in these areas. Throughout the course of their history, the majority of robots used in the medical field have been put to work in specific fields such as robot-assisted surgery. In spite of this, recent developments have resulted in a major expansion of the potential applications of artificial intelligence (AI), machine learning, and robotics in the field of healthcare. An in-depth analysis of the various environmental considerations that are relevant to the application of robots in the field of medical care is presented in the following paragraphs.

Surgical Robotics: One of the most well-known applications of robotics in the field of health is the utilisation of surgical robots, which are utilised to carry out minimally invasive procedures with a high degree of precision and accuracy. This application of robotics is one of the most commonly recognised applications of robotics in the field of medicine. Surgeons utilise these robotic devices in order to improve the quality of their work, reduce the amount of invasiveness involved in procedures, and shorten the amount of time it takes for patients to recuperate.

Telemedicine and Remote Monitoring: The practise of telemedicine, in which medical professionals are able to examine patients without physically being present at their bedsides, is made possible by the utilisation of robots that are equipped with cameras and sensors. In addition to this, they provide a hand in the continuous monitoring of patients by collecting important physiological indicators and communicating this data to medical professionals.

Rehabilitation: There has been a discernible uptick in the number of programmes that include robotic assistance for physical therapy and rehabilitation. Physical therapists play an essential position in the process of helping patients regain their mobility and strength, which is especially important in the aftermath of traumatic events such as injuries, strokes, or surgical procedures. These robotic systems are able to provide exercise regimens that are not only consistent but also tailored to the specific requirements of each unique user.

Medication Management: The distribution of medication and its subsequent administration are both significantly aided by the utilisation of robotic devices. Professionals in the healthcare industry have the ability to assure that patients will receive the correct dosages of their medications at the proper intervals, hence reducing the likelihood that mistakes will be made.

Mental Health and Social Support: Robots have the potential to act as companions and provide therapeutic support to people who are in need of it in the field of mental health. Healthcare providers are expected to take an active part in conversations, offer folks help with empathy, and push patients to adhere to the therapeutic regimens they have been prescribed.

Laboratory and Diagnostic Robots: It has been demonstrated that automating testing procedures in laboratories can both improve the testing process's overall efficiency while simultaneously lowering the number of opportunities for human error. Sample processing, test analysis, and data management are only some of the activities that can be successfully carried out by robotic systems, which also have the capacity to carry out a broad variety of other jobs.

Patient Transport and Logistics: Robotic systems are used in hospitals and other healthcare institutions to ensure the safe and efficient movement of patients, supplies, and equipment. Robotic systems can also be used to diagnose and treat patients. The application of this strategy both lessens the possibility of contamination from one source to another and increases the effectiveness of business operations.

Emergency Response: Robots can be used in situations such as crisis management and first response to get access to areas that are difficult to reach, analyse the situation, and provide emergency teams with up-to-date information. This can be accomplished through the utilisation of robots.

Patient Education: Patients can benefit from the interactive capabilities of robots by learning more about their medical conditions, the various treatment options that are accessible, and the choices they can make regarding their lifestyle. The ability to react to questions and provide pertinent information in a way that is both simple to understand and patient-friendly is a skill that is essential for individuals working in the healthcare industry.

The inclusion of robots into the field of medical care is driven by the aim to improve healthcare outcomes, enhance the experience of patients, and maximise the efficiency of healthcare operations. Nevertheless, there are also substantial obstacles that arise as a result of this issue. These challenges include the necessity for sufficient training, the maintenance of the human element in healthcare, and the consequences of regulatory and ethical questions. It is projected that, as technology continues to advance, there will be a marked uptick in the use of robots in the field of medical care, which will have a revolutionary effect.

SIMULATIONS OF ROBOTIC SYSTEMS IN VIRTUAL ENVIRONMENTS

The utilisation of virtual environments for the purpose of simulating robotic systems is of the utmost importance in the research and development activities being carried out in the field of robotics in the modern day. These facilities provide a controlled environment that is also secure for the purpose of testing, improving, and evaluating robotic systems and the software that goes with them. Engineers, researchers, and developers can examine various aspects of robotics more easily with the use of virtual environments, which serve as digital representations of the physical world and are used in the simulations. This eliminates the need for expensive physical prototypes and makes it possible to investigate both individual components and complex systems. Individuals have the capacity to build and manipulate a diverse range of scenarios within these virtual settings, including a variety of landscapes, items that are constantly changing, and labor-intensive jobs. This makes comprehensive testing and training possibilities much easier to access. The following domains benefit greatly from the utilisation of simulations: Engineers and developers make use of virtual environments for the purpose of designing, prototyping, and iterating robotic system designs. In order to improve the functionality of robots, researchers are able to conduct experiments using a variety of robot configurations, sensor arrangements, and algorithmic approaches.

Virtual environments offer a way to train and assess autonomous robots without having to expose the robots to the risks that are present in the real world. For instance, self-driving cars can be put through rigorous testing in simulated urban environments before being used on actual roads. This can happen before the cars are put into use. Researchers are able to efficiently validate complex algorithms such as those linked to computer vision, path planning, and machine learning within regulated and repeatable settings. This technique helps to ensure that robots are capable of successfully comprehending sensory information and making decisions based on that knowledge. When it comes to the preliminary assessment and risk assessment of deploying robots in high-stakes environments, such as nuclear facilities or disaster zones, simulations are an extremely important part of the process. The protection of human lives and the preservation of valuable resources can be significantly aided by the analysis of potential threats in virtual environments and the implementation of appropriate countermeasures. Building and maintaining actual prototypes can be an expensive and time-consuming procedure that can also entail major costs. It has been discovered that substituting digital simulations for physical testing in virtual settings can result in significant cost savings. This can be accomplished by using virtual environments. The ability to scale is one of the benefits that virtual environments offer.

Researchers are able to examine a wide variety of applications for robotics because of their capacity to simulate a wide variety of scenarios. These scenarios range from small-scale robotic swarms to large-scale smart cities. The study of human-robot interaction also makes use of simulations in a number of different ways. Researchers are able to optimise robot behaviour and usability across a wide variety of settings by simulating human-robot interactions and using the data from such simulations. It is essential for robots to demonstrate predictable conduct in order for them to be deployed in real-world environments. The smooth adaptation of robots to unforeseen conditions in the actual world can be greatly aided by the use of virtual simulations, which play an important part in this process. In light of the aforementioned circumstances, it is clear that simulations are of critical significance to the development of the robotics discipline. Researchers and engineers have the ability, with the help of virtual environments, to speed up the development of cutting-edge robotic systems, reduce the risk of potential hazards, and ultimately improve the capabilities, efficiency, and safety of robots used in a wide variety of applications.

AUTOMATION OF ROBOTIC PROCESSES

The phenomena of automating robotic processes, also known as Robotic Process Automation (RPA), is a rapidly increasing technology-oriented paradigm within the modern sphere of business. RPA is an abbreviation for the acronym "Robotic Process Automation." The activity of utilising software robots, sometimes known as "bots," to mechanise recurrent responsibilities and procedures in an organisational environment is known as "bot farming." Input of data, processing of invoices, communications with customers, and analysis of data are some of the many tasks included in the range of responsibilities. Understanding the setting in which the automation of robotic processes occurs can be enhanced by taking into consideration the following three crucial points: In the modern business climate, which is marked by fierce rivalry, businesses are constantly working to improve their efficiency, limit their operational expenses, and maximise their production. The automation of robotic processes is a solution that satisfies the demand that was indicated above. This solution provides a way to optimise operations and dedicate human resources to occupations that require strategic thinking and creativity.

Rise of Digital Transformation: As a consequence of the introduction of digital transformation across a variety of sectors, businesses have found themselves confronted with a substantial influx of data and an increased incidence of processes that are repetitive. Robotic Process Automation, also known as RPA, is an essential component in efficiently capitalising on the potential of data and integrating a wide variety of digital systems without a hitch. This is of utmost importance in light of the movement of businesses towards utilising cloud-based services, the Internet of Things (IoT), and big data analytics. Robotic process automation, often known as RPA, gives businesses the ability to improve the overall quality and effectiveness of the client services they provide by facilitating the delivery of services in a more timely and efficient manner. Robotic process automation (RPA), when used to the financial services sector, has the ability to hasten the procedures for the approval of loans and increase the effectiveness of dispute resolution. As a direct consequence of this, increased levels of customer satisfaction and brand loyalty are generated.

Accuracy and Compliance: Errors caused by humans can have major effects on finances and can lead to a violation of regulations if they are not caught in time. Robotic Process Automation, abbreviated to RPA, ensures a greater level of precision and uniformity in work by virtue of the fact that automated bots carry out identical tasks with very few variations. Additionally, it is feasible to train these bots in a way that ensures strict adherence to certain compliance and regulatory criteria. This can be accomplished through the use of specific programming. Functionalities related to AI and machine learning are regularly incorporated into robotic process automation (RPA) systems in order to improve their performance. This functionality provides the artificial intelligence agents with the ability to learn knowledge from datasets, modify their behaviour in response to changing situations, and carry out activities that require the processing of unorganised information or the making of judgements. The term "Intelligent Process Automation" (IPA) is also used to refer to this occurrence on occasion.

Employee Empowerment: Employees are able to allocate their time and efforts to work that provides greater value, needing creativity, decision-making, and human connection, thanks to the implementation of automation for occupations that are of a repetitive nature and have the potential to be done again. This phenomenon not only improves the level of happiness enjoyed by workers, but it also enables businesses to maximise the potential of their human capital.

Industry Applications: Robotic process automation, often known as RPA, can be utilised in virtually any sector of the economy. This technology is useful in many different fields, including banking, healthcare, manufacturing, customer service, supply chain management, and a great many more. Robotic process automation (RPA) has been put to use in a wide variety of industries, which is a testament to its inherent versatility. The field of robotic process automation, also known as RPA, is always undergoing new developments and expanding its capabilities. The development of new technologies paves the way for the introduction of automation capabilities that are progressively more intricate and varied. Companies that want to keep their advantage in the marketplace absolutely need to make it a priority to remain current on the most recent developments in robotic process automation (RPA).

Robotic process automation (RPA) presents a number of challenges in addition to offering a variety of benefits. These include worries about the security of data, the loss of employment opportunities, and the requirement for appropriate governance. The application of automation across a variety of industries calls for thoughtful deliberation and investigation due to the ethical considerations that are raised by this trend. The imperative to boost efficiency, precision, and productivity as a response to the process of digital transformation is the driving force behind the rationale behind the automation of robotic processes. This change is being driven by digital technology. The implementation of this technology offers businesses operating in a wide variety of markets a plethora of benefits that can help them compete more effectively. Nevertheless, it also raises a number of issues, including ethical concerns, which demand careful management on the part of the responsible party.

CONCLUSION

The emergence of AI has initiated a paradigm shift in the field of robotics. This abstract elucidates the significant influence of AI on the domain of robotics, emphasising the noteworthy advancements and practical implementations that have arisen. AI-driven robots currently demonstrate improved abilities in perception, decision-making, and adaptability, hence enabling them to thrive in a wide range of jobs across various industries such as manufacturing, healthcare, space exploration, and autonomous vehicles. By utilising machine learning techniques, such as deep learning and reinforcement learning, these robotic systems are capable of acquiring novel abilities, consistently enhancing their efficiency, and intelligently interacting with their environment. Nevertheless, it is important to note that this abstract also emphasises the presence of ethical considerations, safety precautions, and societal repercussions associated with this evolution. The utilisation of artificially intelligent robots in many industries gives rise to inquiries regarding their implications on labour, privacy, and security. It is of utmost importance to address these problems and prioritise responsible development of AI and robotics, given the ongoing intersection of these fields. Looking towards the future, the continuous progress of AI presents significant possibilities for the merging of AI and robotics, which has the potential to revolutionise automation and reshape the dynamics of collaboration between humans and machines. The symbiotic association between AI and robots is anticipated to profoundly transform several industries, enhance operational effectiveness, and facilitate the emergence of novel prospects for innovation. As we find ourselves on the verge of this promising frontier, it is imperative to effectively address the obstacles and capitalise on the potential advantages presented by this integration. This will ultimately contribute to a future characterised by a symbiotic and efficient collaboration between humans and machines. As AI and robotics continue to merge, it will be crucial to conduct interdisciplinary research, collaborate, and prioritise ethical concerns if we want to create a future where AI-powered robots positively impact society.

REFERENCES

Bartneck, C., Lütge, C., Wagner, A., & Welsh, S. (2021). *An introduction to ethics in robotics and AI.* Springer Nature. doi:10.1007/978-3-030-51110-4

Belk, R. (2021). Ethical issues in service robotics and artificial intelligence. *Service Industries Journal, 41*(13-14), 860–876. doi:10.1080/02642069.2020.1727892

Gaur, L., Afaq, A., Singh, G., & Dwivedi, Y. K. (2021). Role of artificial intelligence and robotics to foster the touchless travel during a pandemic: A review and research agenda. *International Journal of Contemporary Hospitality Management, 33*(11), 4079–4098. doi:10.1108/IJCHM-11-2020-1246

Gurjeet, S. (2022). *Vijay, Kumar, Banga., Thaweesak, Yingthawornsuk.* Artificial Intelligence and Industrial Robot., doi:10.1109/SITIS57111.2022.00098

Jonathan, A. (2023). *Development of an AI-Robotics 3D Printed Circle of Command for Enhancing Accessibility and Mobility in Individuals with Mobility Issues.* IEEE. doi:10.5121/csit.2023.130910

Mir, U. B., Sharma, S., Kar, A. K., & Gupta, M. P. (2020). Critical success factors for integrating artificial intelligence and robotics. *Digital Policy. Regulation & Governance, 22*(4), 307–331. doi:10.1108/DPRG-03-2020-0032

Nagadevi, D. (2021). *A Comprehensive Study on Artificial Intelligence and Robotics for Machine Intelligence.* Springer. doi:10.4018/978-1-7998-7701-1.ch011

Nam, K., Dutt, C. S., Chathoth, P., Daghfous, A., & Khan, M. S. (2021). The adoption of artificial intelligence and robotics in the hotel industry: Prospects and challenges. *Electronic Markets, 31*(3), 553–574. doi:10.1007/s12525-020-00442-3

Nehal, R. (2023). AI in Robotics: Advancements, Applications and Challenges. *Journal of Information Technology and Digital World.* doi:10.36548/jitdw.2023.2.009

Oksanen, A., Savela, N., Latikka, R., & Koivula, A. (2020). Trust toward robots and artificial intelligence: An experimental approach to human–technology interactions online. *Frontiers in Psychology, 11*, 568256. doi:10.3389/fpsyg.2020.568256 PMID:33343447

Shaukat, K., Iqbal, F., Alam, T. M., Aujla, G. K., Devnath, L., Khan, A. G., & Rubab, A. (2020). The impact of artificial intelligence and robotics on the future employment opportunities. *Trends in Computer Science and Information Technology, 5*(1), 50-54.

Yinong, C. (2021). Technologies Supporting Artificial Intelligence and Robotics Application Development. doi:10.37965/jait.2020.0065

Chapter 10
Synergistic Swarm:
Multi-Robot Systems in Healthcare

Jaspreet Kaur
Chandigarh University, India

ABSTRACT

The "synergistic swarm" investigates the incorporation of multi-robot systems in healthcare, introducing a fundamental change in patient care and medical operations. This abstract emphasises the collaborative synergy achieved by intelligently coordinating several robotic entities, resulting in improved efficiency, precision, and adaptability in healthcare environments. By utilising cutting-edge technology like artificial intelligence, robotics, and sensor networks, the system seeks to enhance many functions, including diagnostics and patient support, to their maximum efficiency. This chapter highlights the significant potential of combining different approaches to healthcare in order to improve the delivery of medical services. This could lead to more effective, patient-focused, and adaptable healthcare robotics in the changing healthcare industry.

INTRODUCTION

Automating daily operations by understanding the existing routines and practices, generally, it refers to software that imitates an online workforce and carries out repetitive tasks and responsibilities, reducing the requirement for social interaction. In today's competitive world of work, everyone strives to outperform their rivals in order to acquire a competitive advantage. Automating repetitive tasks with Multi-Robots/ Synergistic swarm (RPA) technology significantly enhances the profitability and efficiency of commercial organizations. Automation allows workers to focus on higher-level tasks, unleash their creativity, and devote themselves to advancing their knowledge and skills in the sector. Every Multi-Robots / Synergistic swarm project consists of three main stages: design, execution, and assessment and tracking (Alsamhi & Lee, 2020).

Multi-Robots / Synergistic swarm is considered a significant technological advancement that is capable, resilient, flexible, and reliable enough to make it feasible for use in major corporations. Experts predict that once Multi-Robots / Synergistic swarm (RPA) becomes stable and reliable in the next decade or

DOI: 10.4018/979-8-3693-1962-8.ch010

two, it will lead to significant improvements in efficiency and effectiveness among the working people in terms of emergence of concept of synergistic swarm or Multi-robotic systems. Multi-Robotic system is an emerging technology, aims to provide a sustainable alternative that reduces costs and shipping times while improving the quality of firm operations It is an innovative technology that is increasingly being adopted in businesses that include monotonous activities. Businesses that employ Multi-Robots / Synergistic swarm experience decreased expenses and enhanced operational efficiency (Cao et al., 2022). Introducing Multi-robots to the medical field is an example of such an endeavor as presented in figure 1 below:

The purpose of this study is to shed light on Multi-robotics' significance in healthcare and to lay the groundwork for integrating various strategies to boost productivity and contentment in the workplace . The healthcare sector's operational efficiency has been greatly improved by the application of technology breakthroughs. By streamlining the processing of massive amounts of data, algorithms for machine learning have benefited the healthcare industry (Kaur, 2024).

As compared to the human mind, Multi-Robots / Synergistic swarm can acquire and recall a lot more quickly. This goes against what the human brain is capable of doing cognitively. Systematic optimization of the distribution of healthcare professionals, hospital beds, and resources is possible with the use of AI. Better operational efficiency and lower costs are the outcomes of this optimization (Cao et al., 2022).

The use of cutting-edge robotic technology has made distant surgical procedures a reality, expanding patients' access to highly specialized medical care. By collecting data on a massive scale, Multi-Robots / Synergistic swarm make it possible to study populations for trends and patterns. Multi-Robots / Synergistic swarm is utilized to do this. The management of healthcare for the general public might stand to benefit greatly from this. According to Troccaz et al., (2019), patients and medical professionals can receive continuous and immediate feedback about the patient's status thanks to robotic technology integrated into rehabilitation and prosthetic devices. Accordingly, it becomes easier to tailor rehabilitation programs to each patient's unique requirements, track their progress, and make adjustments as needed (Das et al., 2015).

According to Kaur (2024), as shown in figure 2 & figure 3 below, Synergistic swarm in multi-robot system is a novel concept in healthcare and has numerous advantages:

According to the aforementioned evaluation, Multi-Robots / Synergistic swarm has the potential to be used in virtually every business in a manner that is both exact and does not involve a change to the existing framework. Control will be accomplished by the utilization of intelligent robotics. In addition to this, it has the ability to optimize data analytic for the numerous companies that provide online resources. Automated processes are expected to have a rapid and consistent rise in popularity in the coming years as an increasing number of businesses become aware of the benefits that Multi-Robots / Synergistic swarm can provide through its implementation (Duggal et al., 2022).

The fields of artificial intelligence and robotics are distinct from one another, despite the fact that both may continue to be utilized independently. Over the course of the next few years, Multi-Robots / Synergistic swarm is anticipated to develop into artificial intelligence. This implies that the technology will encompass all aspects of artificial intelligence. By integrating machine learning with the support of staff and instruments, businesses, particularly those in the healthcare industry, are quickly realizing that this combination will lead to full and advanced personnel by boosting the efficacy and effectiveness of the business as a whole, which will eventually serve its stakeholders. The online evolution that involves artificial intelligence analysis is anticipated to be a further significant transformation.

Figure 1. Present state of healthcare and the requirement of multi- robotics

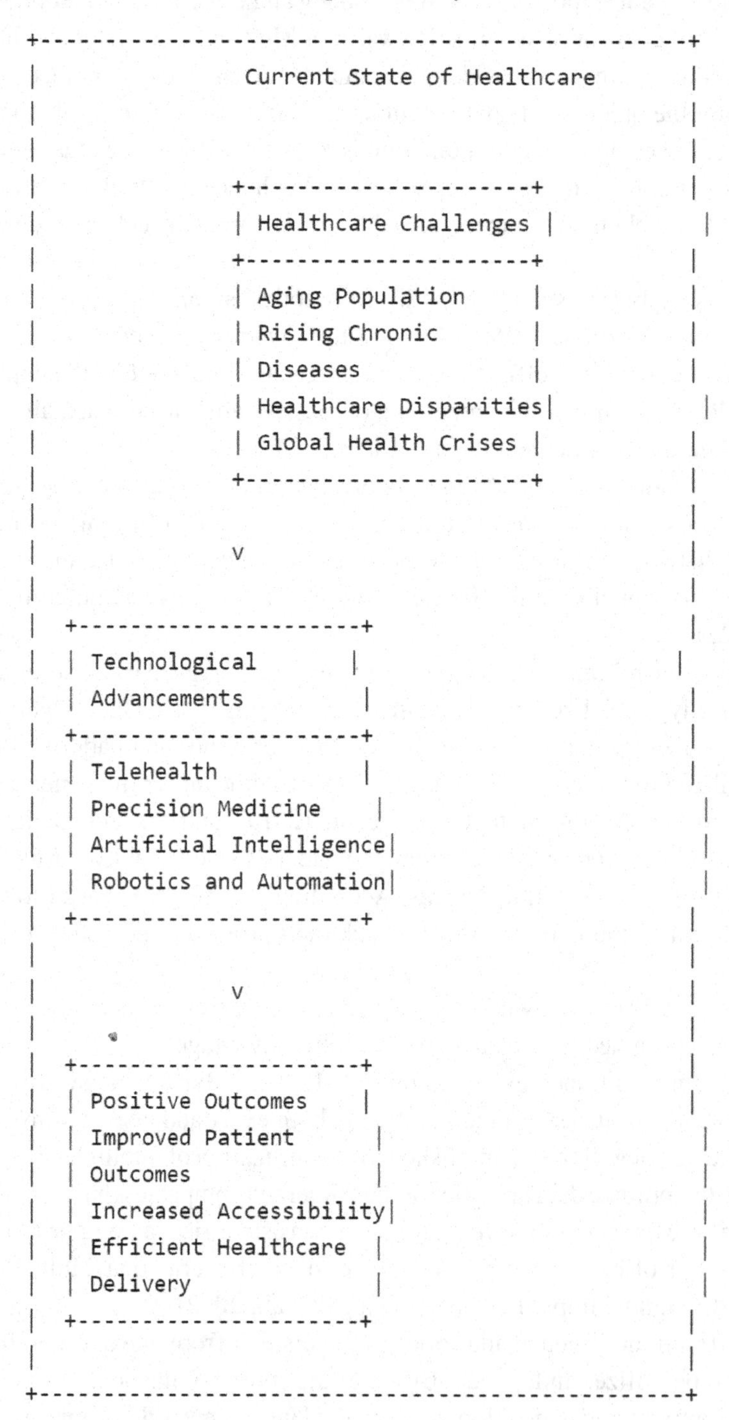

This is because employees who are equipped with automated tools will be able to construct a virtual healthcare staff that encompasses all aspects of monitoring and increases profitability across key domains (Fisher et al., 2017).

Figure 2. Synergistic swarm in multi-robot systems in healthcare

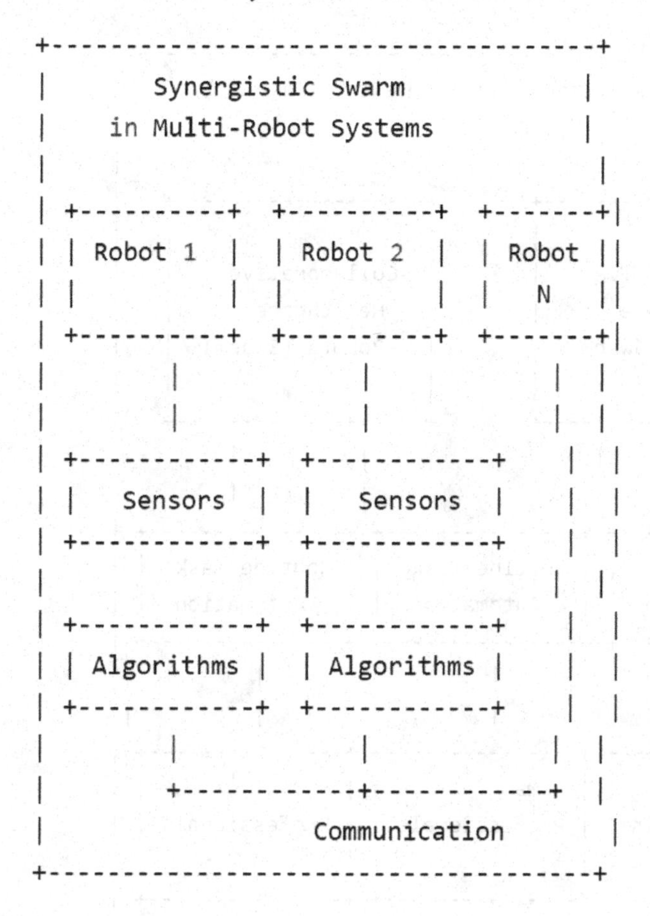

In the not too distant future, Multi-Robots / Synergistic swarm is going to be utilized extensively in a wide range of healthcare-related fields, such as the manufacturing of medical equipment, the analysis of medical data, and the administration of healthcare compliance. In the not too distant future, the extensive information and processing that the Administration is responsible for will be carried out automatically. To be more specific, the installation of Multi-Robots / Synergistic swarm will have a significant impact on the ensuing healthcare programs. It is the core technology that drives the expansion of drug research and development, which is universally accepted as an essential advance in healthcare. It acts as the driving force behind the growth of technology. It typically takes the form of algorithms that are designed for learning (Foroutannia et al., 2021; Kaur, 2024).

By utilizing unique functions and functionalities, health-care automation technologies have the potential to enhance the efficiency, effectiveness, precision, and safety of the delivery of medical services. This is accomplished through the utilization of novel functions and capabilities. The expanding field of artificial intelligence is likely to result in the introduction of a whole new facet to the field of machine learning. As is anticipated, the utilization of machines and artificial intelligence would not only expedite the operations but also greatly enhance the level of safety throughout the process. Furthermore, statistical analysis and developments in software and physical components are likely to have the effect of expanding the scope of applications for robotics in additional medical settings.

Figure 3. Uses of multi-robot system/synergistic swarm

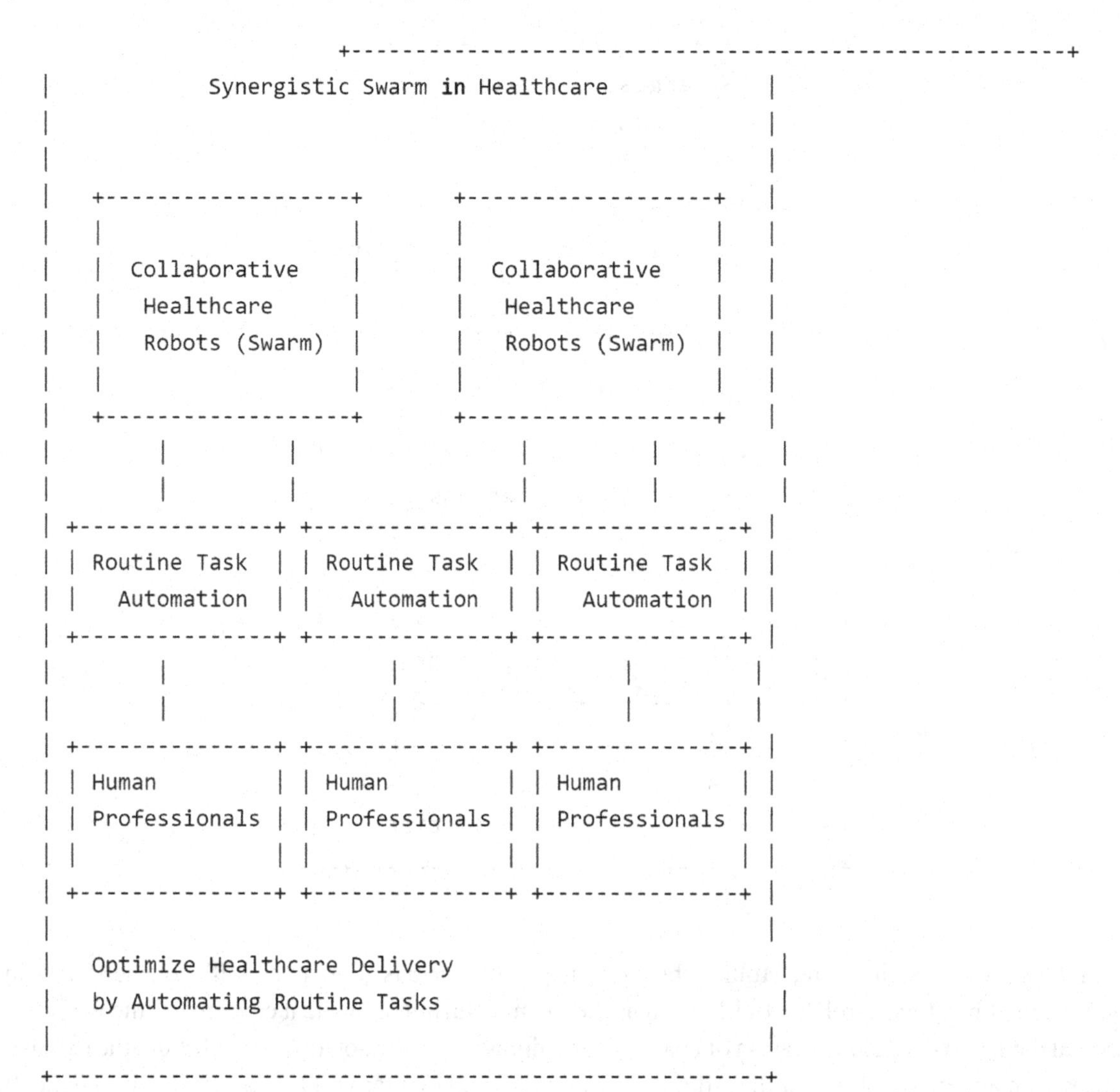

As a result of investments and collaborations between digital companies and medical specialists, the demand from customers for medical robotics is expected to increase even further (Haken & Levi, 2012; Fosch-Villaronga & Drukarch, 2021).

Despite the fact that earlier attempts to make suggestions for examination and treatment were already challenging, experts believe that the installation of Multi-Robots / Synergistic swarm will eventually become proficient in such an area as well. As a result of the rapid advancements that have been made in artificial intelligence related to computed tomography processing, it is becoming increasingly conceivable that the bulk of radio-logical and pathological imaging will soon be examined by a computer. The use of voice and optical character recognition for a variety of reasons, such as therapeutic exchanges and the transcription of diagnostic records, is going to become increasingly frequent (Holland et al., 2021).

Challenges to be Faced by Multi-Robots / Synergistic Swarm

The most significant obstacle that Multi-Robots / Synergistic swarm faces in the health care industry is merely assessing whether or not the technology advancements would be capable of being actually beneficial, rather than ensuring that they will be accepted for diagnostic reasons. In order for Multi-Robots / Synergistic swarm to be widely adopted, it is necessary for it to be endorsed by regulatory bodies, incorporated into electronic health records, sufficiently streamlined so that it is designed to imitate function similarly, the training of healthcare professionals, adequate compensation from either privately funded or publicly funded institutions, and improvement over the course of its use. It is possible that these challenges will be handled in the future; nevertheless, it will take far more time than was originally envisaged for such technology to improve under these circumstances (Humaidi et al., 2020).

As a result, experts estimate that the use of Multi-Robots / Synergistic swarm in therapeutic settings will be restricted during the next twenty years. It is important to emphasize that the introduction of Multi-Robots / Synergistic swarm will not significantly replace physicians in the treatment of patients; rather, it will merely support them in their work. It is possible that in the future, modern physicians may gravitate toward responsibilities and work arrangements that make use of special mental talents such as compassion, understanding, and motivation. It is possible that in the future, these health professionals who refuse to work together with machine learning will be the only ones who are forced to give up their own professions (Karpov & Tarassov, 2017; Kaur, 2024; Kaur & Arora, 2023).

SIGNIFICANCE OF MULTI-ROBOTICS/SYNERGISTIC SWARM

Improved patient diagnosis and remote patient monitoring are two areas where Multi-Robots / Synergistic swarm has shown to be an invaluable asset to healthcare operations. Among its many contributions to healthcare are the following: better health outcomes, more precise patient diagnoses, more effective treatment regimens, and more efficient administration of healthcare facilities (Kernbach, 2011; Lestingi et al., 2022).

By analyzing X-rays, CT images, and MRI's, diagnostic imaging—which includes Multi-Robots / Synergistic swarm systems—assists radiologists in their work. Analyzing these medical images accomplishes this. Radiologists can improve the diagnosis procedure with the help of these technologies by spotting abnormalities more easily. Robots can spot irregularities and minor patterns that a human eye might miss in the early stages of an investigation. Better early detection is made possible by this. Neurological problems like Parkinson's disease and cancer can now be detected much more quickly. Many systematic processes and procedures are followed when new medications are being created (Li et al., 2021).

Our goal in implementing these measures is to make the process of developing and testing novel pharmaceutical substances much faster. In order to find new treatments and evaluate their effectiveness, Multi-Robots / Synergistic swarm-can analyze massive amounts of data, including genetic information. The data analysis procedure does this. The time and money needed for drug trials is thus significantly reduced. When it comes to clinical studies, Multi-Robots / Synergistic swarm-also makes participant selection much easier. To do this, we look at the participants' personality qualities to see if they are a good fit for the study. According to Troccaz et al., (2019)., this means that the experiment has a better chance of succeeding. Robotic technologies help medical professionals make more educated diagnosis and treatment processes by expediting data analysis and providing decision support (Ma et al., 2015; Moura, 2012).

Individualized treatment regimens are made feasible by Robotics' utilization of personal genetic information. This enables for tailored treatment regimens and the capacity to forecast the effectiveness of specific treatments, while also decreasing the risk of unwanted responses. Customizing medications to fit with the specific demands of individual patients can considerably boost the success of treatment optimization. Previous studies (Rizk et al., 2019; Sahu et al., 2022) indicates that Multi-Robots / Synergistic swarm-can increase the quality of procedures in the medical industry-also. Despite the fact that the Synergistic Swarm model has a great deal of promise, there are ethical and regulatory difficulties associated with doing so as depicted in figure 4 below:

The Synergistic Swarm idea, which is used in multi-robot systems, is a revolutionist conceptualization that has the prospective to modify a variety of fields of study as presented in figure 5 below:

- Multi-Robots / Synergistic swarm-enables continual professional education and career growth for healthcare practitioners by leveraging telemedicine and telehealth technology. Telemedicine is the transmission of healthcare information and delivery of medical care to patients who are geographically remote. It does not belong to the realm of technology or represent a novel or revolutionary area of medicine. Telemedicine is the employment of technology to enable the interchange of data, speech, and video between patients and healthcare practitioners during their visits. It enables the supply of healthcare services to distant locations and fosters fair access to medical treatment while maintaining a peaceful interaction between urban and rural inhabitants (ŞEN et al., 2022).
- The implementation of Multi-Robots / Synergistic swarm in the healthcare industry aids in overcoming the issues arising from the geographical distance between healthcare practitioners. The recruitment and retention of healthcare workers in remote regions can offer major obstacles. However, these issues can be overcome by applying multi-robotics, as revealed by numerous research (Kaur, 2024).
- Multi-Robotic surgical aid technology can boost the professional development of healthcare personnel by facilitating the transfer of learning data to remote places and enhancing the sharing of information and knowledge in the field of surgical support. Therefore, it is crucial to determine

Figure 4. Characteristic of synergistic swarm model

Promising Aspects of Synergistic Swarm Model

- High Efficiency in Task Execution
- Improved Problem Solving
- Adaptability to Dynamic Environments
- Increased Robustness and Resilience

Ethical and Regulatory Difficulties

- Lack of Transparent Decision-Making
- Privacy and Security Concerns
- Accountability in Autonomous Systems
- Potential Job Displacement

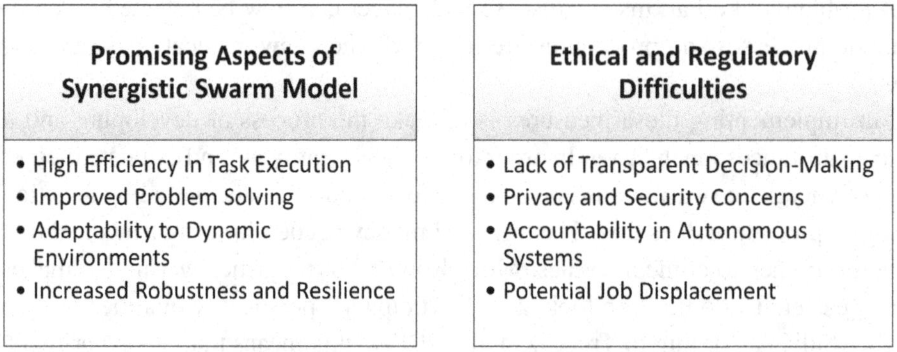

Figure 5. Potential impact across various fields

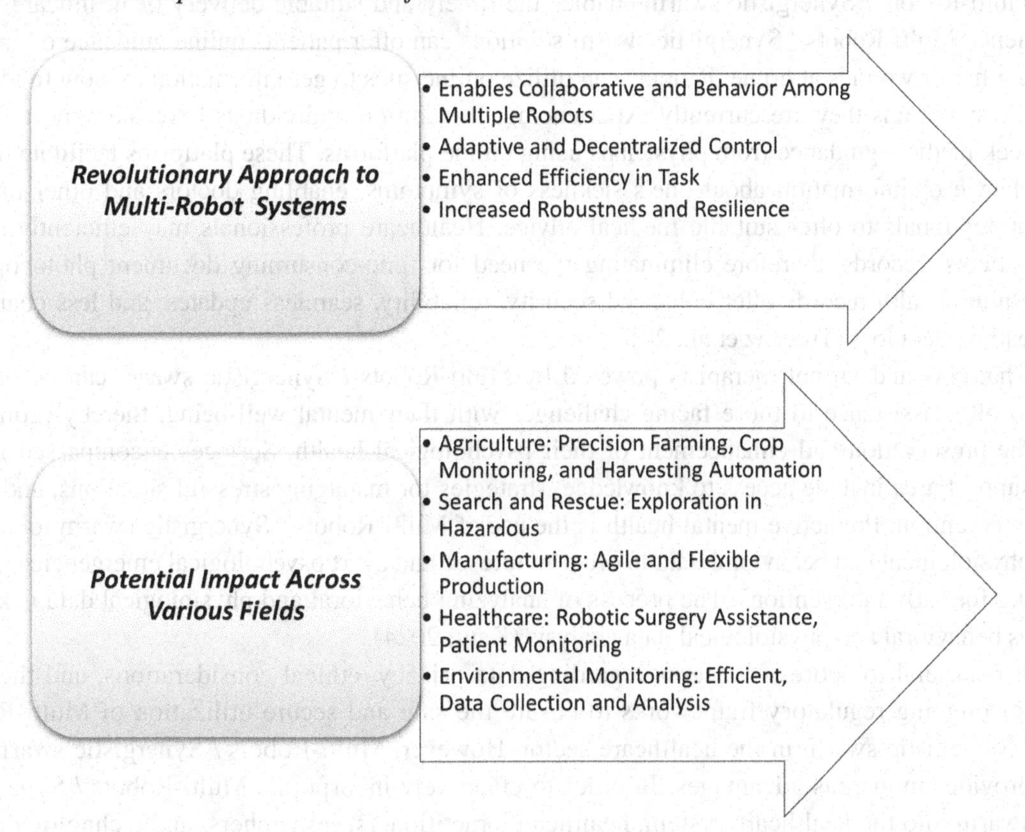

the most effective use of robots to boost the knowledge and learning of healthcare professionals, so supplying them with the necessary foundation for their professional development .

- When it comes to fraud protection, algorithms driven by Multi- Robotics evaluate payment records to discover any discrepancies that may suggest fraudulent activity. Early detection of fraudulent activity is attainable due to this. Implementing Multi-Robots / Synergistic swarm boosts cybersecurity measures, thereby enhancing the protecting of patient data and healthcare information systems (Shallal et al., 2020).
- The utilization of Multi-Robots / Synergistic swarm enables the use of wearable technology, such as trackers, that can continually monitor vital signs and capture real-time data. This data can be promptly communicated to the physician, alerting them to any potential difficulties that may arise. Remote monitoring has the potential to provide major benefits to those with chronic diseases. Some advantages of this technology include the ability to proactively manage symptoms and a decrease in the need for frequent hospital visits (Sugiyama et al., 2010; Kaur & Arora, 2023).
- Assistance in Medical Decision-Making: Robotics provides support to healthcare professionals in determining the best course of action for diagnosing and treating patients. It achieves this by providing them with evidence-based information. To enhance patient safety, it is advisable to utilize the alerts and notifications generated by clinical decision support systems. It is crucial for these cautions and warnings to specifically address the potential occurrence of pharmacological mistakes or interactions (Kaur, 2024).

- Multi-Robots / Synergistic swarm-enables the timely and suitable delivery of healthcare to patients. Multi-Robots / Synergistic swarm-solutions can offer patients online guidance on managing minor wounds at home. Patients can utilize online sites to get information on how to identify the symptoms they are currently experiencing. In addition, individuals have the opportunity to seek medical guidance from physicians using online platforms. These platforms facilitate the exchange of information about one's sickness or symptoms, enabling doctors and other medical professionals to offer suitable medical advice. Healthcare professionals may efficiently access patients' records, therefore eliminating the need for time-consuming document photocopying. Online health records offer enhanced security, reliability, seamless updates, and less chance of patient data loss (Troccaz et al., 2019).
- Chat-bots and virtual therapists powered by Multi-Robots / Synergistic swarm can be utilized to offer assistance to those facing challenges with their mental well-being, thereby promoting the preservation and enhancement of their psychological health. Services encompassed in this support area include access to knowledge, strategies for managing stressful situations, and crisis intervention. Predictive mental health is the use of Multi-Robots / Synergistic swarm to analyze physiological and behavioral data in order to foresee and avert psychological emergencies, allowing for early interventions. The process of analyzing behavioral and physiological data is known as behavioral and physiological data analysis (Kaur, 2024).
- It is crucial to address concerns regarding data privacy, ethical considerations, and the need for ongoing regulatory frameworks to ensure the safe and secure utilization of Multi-Robots / Synergistic swarm in the healthcare sector. However, Multi-Robots / Synergistic swarm still provides numerous advantages. In order to effectively incorporate Multi-Robots / Synergistic swarm into the healthcare system, healthcare practitioners, researchers, and technology developers must collaborate. Population health management can be achieved by employing data analytic to identify patterns and trends found among populations. To accomplish this objective, it is important to carry out a more extensive examination of the data generated by Robotics. This has the potential to significantly assist in the administration of healthcare for the overall population (Troccaz et al., 2019).

Regarding the implementation of measures to improve the health of a population, the healthcare industry can effectively manage its supply chain by utilizing Multi-Robots / Synergistic swarm-devices. These devices can monitor and regulate the storage conditions of pharmaceuticals, immunizations, and other essential supplies. The prevention of spoilage, reduction of waste, and guarantee of essential commodity supply are achieved as a consequence of this. Multi-Robots / Synergistic swarm can be employed in healthcare facilities to monitor and maintain cleanliness, as well as control infections. This utilization could be advantageous for all parties involved. Intelligent sensors possess the capability to monitor the quality of air, assess the extent to which individuals comply with hand hygiene practices, and detect upcoming epidemics. This is achieved by the examination of patterns in patient symptoms or diseases. Integrating biometric authentication methods with Multi-Robots / Synergistic swarm-technology has the potential to enhance the safety and efficiency of healthcare services. This was deliberated within the framework of authentication and security. As a result, this contributes to safeguarding patients' privacy and ensuring compliance with legislation governing the protection of sensitive information. Restricted staff members will be granted exclusive access to patients' personal information (Sugiyama et al., 2010; Kaur & Arora, 2023).

Robotic surgical equipment, powered by Multi-Robots, can swiftly and effectively support surgeons by providing them with up-to-date and valuable information during surgical procedures. The surgeons can acquire this information. These systems may include feedback mechanisms and sensors to optimize surgical outcomes, enhance precision, and minimize errors. By employing Multi Robots, it is possible to carry out the identification and surveillance of indicators of mental health. Here, we can find signs such as stress levels, sleep habits, and mood disorders. Potential exists for employing sensors, adaptations, wearable devices, and smartphone applications. All of them are potential options. Sharing data with healthcare practitioners can facilitate the adoption of personalized psychological treatments, expediting the process (Kaur, 2024). To encourage patient engagement, connected devices in Multi-Robots / Synergistic swarm-can be used to provide individualized health education, medication reminders, and lifestyle counseling opportunities.

In the field of healthcare, the implementation of Multi-Robots / Synergistic swarm (RPA) raises a variety of ethical concerns that need to be taken into consideration. Throughout the course of history, the vast majority of decisions concerning medical treatment were made decisions by individuals. Consequently, the utilization of intelligent technology to either produce treatment decisions or provide support in making them has developed as a result of this (Sugiyama et al., 2010; Kaur & Arora, 2023).

ETHICAL CONCERNS AND CONSIDERATIONS

There are concerns regarding the ethical standards and rules of communication, as well as the preservation of sensitive information, truthfulness, and agreement. Considering the rapid pace at which technological advancements are currently taking place, finding a solution to the issue of openness can be an activity that is fraught with an incredible amount of difficulty. Due to the vast complexity of certain systems, it can be quite challenging to comprehend or interpret them. This is especially true for systems that incorporate deep learning in visual analytic. In the event that patients are presented with information regarding the function that a graphic plays in the process of cancer diagnosis, it is quite probable that they will be interested in acquiring an understanding of the factors that contribute to the diagnosis of their cancer. The fact that it is difficult to articulate the context is something that may come as a surprise to medical professionals who are normally well-versed in the inner-workings of data augmentation .

When it comes to evaluating and treating patients, the utilization of Multi-Robots / Synergistic swarm is prone to errors, which can make the process of holding them accountable for carelessness more difficult. Patients can be evaluated and treated by Multi-Robots / Synergistic swarm. When it comes to providing patients with information that they would rather acquire from a compassionate physician, there will be situations in the not too distant future in which Multi-Robots / Synergistic swarm will fulfill the job of presenting patients with that information. Despite the fact that gender and racial origin are not the actual causes of the disease, it is conceivable for medical artificial intelligence to exhibit computational bias. This might result in larger disease probability estimates based on these factors, despite the fact that these factors are not the actual causes of the disease.

Using the information that is provided by the individual who is ill, an automated system is able to reliably diagnose illnesses using the information that is provided by the one who is ill.In addition to this, it has the potential to offer additional assistance after the completion of therapy. A number of automated algorithms have been developed for the purpose of supporting individuals in performing essential duties, such as the administration of their prescribed medications, the consumption of their nourishment, and the planning of their diet. An individual who has been discharged from the hospital can receive assistance from a caregiver in their recuperation at home after being discharged from the hospital. In light of this, any data or information

The application of an automated system has the ability to control and monitor seventeen distinct health illnesses in an effective manner. These disorders include blood pressure and cardiovascular health issues. According to Sugiyama et al., (2010)., it is feasible that this may be advantageous for the health of an individual because it would enable them to quickly get in touch with their physicians in the event that they experience changes that they had not anticipated.

A variety of advancements in the fields of ethics, medicine, professional practice, and scientific research are anticipated to be brought about as a consequence of the implementation of Multi-Robots / Synergistic swarm (RPA) in healthcare facilities. The construction of protocols for monitoring big circumstances, responding in a responsible manner, and building management systems to prevent adverse effects is an imperative necessity for health systems, as well as for politicians and organizations within the government. This is a necessity that cannot be avoided. Due to the fact that this technology will have a large and long-lasting impact on mankind, it will entail the design of regulations that are severe and will demand constant attention for several generations.

CONCLUSION

In the field of medicine, the use of robots that are independently operated is becoming increasingly common. Medical institutions are confronted with a multitude of demanding operations and stringent criteria, such as the organization of patients and the processing of reimbursements. The consequence of this is a string of discrepancies, increased operational expenses, and procedures that move at a snail's pace. Through the utilization of robotics and machine learning, medical practitioners are able to efficiently address difficulties in the healthcare industry.

Surgical procedures, the overall experience of patients, and the efficiency of medical institutions are all areas that need to be improved, and there are three primary issues that need to be addressed in order to accomplish these goals. Through the utilization of technology and the implementation of effective automated processes, such as Multi-Robots / Synergistic swarm, medical professionals and nurses are able to relieve themselves of the time-consuming and costly responsibilities associated with keeping up with technological advancements. It makes it possible for businesses to concentrate their resources on providing treatment in a more efficient manner.

By utilizing workstation automation to conduct a wide variety of routine duties, ranging from simple to complex, Multi-Robots / Synergistic swarm can be of assistance to medical institutions. This is especially prevalent in situations where enterprises are expanding and handling big customer lists. It is projected that the deployment of Multi-Robots / Synergistic swarm would assist businesses in generating income from a variety of operations that take place on the server side. Also, it is anticipated that Multi-Robots / Synergistic swarm would make it possible to manage huge numbers of patients in healthcare facilities and also supply knowledge about upgraded amenities that will be available in the future.The creation of algorithms that can be utilized for a variety of purposes is becoming an important priority for several suppliers.

Both consumers and medical facilities have made investments in the adoption of automated process solutions. These investments are being made in order to improve the efficiency and lifespan of organizational structures, as well as to increase their market share. In the field of medicine, the development of autonomous robots will be facilitated by the increasing application of cutting-edge technology such as artificial intelligence, data mining, and the internet of things. The use of Multi-Robots / Synergistic swarm is becoming increasingly prevalent in medical facilities as a means of simplifying operations and managing big datasets. Through the utilization of machine learning, organizations are able to independently improve their business procedures and practices.

REFERENCES

Alsamhi, S. H., & Lee, B. (2020). Blockchain-empowered multi-robot collaboration to fight COVID-19 and future pandemics. *IEEE Access : Practical Innovations, Open Solutions*, *9*, 44173–44197. doi:10.1109/ACCESS.2020.3032450 PMID:34786312

Cao, K., Chen, Y., Dang, H., Gao, S., & Yan, K. (2022, July). Multi-robot Coverage System Based on Health Optimization Management Algorithm. In *Proceedings of 2021 5th Chinese Conference on Swarm Intelligence and Cooperative Control* (pp. 902-911). Singapore: Springer Nature Singapore.

Cao, K., Chen, Y., Gao, S., Zhang, H., & Dang, H. (2022). Multi-Robot Formation Control Based on CVT Algorithm and Health Optimization Management. *Applied Sciences (Basel, Switzerland)*, *12*(2), 755. doi:10.3390/app12020755

Das, G. P., McGinnity, T. M., Coleman, S. A., & Behera, L. (2015). A distributed task allocation algorithm for a multi-robot system in healthcare facilities. *Journal of Intelligent & Robotic Systems*, *80*(1), 33–58. doi:10.1007/s10846-014-0154-2

Duggal, A. S., Malik, P. K., Gehlot, A., Singh, R., Gaba, G. S., Masud, M., & Al-Amri, J. F. (2022). A sequential roadmap to Industry 6.0: Exploring future manufacturing trends. *IET Communications*, *16*(5), 521–531. doi:10.1049/cmu2.12284

Foroutannia, A., Shoryabi, M., Anaraki, A. A., & Rowhanimanesh, A. (2021, March). SIN: A Programmable Platform for Swarm Robotics. In *2021 26th International Computer Conference, Computer Society of Iran (CSICC)* (pp. 1-5). IEEE. 10.1109/CSICC52343.2021.9420596

Fosch-Villaronga, E., & Drukarch, H. (2021). On Healthcare Robots: Concepts, definitions, and considerations for healthcare robot governance. arXiv preprint arXiv:2106.03468.

Haken, H., & Levi, P. (2012). *Synergetic agents: From multi-robot systems to molecular robotics*. John Wiley & Sons. doi:10.1002/9783527659524

Holland, J., Kingston, L., McCarthy, C., Armstrong, E., O'Dwyer, P., Merz, F., & McConnell, M. (2021). Service robots in the healthcare sector. *Robotics (Basel, Switzerland)*, *10*(1), 47. doi:10.3390/robotics10010047

Humaidi, A. J., Ibraheem, I. K., Azar, A. T., & Sadiq, M. E. (2020). A new adaptive synergetic control design for single link robot arm actuated by pneumatic muscles. *Entropy (Basel, Switzerland)*, *22*(7), 723. doi:10.3390/e22070723 PMID:33286496

Karpov, V. E., & Tarassov, V. B. (2017, September). Synergetic artificial intelligence and social robotics. In *International Conference on Intelligent Information Technologies for Industry* (pp. 3-15). Cham: Springer International Publishing.

Kaur, J. (2024). Green Finance 2.0: Pioneering Pathways for Sustainable Development and Health Through Future Trends and Innovations. In Sustainable Investments in Green Finance (pp. 294-319). IGI Global.

Kaur, J. (2024). Fueling Healthcare Transformation: The Nexus of Startups, Venture Capital, and Innovation. In Fostering Innovation in Venture Capital and Startup Ecosystems (pp. 327-351). IGI Global.

Kaur, J. (2024). Towards a Sustainable Triad: Uniting Energy Management Systems, Smart Cities, and Green Healthcare for a Greener Future. In Emerging Materials, Technologies, and Solutions for Energy Harvesting (pp. 258-285). IGI Global.

Kaur, J., & Arora, R. (2023, June). Exploring the dimensionality of employee silence in healthcare sector. In AIP conference proceedings (Vol. 2782, No. 1). AIP Publishing. doi:10.1063/5.0154178

Kaur, J., & Arora, R. (2023, June). Exploring the impact of employee silence in private hospitals-a structural equation modeling approach. In AIP Conference Proceedings (Vol. 2782, No. 1). AIP Publishing. doi:10.1063/5.0154173

Kernbach, S. (2011). Robot companions: Technology for humans. arXiv preprint arXiv:1111.5207.

Lestingi, L., Sbrolli, C., Scarmozzino, P., Romeo, G., Bersani, M. M., & Rossi, M. (2022, May). Formal modeling and verification of multi-robot interactive scenarios in service settings. In *Proceedings of the IEEE/ACM 10th International Conference on Formal Methods in Software Engineering* (pp. 80-90). IEEE. 10.1145/3524482.3527653

Li, Y., Jiao, X. Y., Sun, B. Q., Zhang, Q. H., & Yang, J. Y. (2021, March). Multi-welfare-robot cooperation framework for multi-task assignment in healthcare facilities based on multi-agent system. In *2021 IEEE International Conference on Intelligence and Safety for Robotics (ISR)* (pp. 413-416). IEEE. 10.1109/ISR50024.2021.9419496

Ma, Y., Zhang, Y., Wan, J., Zhang, D., & Pan, N. (2015). Robot and cloud-assisted multi-modal healthcare system. *Cluster Computing*, *18*(3), 1295–1306. doi:10.1007/s10586-015-0453-9

Moura, V. (2012). Magnetically Actuated Multiscale Medical Robots. In *IROS 2012 Full-day Workshop* (p. 48). IEEE.

Rizk, Y., Awad, M., & Tunstel, E. W. (2019). Cooperative heterogeneous multi-robot systems: A survey. *ACM Computing Surveys*, *52*(2), 1–31. doi:10.1145/3303848

Sahu, B., Das, P. K., Kabat, M. R., & Kumar, R. (2022). Prevention of Covid-19 affected patient using multi robot cooperation and Q-learning approach: A solution. *Quality & Quantity*, *56*(2), 793–821. doi:10.1007/s11135-021-01155-1 PMID:33972809

Şen, M. O., Okumuş, F., & Kocamaz, F.ŞEN. (2022). Application of blockchain powered mobile robots in healthcare: Use cases, research challenges and future trends. *Türk Doğa ve Fen Dergisi*, *11*(2), 27–35. doi:10.46810/tdfd.1017499

Shallal, A. H., Ucan, O. N., Humaidi, A. J., & Bayat, O. (2020). Multi-robot systems formation control with maneuvring target in system applicable in the hospitality and care-health industry of medical internet of things. *Journal of Medical Imaging and Health Informatics*, *10*(1), 268–278. doi:10.1166/jmihi.2020.2840

Sugiyama, O., Shinozawa, K., Akimoto, T., & Hagita, N. (2010). Case study of a multi-robot healthcare system: Effects of docking and metaphor on persuasion. In *Social Robotics: Second International Conference on Social Robotics*. Singapore.

Troccaz, J., Dagnino, G., & Yang, G. Z. (2019). Frontiers of medical robotics: From concept to systems to clinical translation. *Annual Review of Biomedical Engineering*, *21*(1), 193–218. doi:10.1146/annurev-bioeng-060418-052502 PMID:30822100

Chapter 11
Cogwheels of Care:
Robotic Marvels in the Hospital Landscape

Jaspreet Kaur
Chandigarh University, India

ABSTRACT

"Cogwheels of Care: Robotic Marvels in the Hospital Landscape" examines the incorporation of sophisticated robotics in contemporary healthcare. This abstract explores the profound influence of robots in hospitals, fundamentally changing patient care and enhancing medical capabilities. The chapter explores the ways in which these advanced robots enhance and simplify many jobs, ranging from surgical procedures to everyday activities, by maximising efficiency and accuracy. It emphasises the interdependent connection between technology and healthcare practitioners, focusing on the ethical considerations and societal consequences of this technological transformation. This research highlights the changing healthcare landscape, where the complex interaction between human expertise and robotic innovation is transforming the principles of compassionate and effective patient-centered care.

INTRODUCTION

Currently, healthcare organizations are facing significant pressure to excel in the market due to intense competition. Healthcare professionals experience the burden of meeting expectations in order to preserve and safeguard their status within the organization. Organizations, in response to intense competition, must implement cost reduction measures, which may include downsizing their workforce and increasing the task of the remaining employees. An excessive workload results in an escalation of work-related stress, job discontent, and demotivate. Furthermore, these employees struggle to achieve a harmonious equilibrium between their professional and personal lives (Etukudoh et al., 2024). Hospitals in India face the dual challenge of coping with a scarcity of healthcare personnel while also managing a surge in patient numbers. Healthcare personnel, including nurses, resident physicians, and specialists, are responsible for attending to patients. Due to a scarcity of staff, they are required to work in unconventional shifts and extended hours. It is observed that

DOI: 10.4018/979-8-3693-1962-8.ch011

these staff experience significant pressure to manage the heightened patient workload, resulting in workplace stress and discontent. Additionally, as a result of irregular and excessive working hours, these employees struggle to achieve a harmonious equilibrium between their professional and home lives. Investing additional time in their professional responsibilities and experiencing work-related exhaustion can disrupt their personal lives. This can diminish the productivity of people in their work. Employees may resign from their employment in severe circumstances of overwhelming strain. This issue is prevalent throughout numerous healthcare organizations. Content and satisfied personnel contribute to a positive and prosperous organization.Employees are crucial to the functioning of any organization. Therefore, it is imperative to take proactive measures to address this issue in order to remain competitive in today's market (Ala'a, 2023).

One example of such an endeavour is the implementation of robotics in the healthcare industry. This can be accomplished by acknowledging and valuing an employee's right to a satisfying existence both within and outside of their job, resulting in advantages for the individual, the organization, and society as a whole. Work-family balance refers to a situation when the demands and expectations of one's work and family roles are in conflict and cannot be reconciled in some way.The integration of robotics in healthcare has a crucial role in enhancing overall pleasure in all aspects of life, including personal well-being, professional success, and family dynamics. It aids in mitigating stress and lowering staff turnover and job-related stress. This study illuminates the significance of Robotics for healthcare and aims to establish a foundation for implementing various methods to enhance employee job happiness and productivity. These kind of activities will assist employees in achieving a harmonious work-life balance and attaining personal contentment, fulfillment, and overall well-being both in their personal lives and within the organization. The use of advances in technology initiatives has led to a substantial improvement in the operational effectiveness of the healthcare industry. The utilization of algorithms for machine learning has been advantageous to the healthcare industry as it has streamlined the task of analyzing vast quantities of data.

Robotics is capable of retaining and acquiring knowledge at a significantly faster rate than the human brain does. This is in comparison to the performance of the human brain. The distribution of medical experts, hospital beds, and resources can be optimized with the use of solutions that are powered by artificial intelligence. This optimization leads to enhanced operational efficiency and decreased expenses. In order to shed light on the benefits that may be acquired by using Artificial Intelligence, Robotics, Robotic Process Automation, and Machine Learning algorithms in the healthcare industry, the objective of this study is to shine light on those benefits. This opens the door for the development of less invasive operations that require shorter recovery times as depicted in figure 1 below.

Through the utilization of cutting-edge robotic equipment, it is now feasible to perform surgical procedures remotely, which in turn expands the availability of specialized medical care. The Robotics generates data that has the potential to be researched on a more extensive scale in order to identify patterns and trends among populations. This is accomplished through the use of the Robotics. There is a possibility that this will prove to be of considerable aid in the management of health care for the general population. When it comes to prosthetics and rehabilitation equipment, the implementation of Robotics technology makes it feasible to provide patients and medical professionals with quick and continuous feedback regarding the patient's status (Aggarwal et al., 2019). Because of this, both the personification of rehabilitation programme and the tracking of changes, as well as the tweaking of these programme to match the specific requirements of each individual patient, are less difficult to do (AlShamsi et al., 2022). Furthermore, the implementation of robots in the sphere of eldercare

Figure 1. Flowchart depicting the utilization of robotics in the healthcare industry

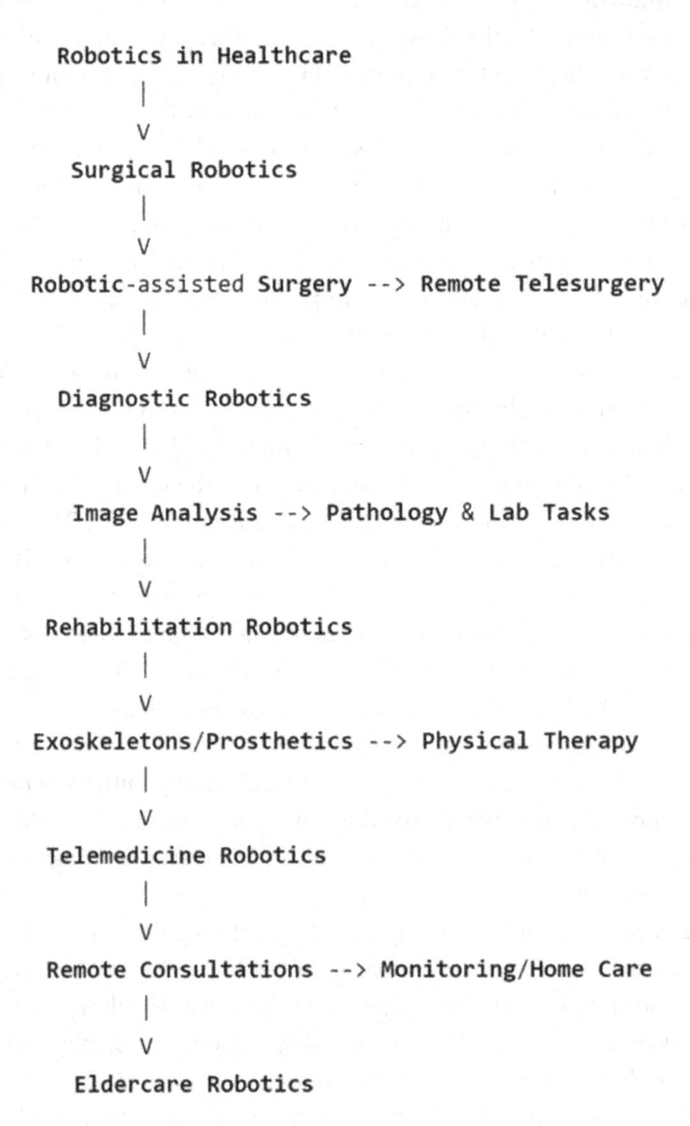

guarantees enhanced mobility and safety for the senior population (AlShamsi et al., 2022) as depicted in figure 2 below:

The purpose of this study is to provide an overview of the research approaches that were utilized in order to find, locate, and analyse papers that were relevant. The research questions that were utilized to investigate the content of the study are summarized in this paragraph, which provides an overview of those questions. Both the criteria that were used to include or exclude content, as well as the method that was used to gather data, will be discussed in this text. In order to complete the evaluation of the pertinent literature, four independent research inquiries were devised. The objective of this evaluation is to provide aid in the process of data extraction by assessing whether or not the papers that were selected address any of the study subjects. This evaluation will be carried out in order to accomplish this purpose. The

Figure 2. Application of robots in the domain of eldercare

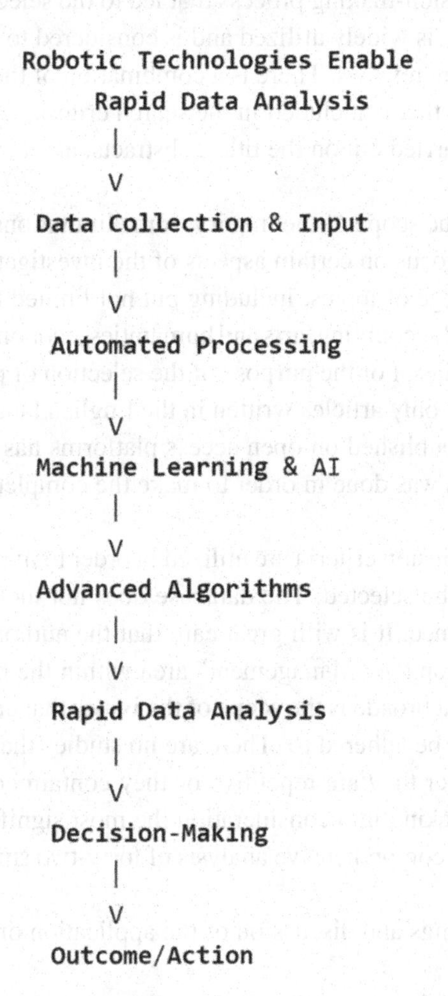

following questions, which together make up the guiding principles, served as the basis for the research that was carried out for this project.

1. To what extent does the implementation of Robotics in the field of healthcare allow for the provision of a wide range of advantageous advantages?
2. What are the many advantages that the Robotics delivers to the field of healthcare as a whole, and to what extent does it bring these advantages?
3. What are the numerous advantages that the implementation of robotic process automation provides to the field of healthcare, and how may these advantages be utilized successfully?
4. When it comes to the healthcare industry, what are the numerous advantages that machine learning algorithms offer? This is the fourth research subject that needs investigating.

It is necessary to make use of the published literature that is accessible through the Scopus database in order to carry out the research. An inquiry is conducted into the Scopus database in order to discover publications that are pertinent to the study that is currently being conducted. There were two distinct elements that had a role in the decision-making process that led to the selection of this particular database. In addition, the Scopus database is widely utilized and is considered to be one of the most comprehensive databases for doing literature reviews. There is a combination of the terms "Artificial Intelligence" or "Robotics" and "Healthcare" that is included in the search criteria. By making use of the string that was specified, the search was carried out on the titles, abstracts, and keywords of the articles that were being considered.

For the aim of establishing the scope of the inquiry, some limitations were placed, meaning that the investigators were obligated to focus on certain aspects of the investigation.

This study covers a wide range of topics, including but not limited to the following: the social sciences, business management and accounting, arts and humanities, economics, econometrics, and finance, as well as multidisciplinary studies. For the purpose of the selection of publications, an extra constraint was imposed; more specifically, only articles written in the English language were taken into consideration. The study that has been published on open access platforms has been chosen with great care in order to extract the results. This was done in order to make the complete document more accessible to the general public.

Additional inclusion and exclusion criteria are utilized in order to make a decision regarding the final collection of research that will be selected. The database does not include any research that has been repeated as a result of its existence. It is with great care that the authors choose the study that will be reviewed, with a specific focus on the "Management" area within the healthcare industry. In addition, the incorporation of research that broadens the scope of the issues that are being investigated is the most important concept that needs to be adhered to. There are no studies that are included because they fall into one of two categories: either they are repetitive or they contain conclusions that are limited to a particular aspect of the issue. Taking into consideration the most significant findings from each of the studies, the authors conducted a comprehensive analysis of forty-two study investigations. This analysis was carried out by the authors.

This study presents the findings and discussion of the application of Robotics(AI) in the healthcare industry.

Robotics(AI) has the potential to be utilized in a wide range of various ways within the field of medical care (Cao & Rogers, 2007). Significant advancements have been made in the areas of diagnosis and therapy over the course of the past few years (Lloyd et al., 2024). Although it is anticipated that Robotics(AI) will serve as a supplement to human work, it is not anticipated that AI will totally replace human employment. Using Robotics(AI) assists with the monitoring of patients and provides individualized support through the analysis of images and the automation of medical equipment (Bogue, 2011).This is accomplished through the use of AI. Further, Robotics(AI) is a factor that contributes to the development of a wide range of industries, such as robotic-assisted surgery, data analysis, research in the pharmaceutical business, and many others. Tasks are completed more quickly and with greater efficiency thanks to artificial intelligence. The process of making decisions regarding patient diagnoses and the process of remotely monitoring patients are both areas in which the utilization of Robotics facilitates both of these processes (Butter et al., 2008).

Robotics makes a contribution to the improvement of healthcare operations in a variety of domains, including the optimization of health facility administration, the correct diagnosis of patients, the development of treatment plans, and the improvement of health outcomes. Furthermore, robotic equipment contribute to the diagnostic procedure by offering support, as depicted in figure 3:

The diagnostic imaging, which includes:

Figure 3. Effect of robotic surgical platforms on both diagnosis and operation

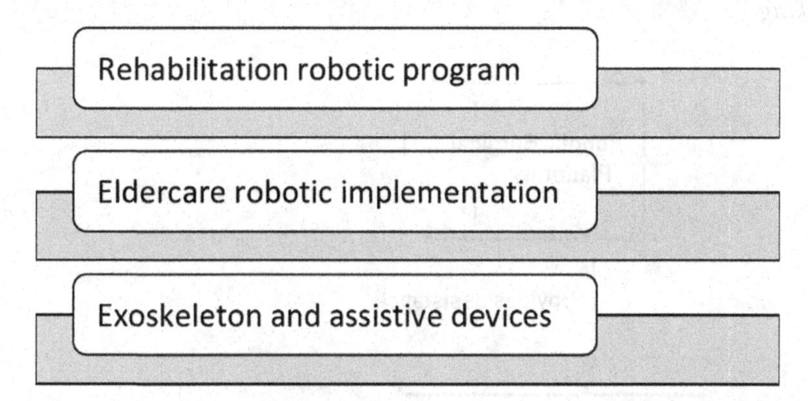

Robotics systems offer support to radiologists working in the field of radiology by analyzing X-rays, CT scans, and MRIs. This is accomplished through the study of these medical images. The use of these systems assists radiologists in identifying abnormalities and contributes to the process of diagnosis.

Early Detection: Robotics has the ability to recognize anomalies and minute patterns that human vision can overlook. This facilitates more accurate early detection. The quick diagnosis of diseases such as cancer and neurological disorders such as Parkinson's disease is made possible as a result of this (Kim et al., 2024). In the process of developing new medications, there are a variety of processes and procedures that are carried out in a methodical manner. These processes and actions are designed to facilitate the development of novel pharmaceutical substances and their subsequent testing.

Robotics has the capability to study enormous amounts of data, including genetic information, in order to discover potential new treatments and to estimate how effective they might be. This is accomplished through the process of data analysis. This consequently results in a significant reduction in the amount of time and money that is required for the study of medications (Khang et al., 2024).

When it comes to clinical trials, Robotics makes the process of selecting trial participants more efficient. This is accomplished by analyzing the characteristics of the individuals to evaluate whether or not they are suitable for the research being conducted. As a consequence of this, the probability that the experiment would be successful is increased (Gonzalo et al., 2024). In addition, robotic technologies facilitate swift data analysis and decision support, aiding medical personnel in making educated diagnosis and treatment processes, as depicted in figure 4 below:

Plans of treatment that are individualized:

The utilization of personal genetic information by Robotics makes it possible to personalize treatment regimens, which in turn makes it possible to forecast the efficacy of specific medications and reduce the possibility of adverse reactions caused by those medications. It is possible to considerably improve the efficacy of treatment optimization through the process of customizing medications to match the specific requirements of each individual patient (Etukudoh et al., 2024). Existing literature suggests that robotics has the ability to enhance the quality of processes in medical field, as depicted in figure 5 below.

Figure 4. Flowchart depicting the process by which robotic technologies facilitate swift data processing and decision-making

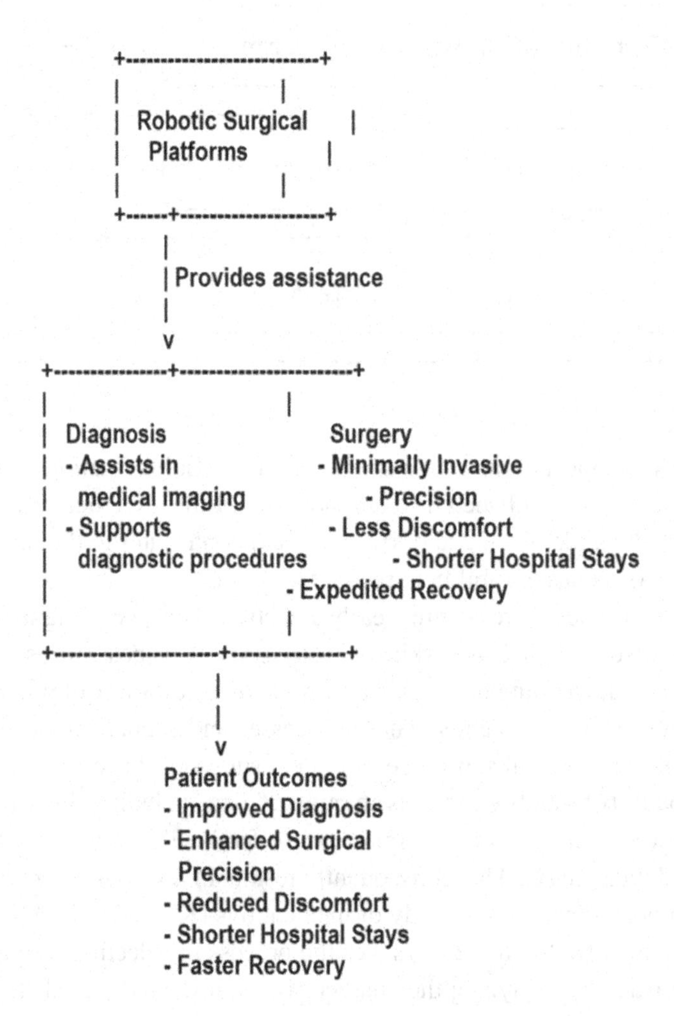

Advanced Analytic

Robotics enables ongoing professional education and advancement of healthcare practitioners through the use of telemedicine and telehealth. Telemedicine is the process of exchanging healthcare data and providing medical treatment to patients who are located far away. It does not fall within the category of technology or constitute a distinct or novel field of medicine. Telemedicine refers to the use of technology to facilitate the interchange of data, voice, and video between patients and healthcare professionals during their interactions. It facilitates the provision of healthcare services to remote regions and promotes equal access to medical care while preserving a balance between urban and rural communities.

The utilization of robotics in the healthcare sector helps to mitigate the challenges caused by the geographical separation of healthcare workers. In rural regions, the task of recruiting and retaining healthcare personnel is frequently challenging.This difficulty can be tackled by employing robot-

Figure 5. Ability of robotics in medical field

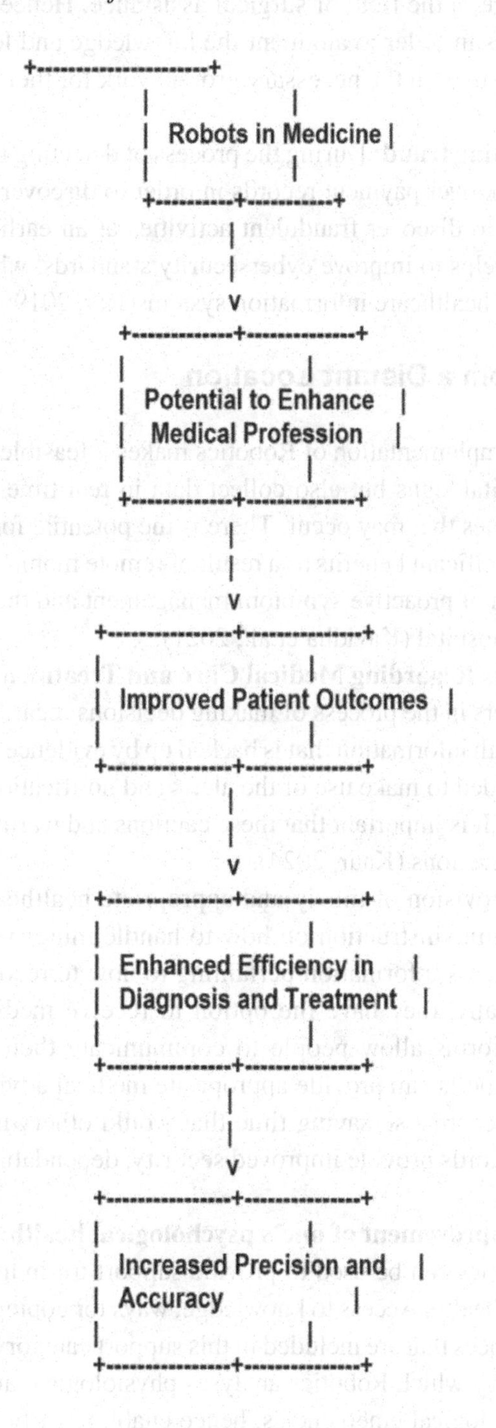

ics (Aggarwal et al., 2019; Cresswell et al., 2018; Elendu et al., 2023; Javaid et al., 2022; Guntur et al., 2019).

Robotic surgical assistance technology has the potential to enhance the professional development of healthcare staff by facilitating the exchange of learning data to rural areas and improving the sharing of information and knowledge in the field of surgical assistance. Hence, it is crucial to ascertain the optimal utilization of robotics in order to augment the knowledge and learning of healthcare professionals, thereby providing them with the necessary groundwork for their professional growth (Joseph et al., 2018).

When it comes to preventing fraud: During the process of detecting fraudulent activity, algorithms powered by Robotics(AI) look over payment records in order to discover any inconsistencies that may have occurred. It is possible to discover fraudulent activities at an earlier stage because to this. The application of Robotics(AI) helps to improve cybersecurity standards, which in turn helps to boost the protection of patient data and healthcare information systems(Kar, 2019; Katevas, 2001).

Observing Patients From a Distant Location

Wearable technology: The implementation of Robotics makes it feasible to wear trackers that not only continuously monitor your vital signs but also collect data in real time and immediately notify your physician of any potential issues that may occur. There is the potential for individuals who suffer from chronic conditions to have significant benefits as a result of remote monitoring. A few examples of these advantages are the facilitation of proactive symptom management and the reduction in the necessity of making frequent trips to the hospital (Kavidha et al., 2021).

Help in Making Decisions Regarding Medical Care and Treatment Provision: Robotics(AI) offers aid to medical practitioners in the process of making decisions regarding how to diagnose and treat patients by supplying them with information that is backed up by evidence. For the purpose of enhancing patient safety, it is recommended to make use of the alerts and notifications that are produced by clinical decision support systems. It is important that these cautions and warnings address the possibility of pharmaceutical errors or interactions (Kaur, 2024).

Robotics facilitates the provision of timely and appropriate healthcare to patients.Robotics tools can provide patients with online instruction on how to handle minor wounds at home. Patients can utilise online resources to access information pertaining to how to recognise the symptoms they are now experiencing. Additionally, they have the option to receive medical advice from doctors via online platforms.These platforms allow people to communicate their illness or symptoms, while doctors and other medical experts can provide appropriate medical advice. The healthcare personnel can easily get the patients' records, so saving time that would otherwise be spent on photocopying documents. Online health records provide improved security, dependability, easy upgrading, and less risk of patient data loss.

The preservation and improvement of one's psychological health: Chat-bots and virtual therapists that are driven by Robotics can be used to provide support for individuals who are experiencing difficulties with their mental health. Access to knowledge, ways for coping with stressful situations, and crisis intervention are all services that are included in this support category. The term "predictive mental health" refers to the process by which Robotics analyses physiological and behavioral data in order to anticipate and prevent psychological emergencies, hence enabling early interventions. This process is referred to as behavioral and physiological data analysis (Kaur, 2024).

It is vital to address issues such as data privacy, ethical considerations, and the demand for continual regulatory frameworks in order to guarantee the safe and secure use of Robotics(AI) in the healthcare industry. This is the case despite the fact that Robotics(AI) offers a variety of benefits. For the purpose of achieving the objective of successfully integrating Robotics(AI) into the ecosystem of healthcare, it will be required for healthcare practitioners, researchers, and technology developers to collaborate with one another. The management of population health can be accomplished through the utilization of data analytic in order to find patterns and trends observed among populations. In order to achieve this goal, it is necessary to conduct a more comprehensive analysis of the data that is produced by Robotics. There is a possibility that this will prove to be of considerable aid in the management of health care for the general population. Specifically with reference to the deployment of preventative measures with the intention of improving the health of a populations (Kaur, 2024).

The supply chain in the healthcare industry can be managed through the implementation of Robotics devices that are able to monitor and regulate the conditions under which pharmaceuticals, immunizations, and other important supplies are stored. It is as a result of this that the prevention of spoilage, the reduction of waste, and the guarantee of the supply of essential commodities are all accomplished (Maalouf et al., 2018).

Robotics have the potential to be utilized in healthcare facilities for the goal of monitoring and maintaining cleanliness as well as infection control. This usage could be beneficial for all parties involved. Intelligent sensors have the ability to monitor air quality, evaluate the degree to which persons adhere to hand hygiene measures, and identify future epidemics. This is accomplished by the analysis of patterns of patient symptoms or diseases. The introduction of biometric authentication methods into Robotics technology has the potential to increase the efficiency of healthcare systems in terms of safety. This was discussed in the context of authentication and security. The result of this is that it contributes to the protection of patients' privacy and assures compliance with the regulations that regulate the protection of sensitive information. Staff members who are restricted in their access will be provided exclusive access to the personal information of patients (Kolpashchikov et al., 2022; Kim et al., 2016).

Current information: During surgical procedures, robotic surgical equipment that is enabled by the Robotics has the ability to provide surgeons with quick and valuable assistance by giving them with information that is current. This information can be obtained by the surgeons. It is possible that these systems will incorporate feedback mechanisms and sensors in order to enhance the outcomes of surgical procedures, as well as to improve precision and cut down on the amount of errors that take place.

Through the utilization of Robotics, it is feasible to perform detection and monitoring of markers of mental health. Stress levels, sleep patterns, and mood problems are some of the indicators that can be found here. There is the potential for the utilization of sensors, modifications, wearable devices, and applications for smartphones. These are all possibilities. It is possible to share the data with healthcare practitioners in order to expedite the process of adopting psychological treatments that are specifically suited to the individual (Kaur, 2024).

Patient Engagement and Education: As a means of encouraging patients to take an active role in their own healthcare, devices that are connected to the Robotics can be utilized to provide patients with individualized health education, medication prompts, and lifestyle counselling opportunities. Patients are motivated to take an active role in the management of their own health as a result of the implementation of this method, which in turn enhances the likelihood that they will adhere to the treatment regimens that have been prescribed to them (Kyrarini et al., 2021; Kaur, 2024).

Robotics is a technology that enables Real-time position Services (RTLS) to monitor the precise location of patients, medical staff, and medical equipment within a healthcare institution or clinic in real time. This is made possible by the Robotics. Consequently, this not only enhances the entire patient experience but also boosts the efficiency of the procedure, which in turn reduces the length of time that patients are required to wait. The integration of block-chain technology with Robotics technology has the potential to increase the dependability and security of health data management. Smart prostheses and rehabilitation equipment are now able to provide fast feedback to both patients and healthcare practitioners thanks to Robotics, which makes this capability possible. Robotics technology is responsible for making this capability a reality. The customization of rehabilitation programme, the monitoring of these programme progress, and the modification of these programme to match the specific requirements of each individual patient are all made simpler as a result of this reason. When it comes to the healthcare industry, it is required to solve issues such as data security, interoperability, and ethical considerations in order to make use of Robotics in a manner that is both suitable and successful.

Challenges

There are a number of challenges that the healthcare business must contend with, some of which include the complexity of procedures, the integration of data from a wide variety of sources, the management of information systems, and laboratory information systems. The software robotics that are currently accessible allow for the extraction of data from a wide number of sources and its subsequent utilization within businesses. Through the use of robotic process automation, there has been a notable reduction in the number of processes that are repetitive. It is possible to relate this to the fact that robots are more efficient than humans in terms of their operational capabilities. There is the opportunity for individuals to commit their time to tasks that require them to engage in direct interaction with other consumers (Morgan et al., 2022; Pee et al., 2019; Narula et al., 2014; Kaur, 2024).When it comes to aspects such as speed, efficiency, quality, and intellect, there is no comparison that can be made between humans and robots. The utilization of robotic process automation will result in a reduction of the expenses that are associated with human resources.

There will be a huge rise in productivity as a consequence of job automation processes. As a consequence of the implementation of automation, the total efficiency of the organization will experience an improved state.

Through the implementation of a digital workplace, the organization will be able to enhance its capacity to optimize and supervise data management, which will result in enhanced efficiency. There will be an increase in the organization's overall efficiency as a consequence of the introduction of robotic process automation as well as other technical advancements. It is currently being worked on to automate the procedures for requesting information regarding the status of claims. The fact that claim requests are handled automatically is one factor that contributes to the overall satisfaction of the target audience. Robotics techniques that enable computers to learn and generate predictions without being explicitly programmed and without the need for human interaction are referred to as "machine learning" techniques.

Predictive analytic is a term that refers to the process of applying machine learning algorithms to the analysis of data that is obtained from a range of health records. This analytic is being conducted with the intention of identifying patterns and trends that can be utilized in the process of formulating policies pertaining to public health. There is not a single piece of technology that does not make a major contribution to the enhancement of organizational efficiency and the facilitation of predictive analytic. By doing so, the organization is able to acquire a more comprehensive understanding of its patients and to make decisions concerning the health of those individuals. It becomes abundantly evident, as a result of the process of analyzing the data, which clients require particular types of treatment. In order to expedite the patient's recovery even further, individualized medication may also be supplied to the person. The implementation of robotic process automation has led to a major improvement in both the speed and quality of surgical treatments. Additionally, it has been responsible for a large reduction in the number of errors that occur during surgical procedures.

CONCLUSION

The application of robotic technology exemplifies the unwavering commitment to improving patient care through innovative advancements in the ever-changing and developing field of healthcare. The adoption of robotics in healthcare facilities has complex repercussions. These technological marvels have demonstrated their capacity to significantly enhance healthcare standards in various domains, such as providing logistical assistance and performing surgical procedures with utmost accuracy. An examination of the relevant facts has revealed the chronological progression of robotics, starting from the initial prototypes and ending with the most advanced systems currently in operation (Ragno et al., 2023; Kaur, 2024; (Sarker et al., 2021). Nevertheless, the research emphasizes that the proper implementation of these technologies, together with their flexibility and training, are crucial factors that could enhance the efficiency of these technologies. This is a finding underlined by the research (Kaur,2024).

This comprehensive analysis leads to the conclusion that robotic technology is poised to play a significant role in the future of healthcare. Based on the investigation's results, this conclusion was reached. Several proposals have been put forward for future research endeavour focused on enhancing existing systems, tackling identified limitations, and managing ethical issues arising from the increasing automation in healthcare settings. These recommendations have been put up in the last few years. In order to ensure the ethical and safe integration of robots into healthcare services, it is crucial for technologists, healthcare practitioners, ethicists, and policymakers to collaborate and set comprehensive guidelines. This will enable the assurance of proper implementation of robots in healthcare services (Soriano et al., 2022; Stahl & Coeckelbergh, 2016).

The future of healthcare will be shaped by the integration of human skill with the advancement of technical innovation. This will be the case as the future of healthcare necessitates the combination of both. The combination of a visually realistic medical environment, along with the seamless integration of human empathy and precise robotic capabilities, has the potential to bring about a new era of patient-centered healthcare. This would represent a substantial advancement in the realm of healthcare. As healthcare becomes more complex, it is crucial to handle the issue with caution to ensure that compassion remains a vital aspect of technological advancements in the healthcare profession. The rationale behind this is that compassion plays a crucial role in the progress of technology innovations in the healthcare sector. In summary, the integration of robotics into healthcare facilities leads to numerous impressive achievements, as well as the potential for confronting challenges. To ensure that healthcare evolves into a future where human creativity and technology proficiency seamlessly merge, it is crucial for the various individuals involved to collaborate and contribute their diverse areas of knowledge. This will enable the path to be directed in order to accomplish the intended result (Vallès-Peris & Domènech, 2023; Vallès-Peris et al., 2021; Yoon & Lee, 2018).

This conclusion provides a succinct evaluation of the significant findings, viewpoints, and suggestions. Additionally, it introduces a thought-provoking topic of the harmonious existence of individuals and technology within the healthcare domain. Robotics is progressively becoming a significant component of the daily tasks performed by healthcare professionals. The significance of technology, particularly Robotics, in the healthcare industry has been widely acknowledged. Health-care professionals must possess computer abilities to efficiently navigate the ever expanding environment.Healthcare workers will dedicate time utilising Robotics tools.This technology would also facilitate the professional development of healthcare personnel in terms of knowledge and lifelong learning by enabling the sharing of educational data to rural areas and enhancing the transfer of data and expertise among faculty, students, and colleagues. Hence, it is crucial to ascertain the utilisation of Robotics in augmenting the comprehension and education of healthcare professionals, thereby providing them with the necessary groundwork for their professional advancement.

REFERENCES

Aggarwal, S., Gupta, D., & Saini, S. (2019, November). A literature survey on robotics in healthcare. In *2019 4th International Conference on Information Systems and Computer Networks (ISCON)* (pp. 55-58). IEEE. 10.1109/ISCON47742.2019.9036253

Ala'a, A. M. (2023). Adoption of Roboticsand Robotics in Healthcare: A Systematic Literature Review. [IJCMIT]. *International Journal of Contemporary Management and Information Technology*, *3*(6), 1–16.

AlShamsi, S., AlSuwaidi, L., & Shaalan, K. (2022). Robotics and AI in Healthcare: A Systematic Review. *Recent Innovations in Roboticsand Smart Applications*, 319-343.

Antony, V. N., Li, M., Lin, S. H., Li, J., & Huang, C. M. (2024). *Social Robots for Sleep Health: A Scoping Review*. arXiv preprint arXiv:2403.04169.

Bakshi, G., Kumar, A., & Puranik, A. N. (2021). Adoption of robotics technology in healthcare sector. In Advances in Communication, Devices and Networking [Singapore: Springer Singapore.]. *Proceedings of ICCDN, 2020*, 405–414.

Bogue, R. (2011). Robots in healthcare. Industrial Robot. *International Journal (Toronto, Ont.)*, *38*(3), 218–223.

Butter, M., Rensma, A., Kalisingh, S., Schoone, M., Leis, M., Gelderblom, G. J., & Korhonen, I. (2008). *Robotics for healthcare*.

Cao, C. G., & Rogers, G. (2007). *Robotics in health care: HF issues in surgery. Handbook of Human Factors and Ergonomics in Health Care and Patient Safety*. Lawrence Earlbaum & Associates.

Cresswell, K., Cunningham-Burley, S., & Sheikh, A. (2018). Health care robotics: Qualitative exploration of key challenges and future directions. *Journal of Medical Internet Research*, *20*(7), e10410. doi:10.2196/10410 PMID:29973336

Elendu, C., Amaechi, D. C., Elendu, T. C., Jingwa, K. A., Okoye, O. K., Okah, M. J., & Alimi, H. A. (2023). Ethical implications of AI and robotics in healthcare. *Revista de Medicina (São Paulo), 102*(50), e36671. PMID:38115340

Etukudoh, N. S., Esame, N. V., Obeta, U. M., Ejinaka, O. R., & Khang, A. (2024). Automations and Robotics Improves Quality Healthcare in the Era of Digital Medical Laboratory. In Computer Vision and AI-Integrated IoT Technologies in the Medical Ecosystem (pp. 419-434). CRC Press. doi:10.1201/9781003429609-24

Gonzalo de Diego, B., González Aguña, A., Fernández Batalla, M., Herrero Jaén, S., Sierra Ortega, A., Barchino Plata, R., & Santamaría García, J. M. (2024, March). Competencies in the Robotics of Care for Nursing Robotics: A Scoping Review. In Healthcare (Vol. 12, No. 6, p. 617). MDPI. doi:10.3390/healthcare12060617

Guntur, S. R., Gorrepati, R. R., & Dirisala, V. R. (2019). Robotics in healthcare: an internet of medical robotic things (IoMRT) perspective. In *Machine learning in bio-signal analysis and diagnostic imaging* (pp. 293–318). Academic Press. doi:10.1016/B978-0-12-816086-2.00012-6

Javaid, M., Haleem, A., Singh, R. P., Rab, S., Suman, R., & Kumar, L. (2022). Utilization of Robotics for Healthcare: A Scoping Review. *Journal of Industrial Integration and Management, 2250015.*

Joseph, A., Christian, B., Abiodun, A. A., & Oyawale, F. (2018). A review on humanoid robotics in healthcare. In *MATEC Web of Conferences* (Vol. 153, p. 02004). EDP Sciences. 10.1051/matecconf/201815302004

Kar, S. (2019, October). Robotics in HealthCare. In *2019 2nd International Conference on Power Energy, Environment and Intelligent Control (PEEIC)* (pp. 78-83). IEEE. 10.1109/PEEIC47157.2019.8976668

Katevas, N. (Ed.). (2001). *Mobile robotics in healthcare* (Vol. 7). IOS Press.

Kaur, J. (2024). Green Finance 2.0: Pioneering Pathways for Sustainable Development and Health Through Future Trends and Innovations. In Sustainable Investments in Green Finance (pp. 294-319). IGI Global.

Kaur, J. (2024). Fueling Healthcare Transformation: The Nexus of Startups, Venture Capital, and Innovation. In Fostering Innovation in Venture Capital and Startup Ecosystems (pp. 327-351). IGI Global.

Kaur, J. (2024). Towards a Sustainable Triad: Uniting Energy Management Systems, Smart Cities, and Green Healthcare for a Greener Future. In Emerging Materials, Technologies, and Solutions for Energy Harvesting (pp. 258-285). IGI Global.

Kaur, J., & Arora, R. (2023, June). Exploring the dimensionality of employee silence in healthcare sector. In AIP conference proceedings (Vol. 2782, No. 1). AIP Publishing. doi:10.1063/5.0154178

Kaur, J., & Arora, R. (2023, June). Exploring the impact of employee silence in private hospitals-a structural equation modeling approach. In AIP Conference Proceedings (Vol. 2782, No. 1). AIP Publishing. doi:10.1063/5.0154173

Kavidha, V., Gayathri, N., & Kumar, S. R. (2021). AI, IoT and robotics in the medical and healthcare field. *AI and IoT-Based Intelligent Automation in Robotics*, 165-187.

Khang, A., Rath, K. C., Anh, P. T. N., Rath, S. K., & Bhattacharya, S. (2024). Quantum-Based Robotics in the High-Tech Healthcare Industry: Innovations and Applications. In Medical Robotics and AI-Assisted Diagnostics for a High-Tech Healthcare Industry (pp. 1-27). IGI Global.

Kim, H., Kwon, D., Son, J., & Choi, J. (2024). A Novel Robotic Healthcare Device for Treating Chronic Venous Insufficiency in People Who Sit for Prolonged Periods. *IEEE Transactions on Medical Robotics and Bionics*, *6*(2), 618–631. doi:10.1109/TMRB.2024.3373909

Kim, J., Gu, G. M., & Heo, P. (2016). Robotics for healthcare. *Biomedical Engineering: Frontier Research and Converging Technologies*, 489-509.

Kolpashchikov, D., Gerget, O., & Meshcheryakov, R. (2022). *Robotics in healthcare. Handbook of Roboticsin Healthcare* (Vol. 2). Practicalities and Prospects.

Kyrarini, M., Lygerakis, F., Rajavenkatanarayanan, A., Sevastopoulos, C., Nambiappan, H. R., Chaitanya, K. K., Babu, A. R., Mathew, J., & Makedon, F. (2021). A survey of robots in healthcare. *Technologies*, *9*(1), 8. doi:10.3390/technologies9010008

Lloyd, P., Dall'Armellina, E., Schneider, J. E., & Valdastri, P. (2024). *Future cardiovascular healthcare via magnetic resonance imaging-driven robotics.*

Maalouf, N., Sidaoui, A., Elhajj, I. H., & Asmar, D. (2018). Robotics in nursing: A scoping review. *Journal of Nursing Scholarship*, *50*(6), 590–600. doi:10.1111/jnu.12424 PMID:30260093

Mois, G., & Beer, J. M. (2020). The role of healthcare robotics in providing support to older adults: A socio-ecological perspective. *Current Geriatrics Reports*, *9*(2), 82–89. doi:10.1007/s13670-020-00314-w PMID:32435576

Morgan, A. A., Abdi, J., Syed, M. A., Kohen, G. E., Barlow, P., & Vizcaychipi, M. P. (2022). Robots in healthcare: A scoping review. *Current Robotics Reports*, *3*(4), 271–280. doi:10.1007/s43154-022-00095-4 PMID:36311256

Narula, A., Narula, N. K., Khanna, S., Narula, R., Narula, J., & Narula, A. (2014). Future prospects of Roboticsin robotics software, a healthcare perspective. *International Journal of Applied Engineering Research: IJAER*, *9*(22), 10271–10280.

Oña, E. D., Garcia-Haro, J. M., Jardón, A., & Balaguer, C. (2019). Robotics in health care: Perspectives of robot-aided interventions in clinical practice for rehabilitation of upper limbs. *Applied Sciences (Basel, Switzerland)*, *9*(13), 2586. doi:10.3390/app9132586

Patel, A. R., Patel, R. S., Singh, N. M., & Kazi, F. S. (2017). Vitality of robotics in healthcare industry: Robotics perspective. *Robotics and big data technologies for next generation healthcare*, 91-109.

Pee, L. G., Pan, S. L., & Cui, L. (2019). Roboticsin healthcare robots: A social informatics study of knowledge embodiment. *Journal of the Association for Information Science and Technology*, *70*(4), 351–369. doi:10.1002/asi.24145

Ragno, L., Borboni, A., Vannetti, F., Amici, C., & Cusano, N. (2023). Application of Social Robots in Healthcare: Review on Characteristics, Requirements, Technical Solutions. *Sensors (Basel)*, *23*(15), 6820. doi:10.3390/s23156820 PMID:37571603

Sarker, S., Jamal, L., Ahmed, S. F., & Irtisam, N. (2021). Robotics and Roboticsin healthcare during COVID-19 pandemic: A systematic review. *Robotics and Autonomous Systems*, *146*, 103902. doi:10.1016/j.robot.2021.103902 PMID:34629751

Silvera-Tawil, D. (2024). Robotics in Healthcare: A Survey. *SN Computer Science*, *5*(1), 189. doi:10.1007/s42979-023-02551-0

Soriano, G. P., Yasuhara, Y., Ito, H., Matsumoto, K., Osaka, K., Kai, Y., & Tanioka, T. (2022, August). Robots and robotics in nursing. In Healthcare (Vol. 10, No. 8, p. 1571). MDPI. doi:10.3390/healthcare10081571

Stahl, B. C., & Coeckelbergh, M. (2016). Ethics of healthcare robotics: Towards responsible research and innovation. *Robotics and Autonomous Systems*, *86*, 152–161. doi:10.1016/j.robot.2016.08.018

Vallès-Peris, N., Barat-Auleda, O., & Domènech, M. (2021). Robots in healthcare? What patients say. *International Journal of Environmental Research and Public Health*, *18*(18), 9933. doi:10.3390/ijerph18189933 PMID:34574861

Vallès-Peris, N., & Domènech, M. (2023). Caring in the in-between: A proposal to introduce responsible AI and robotics to healthcare. *AI & Society*, *38*(4), 1685–1695. doi:10.1007/s00146-021-01330-w

Yoon, S. N., & Lee, D. (2018). Roboticsand robots in healthcare: What are the success factors for technology-based service encounters? *International Journal of Healthcare Management*.

Chapter 12
Bio-Inspired Nanorobots for Cancer Diagnosis and Therapy

Anshit Mukherjee
https://orcid.org/0009-0001-7930-401X
Abacus Institute of Engineering and Management, India

Gunjan Mukherjee
https://orcid.org/0000-0002-3959-3718
Brainware University, India

ABSTRACT

Cancer is one of the most serious threats to human health and life. Despite the advances in conventional therapies, such as surgery, chemotherapy, radiotherapy, and immunotherapy, there are still many challenges and limitations in achieving effective and precise cancer treatment. Nanorobots, inspired by natural biological nanomachines, offer a promising alternative for cancer diagnosis and therapy. Nanorobots are nanoscale devices that can perform various tasks under the guidance of external stimuli, such as magnetic fields, light, ultrasound, or chemical gradients. Nanorobots can be designed to target specific cancer cells or tissues, deliver drugs or genes, sense tumor biomarkers, perform minimally invasive surgery, or combine multiple functions for comprehensive treatment. In this chapter, the authors review the recent progress and applications of bio-inspired nanorobots for cancer diagnosis and therapy, with a focus on magnetic field-driven nanorobots. They also discuss the challenges and future perspectives of nanorobots in clinical translation.

INTRODUCTION

Bio-inspired nanorobots are a type of nanomedicine that mimics the natural, physical, and chemical properties of biological systems or processes. They have great potential for cancer diagnosis and therapy, as they can access remote and hard-to-reach body regions, perform various medical tasks, and improve the efficiency and effectiveness of treatment with reduced toxicity and side effects. Bio-inspired nanorobots can be designed and constructed using different materials, such as metals,

DOI: 10.4018/979-8-3693-1962-8.ch012

polymers, lipids, proteins, DNA, or cells. Depending on the material (Jun-Bing Fan,2022), they can have different shapes, sizes, functions, and biocompatibility. Some examples of bio-inspired nanorobots are magnetic nanorobots, DNA nanorobots, cell-based nanorobots, etc. Bio-inspired nanorobots can be driven by various methods, such as magnetic fields, light, ultrasound, chemical reactions, or biological motors. These methods(Xiangyi Kong,2023) can provide different levels of control, speed, and power for the nanorobots. Some examples of driving methods are water propulsion, enzyme catalysis, bacterial flagella, etc. Bio-inspired nanorobots can be applied for various purposes in cancer diagnosis and therapy, such as drug delivery, tumor sensing and diagnosis, targeted therapy, minimally invasive surgery, and immunotherapy. These applications can enhance the accuracy, sensitivity, specificity, and efficacy of cancer treatment. Some examples of applications are tumor targeting, tumor imaging, tumor ablation, tumor surgery, etc. Bio-inspired nanorobots are a promising and emerging field of research that could revolutionize cancer treatment in the near future. However, there are still many challenges and opportunities for further development, such as improving the biostability, biodegradability, biosafety, and bioethics of nanorobots; optimizing the design, fabrication, and integration of nanorobots; exploring new materials, driving methods, and functions of nanorobots; and validating the clinical feasibility and effectiveness of nanorobots(Lianging Liu, 2022). They have emerged as a promising platform for cancer diagnosis and therapy, owing to their advantages of high specificity, low toxicity, and self-assembly. In this book chapter, we will review the recent advances in the design, fabrication, and application of bio-inspired nanorobots for cancer detection and treatment. We also discuss the challenges and opportunities for the future development of bio-inspired nanorobots in the field of nanomedicine. Some of the possible key points we are going to discuss this chapter are:

- The advantages and disadvantages of different materials, shapes, sizes, and biocompatibility of bio-inspired nanorobots.
- Framework of bio-inspired nanorobots that can mimic the natural functions of biological entities, such as bacteria, viruses, cells, and enzymes, for cancer detection and therapy.
- How can bio-inspired nanorobots be driven by different methods, such as magnetic fields, light, ultrasound, chemical reactions, or biological motors? What are the trade-offs between control, speed, and power of these methods?
- How can bio-inspired nanorobots be applied for various purposes in cancer diagnosis and therapy, such as drug delivery, tumor sensing and diagnosis, targeted therapy, minimally invasive surgery, and immunotherapy? What are the benefits and limitations of these applications?
- How can bio-inspired nanorobots improve the accuracy, sensitivity, specificity, and efficacy of cancer treatment with reduced toxicity and side effects?

REVIEW

Bio-inspired nanorobots are nanoscale devices that mimic the functions and behaviors of natural biological systems, such as bacteria, viruses, enzymes, and cells. They can perform various tasks, such as sensing, targeting, delivery, diagnosis, and therapy, in complex biological environments. Bio-inspired nanorobots have great potential for cancer diagnosis and therapy, as they can overcome some of the limitations of conventional methods, such as low specificity, high toxicity, and drug resistance.

One of the main challenges in designing bio-inspired nanorobots is to achieve autonomous and controllable motion in biological fluids. Various propulsion mechanisms have been explored, such as chemical, magnetic, acoustic, optical, and electrical. Chemical propulsion is based on the catalytic decomposition of fuel molecules by nanomotors, which generates thrust from the reaction products. Magnetic propulsion relies on the application of external magnetic fields to manipulate the orientation and movement of magnetic nanomotors. Acoustic propulsion utilizes ultrasound waves to induce acoustic streaming around nanomotors, which results in fluid flow and propulsion. Optical propulsion exploits light energy to generate thermal gradients or optical forces that drive the motion of nanomotors. Electrical propulsion employs electric fields to induce electrokinetic effects, such as electrophoresis and electro-osmosis, that propel nanomotors.

Another challenge in designing bio-inspired nanorobots is to achieve specific and efficient targeting of cancer cells or tissues. Various targeting strategies have been developed, such as passive targeting, active targeting, and stimuli-responsive targeting. Passive targeting exploits the enhanced permeability and retention (EPR) effect of tumors, which allows nanomotors to accumulate in tumor sites due to their leaky vasculature and impaired lymphatic drainage. Active targeting involves the functionalization of nanomotors with ligands that can bind to receptors or antigens overexpressed on cancer cells or tissues. Stimuli-responsive targeting utilizes external or internal stimuli, such as pH, temperature, enzymes, or redox potential, to trigger the release or activation of nanomotors at tumor sites.

Bio-inspired nanorobots can perform various functions for cancer diagnosis and therapy, such as drug delivery, tumor sensing and diagnosis, targeted therapy, minimally invasive surgery, and other comprehensive treatments. Drug delivery is one of the most common applications of bio-inspired nanorobots, which can transport drugs or other therapeutic agents to tumor sites with high efficiency and specificity. Tumor sensing and diagnosis is another important application of bio-inspired nanorobots, which can detect tumor biomarkers or microenvironmental changes with high sensitivity and selectivity. Targeted therapy is a promising application of bio-inspired nanorobots, which can destroy cancer cells or tissues by physical or chemical means, such as mechanical drilling, photothermal ablation, or gene silencing. Minimally invasive surgery is an emerging application of bio-inspired nanorobots, which can perform surgical tasks at the microscale or nanoscale level with minimal damage to healthy tissues. Other comprehensive treatments are also possible with bio-inspired nanorobots, which can integrate multiple functions into one platform for synergistic effects.

The Table -1 summarizes some of the recent research work on bio-inspired nanorobots for cancer diagnosis and therapy:

FRAMEWORK OF BIO-INSPIRED NANOROBOTS THAT CAN MIMIC THE NATURAL FUNCTIONS OF BIOLOGICAL ENTITIES, SUCH AS BACTERIA, VIRUSES, CELLS, AND ENZYMES, FOR CANCER DETECTION AND THERAPY

The nanorobot based testing process has taken place in numberof steps. Figure 1 presents the detailed steps testing starting from the Material selection process.

One possible way is as follows (Calvalcanti, 2008):

Table 1. Review works of recent research works

Researcher's Name	Methodologies Used	Accuracy in Result
(Kong et al.,2023)	Nanosubmarines powered by chemical fuels for cancer treatment	Achieved autonomous navigation in blood vessels and selective destruction of cancer cells with high speed and precision
(Wang et al.,2020)	AuNPs-based micro/nanomotors for cancer-targeted delivery, diagnosis and imaging-guided therapy	Achieved efficient delivery of drugs or imaging agents to tumor sites with enhanced therapeutic efficacy or imaging contrast
(Li et al.,2020)	DNA origami-based photothermal motors for tumor ablation	Achieved light-driven motion in biological fluids and effective ablation of tumor cells with low laser power
(Li et al.,2019)	DNA origami-based gene motors for RNA interference therapy	Achieved electric field-driven motion in biological fluids and efficient delivery of siRNA to tumor cells with high gene silencing efficiency
(Wang et al.,2019)	Enzyme-powered micromotors for tumor microenvironment modulation	Achieved urease-driven motion in biological fluids and effective modulation of tumor acidity with enhanced drug efficacy
(Liu et al.,2018)	Magnetic helical microrobots for minimally invasive surgery	Achieved magnetic field-driven motion in biological fluids and precise manipulation of microscale objects with high dexterity
(Gao et al.,2017)	Ultrasound-powered nanorobots for tumor penetration and therapy	Achieved ultrasound-driven motion in biological fluids and deep penetration into tumor tissues with improved drug delivery
(Medina-Sánchez et al.,2017)	Magnetically actuated sperm-hybrid micromotors for drug delivery	Achieved magnetic field-driven motion in biological fluids and targeted delivery of drugs to tumor sites with high biocompatibility
(Esteban-Fernández de Ávila et al.,2017)	Bacteria-powered microrobots for tumor targeting and therapy	Achieved bacteria-driven motion in biological fluids and selective accumulation at tumor sites with enhanced drug efficacy
(Li et al.,2016)	Catalytic nanomotors for tumor sensing and diagnosis	Achieved catalytic motion in biological fluids and rapid detection of tumor biomarkers with high sensitivity and specificity

Figure 1. The steps n nanorobot testing from the material selection process

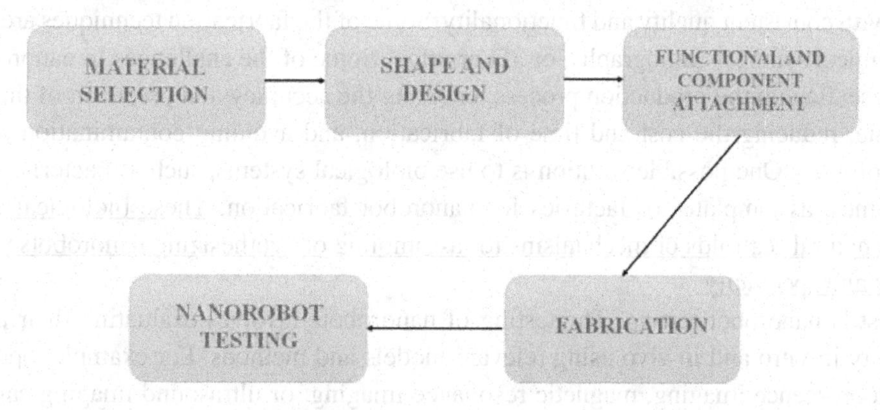

- First step is material selection. The material for the nanorobot body should be
- biocompatible, stable, and functionalizable. Biocompatibility means that the material should not cause adverse reactions or toxicity in the biological environment. Stability means that the material should not degrade or lose its properties over time. Functionalizability means that the material should allow the attachment of functional components, such as sensors, actuators, or drugs. Some of the challenges in material selection are finding materials that meet all these criteria, ensuring the quality and purity of the materials, and controlling the size and shape of the materials. One possible solution is to use natural biomolecules, such as DNA,

proteins, or lipids, as building blocks for the nanorobot body. These biomolecules are inherently biocompatible, stable under certain conditions, and functionalizable through molecular interactions .

- Second step is shape and size design. The shape and size of the nanorobot should be designed according to the desired function and target. For example, spherical nanorobots can be used for drug delivery, rod-shaped nanorobots can be used for cell penetration, or helical nanorobots can be used for propulsion. Some of the challenges in shape and size design are optimizing the geometry and dimensions of the nanorobot for maximum performance, minimizing the drag and friction forces on the nanorobot, and avoiding aggregation or clearance of the nanorobot by the immune system. One possible solution is to use computer-aided design (CAD) tools to model and simulate the nanorobot shape and size under different conditions and scenarios .
- Third step is functional component attachment.The functional components are the parts of the nanorobot that enable sensing, actuation, communication, or drug release. These components can be derived from natural biomolecules, such as antibodies, enzymes, DNA, or peptides, or from synthetic materials, such as magnetic nanoparticles, quantum dots, or organic molecules. Some of the challenges in functional component attachment are selecting the appropriate components for the specific function and target, ensuring the compatibility and functionality of the components with the nanorobot body and the biological environment, and controlling the number and location of the components on the nanorobot surface. One possible solution is to use self-assembly methods to attach the functional components to the nanorobot body through specific molecular interactions, such as DNA hybridization, antigen-antibody binding, or enzyme-substrate recognition .
- Fourth step is fabrication. The fabrication of nanorobots involves creating large quantities of nanorobots with consistent quality and functionality. Some of the fabrication techniques are self-assembly, chemical synthesis,lithography, or 3D printing. Some of the challenges in nanorobot fabrication are scaling up the production process, ensuring the accuracy and precision of the fabrication technique, reducing the cost and time of fabrication, and avoiding contamination or defects in the nanorobots. One possible solution is to use biological systems, such as bacteria,viruses,cells, or enzymes, as templates or factories for nanorobot fabrication. These biological systems can provide natural scaffolds or mechanisms for assembling or synthesizing nanorobots with high efficiency and specificity .
- Fifth test is nanorobot testing. The testing of nanorobots involves evaluating their performance and safety in vitro and in vivo using relevant models and methods. For example, optical microscopy, fluorescence imaging, magnetic resonance imaging, or ultrasound imaging can be used to monitor the location and movement of the nanorobots. Cell culture, animal models, or human trials can be used to evaluate the efficacy and toxicity of the nanorobots. Some of the challenges in nanorobot testing are measuring and analyzing the behavior and interactions of nanorobots at different scales and levels, ensuring the validity and reliability of the testing models and methods, addressing ethical issues related to animal or human experiments, and complying with regulatory standards and guidelines for clinical applications.

HOW BIO-INSPIRED NANOROBOTS CAN BE USED FOR CANCER DETECTION?

They can be used for cancer detection by performing tasks such as(Wang,2019):

- Navigating through the blood vessels and tissues to reach the tumor site.
- Recognizing and binding to specific cancer cells or markers using molecular recognition or targeting strategies.
- Transmitting signals or data to external devices or receivers using optical, magnetic, electrical, or acoustic methods.
- Delivering contrast agents or tracers to enhance the visibility of cancer cells under various imaging modalities, such as MRI, PET, CT, or ultrasound.
- Collecting samples or biopsies from the tumor tissue for further analysis or diagnosis.

However, using bio-inspired nanorobots for cancer detection is not easy. There are many challenges and limitations that need to be solved, such as:

- Biocompatibility: How to make sure that the nanorobots do not harm the healthy cells or tissues in the body?
- Safety: How to prevent the nanorobots from causing unwanted side effects, such as inflammation, infection, or immune response?
- Specificity: How to ensure that the nanorobots only target the cancer cells or markers and not the normal ones?
- Stability: How to protect the nanorobots from degradation or damage by the body's enzymes or fluids?
- Scalability: How to produce and deliver enough nanorobots for effective cancer detection?

One possible solution to overcome these challenges is to use living cells or biological materials as building blocks for nanorobots. This way, the nanorobots can mimic natural functions and behaviors and avoid being rejected by the body. For example, some researchers have used bacteria as nanorobots that can sense and respond to oxygen levels in the tumor microenvironment. Other researchers have used DNA as nanorobots that can self-assemble into various shapes and perform logic operations. These are some of the examples of how bio-inspired nanorobots can be improved by using living cells or biological materials. However, ethical, social, and regulatory issues also need to be considered before nanorobots can be widely used in clinical settings.

HOW CAN BIO-INSPIRED NANOROBOTS BE DRIVEN BY DIFFERENT METHODS, SUCH AS MAGNETIC FIELDS, LIGHT, ULTRASOUND, CHEMICAL REACTIONS, OR BIOLOGICAL MOTORS? WHAT ARE THE TRADE-OFFS BETWEEN CONTROL, SPEED, AND POWER OF THESE METHODS? WHAT ARE THE BENEFITS AND LIMITATIONS OF THESE APPLICATIONS?

Bio-inspired nanorobots can be driven by different methods depending on their design, size, and function. Some of the common methods are:

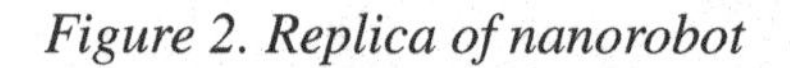
Figure 2. Replica of nanorobot

- Magnetic fields: This method uses an external magnetic field to generate torque and force on the nanorobots that have magnetic materials or components(Yan, 2019). The advantages of this method are that it is non-invasive, wireless, and can be controlled remotely and precisely. The disadvantages are that it requires a strong and uniform magnetic field, which may interfere with other devices or biological tissues, and that it may cause heating or damage to the nanorobots or the surrounding environment .
- Light: This method uses light to activate or power the nanorobots that have light-sensitive materials or components, such as photochromic molecules, photovoltaic cells, or optical fibers. The advantages of this method are that it is wireless, biocompatible, and can be modulated by changing

the wavelength, intensity, or direction of the light(Martel,2009). The disadvantages are that it may have limited penetration depth in biological tissues, which may reduce the range and efficiency of the nanorobots, and that it may cause photodamage or photobleaching to the nanorobots or the surrounding environment .

- Ultrasound: This method uses ultrasound waves to generate acoustic radiation force or streaming on the nanorobots that have acoustic materials or components, such as piezoelectric crystals, microbubbles, or microcapsules. The advantages of this method are that it is wireless, biocompatible, and can be focused and steered by changing the frequency, amplitude, or phase of the ultrasound waves. The disadvantages are that it may have limited resolution and accuracy in controlling the nanorobots, and that it may cause cavitation or heating to the nanorobots or the surrounding environment .
- Chemical reactions: This method uses chemical reactions to generate propulsion or power for the nanorobots that have catalytic materials or components, such as metal nanoparticles, enzymes, or fuel molecules. The advantages of this method are that it is self-powered, autonomous, and can be triggered by specific stimuli, such as pH, temperature, or biomolecules. The disadvantages are that it may have limited control and directionality of the nanorobots, and that it may cause toxicity or depletion of the nanorobots or the surrounding environment .
- Biological motors: This method uses biological motors to drive the nanorobots that have biological materials or components, such as flagella, cilia, or muscle cells. The advantages of this method are that it is biocompatible, bio functional, and can be integrated with natural systems. The disadvantages are that it may have limited stability and durability of the nanorobots, and that it may cause immune response or infection to the nanorobots or the surrounding environment .

The trade-offs between control, speed, and power of these methods depend on various factors, such as the design parameters of the nanorobots, the properties of the external fields or stimuli, and the characteristics of the operating environment. Generally speaking, magnetic fields and light offer high control and speed but low power; ultrasound and chemical reactions offer moderate control and speed but high power; biological motors offer low control and speed but moderate power . However, these trade-offs are not fixed and can be optimized by combining different methods or improving the performance of the nanorobots.

HOW CAN BIO-INSPIRED NANOROBOTS BE APPLIED FOR VARIOUS PURPOSES IN CANCER DIAGNOSIS AND THERAPY, SUCH AS DRUG DELIVERY, TUMOR SENSING AND DIAGNOSIS, TARGETED THERAPY, MINIMALLY INVASIVE SURGERY, AND IMMUNOTHERAPY?

Bio-inspired nanorobots are nanoscale devices that mimic the behavior and functions of natural biological systems, such as bacteria, viruses, enzymes, and cells. They can be applied for various purposes in cancer diagnosis and therapy, such as drug delivery, tumor sensing and diagnosis, targeted therapy, minimally invasive surgery, and immunotherapy. Here are some examples of how bio-inspired nanorobots can work in each of these applications:

Figure 3. Application of nanorobot in the cancer diagnosis

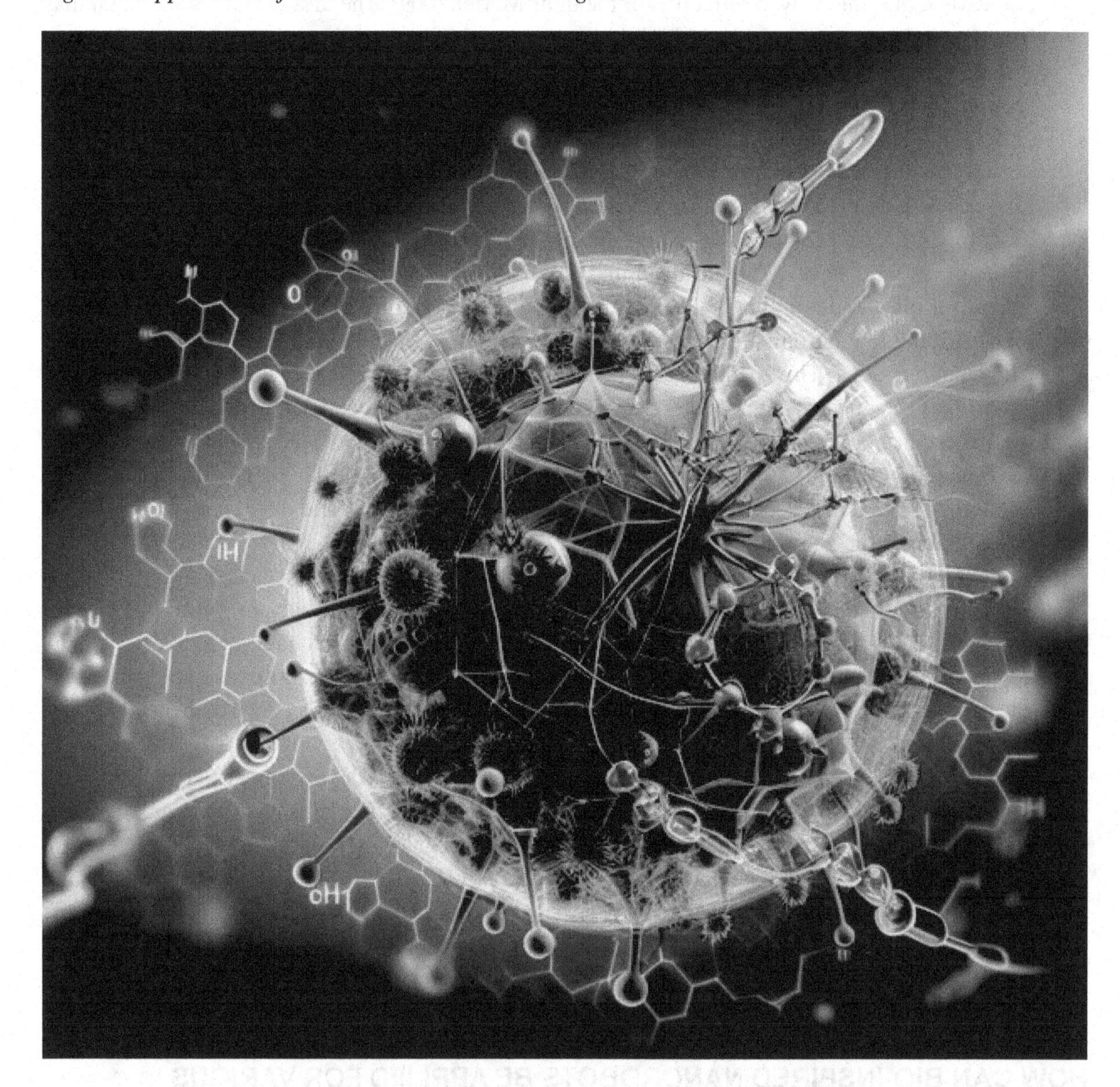

- Drug delivery: Bio-inspired nanorobots can carry anticancer drugs and deliver them to the tumor cells with high precision and efficiency. For example, researchers have invented self-propelling nanorobots that can precisely target and deliver an anticancer drug to human colon cancer cells. The nanorobots are made of magnesium nanoparticles coated with a biocompatible polymer and a layer of red blood cell membranes. The magnesium nanoparticles react with water to produce hydrogen bubbles, which propel the nanorobots toward the tumor cells (Banerjee, 2020). The red blood cell membranes help the nanorobots evade the immune system and recognize the tumor cells by binding to their surface receptors. The nanorobots release the drug slowly at the target cancer cells, enabling the drug to kill cancer cells at a low dose, sparing healthy cells.

The benefit of using bio-inspired nanorobots for drug delivery is that they can enhance the specificity, efficiency, and safety of anticancer drugs. They can target tumor cells with high accuracy, deliver drugs at a controlled rate and dose, and avoid harming healthy cells or causing systemic toxicity. The limitation of this process is that it may face challenges such as immune clearance, biofouling, biodegradation, and drug resistance. The nanorobots may be recognized and eliminated by the immune system, or they may lose their functionality due to the accumulation of biological molecules or the degradation of their materials. The tumor cells may also develop resistance to the drugs delivered by the nanorobots, reducing their effectiveness.

- Tumor sensing and diagnosis: Bio-inspired nanorobots can sense and detect tumor cells in the blood or tissue samples of cancer patients. For example, researchers have designed efficient, light-emitting magnetic nanorobots that can capture circulating tumor cells (CTCs) in the blood samples of cancer patients. The nanorobots are made of iron oxide nanoparticles coated with a fluorescent polymer and antibodies that recognize CTCs. The iron oxide nanoparticles allow the nanorobots to be controlled by a magnetic field and to be separated from the blood sample. The fluorescent polymer emits light when excited by a laser, making the nanorobots visible under a microscope. The antibodies help the nanorobots capture CTCs by binding to their surface antigens. The captured CTCs can be used for further analysis and diagnosis.

The benefit of using bio-inspired nanorobots for tumor sensing and diagnosis is that they can improve the sensitivity, specificity, and speed of cancer detection. They can capture and identify tumor cells or biomarkers in blood or tissue samples with high affinity and selectivity. They can also provide real-time and non-invasive imaging of tumor location and progression. The limitation of this process is that it may encounter difficulties such as low signal-to-noise ratio, interference from background substances, and false-positive or false-negative results. The nanorobots may not be able to generate enough signal or contrast to distinguish tumor cells or biomarkers from normal cells or molecules. They may also be affected by other substances in the biological environment that may interfere with their sensing or imaging functions. They may also fail to capture or detect some tumor cells or biomarkers due to their heterogeneity or variability.

- Targeted therapy: Bio-inspired nanorobots can deliver therapeutic agents or signals to specific tumor cells or tissues, enhancing the efficacy and reducing the side effects of cancer treatment. For example, researchers have developed DNA-based nanorobots that can deliver thrombin molecules to tumor-associated blood vessels, inducing blood clotting and cutting off the blood supply to the tumor (Wavhale,2021). The nanorobots are made of DNA origami structures that fold into hollow tubes with two open ends. One end is closed by a DNA aptamer that binds to nucleolin, a protein overexpressed on the surface of tumor-associated endothelial cells. The other end is loaded with thrombin molecules that are released when the aptamer binds to nucleolin. The thrombin molecules trigger coagulation cascade in the blood vessels, leading to tumor necrosis.

The benefit of using bio-inspired nanorobots for targeted therapy is that they can induce tumor cell death or inhibit tumor growth by delivering therapeutic agents or signals to specific tumor cells or tissues. They can overcome the limitations of conventional therapies such as low efficacy, high toxicity, and adverse effects. They can also achieve synergistic effects by combining different therapeutic modalities such as chemotherapy, radiotherapy, gene therapy, immunotherapy, and phototherapy. The limitation of this process is that it may face challenges such as poor penetration, low retention, and undesired side effects. The nanorobots may not be able to reach deep-seated tumors or diffuse tumors due to the physical barriers or physiological factors in the body. They may also be quickly cleared from the circulation or the tumor site due to their size, shape, charge, or surface properties. They may also cause unwanted side effects such as inflammation, infection, or immunogenicity due to their interaction with the biological system..

- Minimally invasive surgery: Bio-inspired nanorobots can perform surgical tasks in hard-to-reach or delicate body regions, reducing the trauma and risk of infection associated with conventional surgery. For example, researchers have created magnetic helical nanorobots that can drill through biological barriers and remove unwanted tissue or foreign objects. The nanorobots are made of nickel–gold–iron hybrid nanoparticles that form helical shapes under a magnetic field. The magnetic field also controls the rotation and direction of the nanorobots, enabling them to drill through mucus layers or cell membranes. The nanorobots can also carry functional molecules or nanoparticles that can degrade or detach unwanted tissue or foreign objects.

The benefit of using bio-inspired nanorobots for minimally invasive surgery is that they can perform surgical tasks in hard-to-reach or delicate body regions, reducing the trauma and risk of infection associated with conventional surgery. They can drill through biological barriers, remove unwanted tissue or foreign objects, repair damaged tissue or organs, and deliver drugs or biomaterials to the surgical site. They can also provide feedback and guidance for the surgeon during the operation. The limitation of this process is that it may encounter difficulties such as navigation, control, communication, and coordination. The nanorobots may not be able to navigate through complex and dynamic environments in the body due to the lack of reliable sensors or actuators. That may also be difficult to control remotely due to the interference from electromagnetic fields or biological signals. They may also have problems with communication and coordination with other nanorobots or external devices due to the limited bandwidth or power.

- Immunotherapy: Bio-inspired nanorobots can modulate the immune system to fight against cancer cells or prevent tumor recurrence. For example, researchers have fabricated biomimetic nanorobots that can activate dendritic cells (DCs), which are key players in initiating adaptive immune responses. The nanorobots are made of mesoporous silica nanoparticles coated with red blood cell membranes and loaded with antigen peptides and adjuvants. The red blood cell membranes help the nanorobots evade the immune system and target DCs by binding to their surface receptors. The antigen peptides and adjuvants stimulate DCs to mature and present antigens to T cells, which then recognize and kill tumor cells.

The benefit of using bio-inspired nanorobots for immunotherapy is that they can modulate the immune system to fight against cancer cells or prevent tumor recurrence. They can activate or suppress immune cells, enhance or inhibit immune responses, and deliver antigens or adjuvants to elicit specific immune reactions. They can also evade or overcome immune evasion mechanisms of tumor cells such as antigen loss, immunosuppression, or tolerance. The limitation of this process is that it may face challenges such as immunogenicity, immunotoxicity, and autoimmunity. The nanorobots may be recognized and attacked by the immune system, reducing their functionality or causing inflammation. They may also trigger excessive or inappropriate immune responses that may damage normal tissues or organs. They may also induce autoimmune diseases by stimulating self-reactive immune cells.

DEVELOP AND EVALUATE NOVEL BIOSENSING CHIPS THAT CAN CAPTURE AND ISOLATE CIRCULATING TUMOR CELLS FROM BLOOD SAMPLES USING NANOROBOTS

One possible way to develop biosensing chips that can capture and isolate CTCs from blood samples using nanorobots is to use magnetic nanowires as the building blocks of the nanorobots. Magnetic nanowires are thin rods of magnetic materials that can be manipulated by external magnetic fields. They can also be functionalized with antibodies or other molecules that can specifically bind to CTCs. By applying a rotating magnetic field, the magnetic nanowires can form helical structures that can propel themselves in fluid and capture CTCs on their surface (Yan, 2019). The captured CTCs can then be isolated from the blood sample by applying a gradient magnetic field that pulls the nanowires out of the fluid.

To evaluate the performance of the biosensing chips, one could measure the capture efficiency, purity, and viability of the CTCs. Capture efficiency is the ratio of the number of captured CTCs to the number of CTCs in the blood sample. Purity is the ratio of the number of captured CTCs to the number of total cells on the chip. Viability is the percentage of captured CTCs that are alive and functional. These metrics can be assessed by using various techniques, such as fluorescence microscopy, flow cytometry, or polymerase chain reaction (PCR).

How can bio-inspired nanorobots improve the accuracy, sensitivity, specificity, and efficacy of cancer treatment with reduced toxicity and side effects?

Bio-inspired nanorobots are nanoscale devices that mimic the structure and function of natural biological systems, such as bacteria, viruses, or cells. They can improve the accuracy, sensitivity, specificity, and efficacy of cancer treatment with reduced toxicity and side effects by performing the following tasks:

- **Drug delivery**: Bio-inspired nanorobots can carry and deliver large amounts of anti-cancer drugs into cancerous cells without harming healthy cells, reducing the side effects related to current therapies such as chemotherapy damage. For example, researchers have designed nanorobots that can self-assemble into helical structures that can propel themselves in fluid and capture circulating tumor cells (CTCs) on their surface. The captured CTCs can then be isolated from the blood sample by applying a gradient magnetic field that pulls the nanorobots out of the fluid.
- **Tumor sensing and diagnosis**: Bio-inspired nanorobots can detect and measure biological signals, such as DNA, proteins, or cells, that are associated with cancer. They can also provide real-time feedback and imaging of the tumor microenvironment. For example, researchers have developed nanorobots that can use DNA origami to fold into tubular structures that can recognize and bind to specific nucleic acid sequences on the surface of cancer cells. The binding triggers a conformational change in the nanorobots that exposes a fluorescent dye for optical detection.
- **Targeted therapy**: Bio-inspired nanorobots can selectively destroy cancer cells by using various mechanisms, such as mechanical drilling, thermal ablation, or gene editing. For example, researchers have used nanorobots to drill into cancer cells, killing them in just 60 seconds. They are now experimenting on micro-organisms and small fish, before moving on to rodents (Martel,

2009). Clinical trials in humans are expected to follow and it is hoped that the results may have the potential to save millions of lives.

- **Minimally invasive surgery**: Bio-inspired nanorobots can perform surgical tasks at the nanoscale, such as cutting, suturing, or removing tissue, with minimal damage to the surrounding healthy tissue. For example, researchers have proposed nanorobots that can use carbon nanotubes as surgical blades to cut through tissue with high precision and low friction.

THE ADVANTAGES AND DISADVANTAGES OF DIFFERENT MATERIALS, SHAPES, SIZES, AND BIOCOMPATIBILITY OF BIO-INSPIRED NANOROBOTS FOR THE TOPIC BIO-INSPIRED NANOROBOTS FOR CANCER DIAGNOSIS AND THERAPY

Some of the advantages and disadvantages of different materials, shapes, sizes, and biocompatibility of bio-inspired nanorobots for cancer diagnosis and therapy (Datta, 2022):

- **Materials**: Different materials can be used to make bio-inspired nanorobots, depending on what they are supposed to do and where they are going to be used. For instance, nanorobots made of metals can be good for optical or thermal applications, such as finding or destroying cancer cells with light or heat. But metal nanorobots may not be very friendly to the body and may cause harm by releasing metal ions or causing oxidative stress. So, metal nanorobots need to be covered with biocompatible materials or made by green methods to make them safer. Another example is nanorobots made of polymers, which can be good for drug delivery applications, as they can hold and release drugs in a controlled way. Polymer nanorobots can also change their behavior according to different stimuli, such as pH, temperature, light, or magnetic fields. But polymer nanorobots may not last long and degrade in biological environments. So polymer nanorobots need to be adjusted for their physical and chemical properties and biodegradability. A third example is nanorobots made of DNA, which can be good for tumor sensing and targeted therapy applications. DNA nanorobots can form various shapes and structures that can recognize and bind to specific DNA sequences on the surface of cancer cells. When the nanorobots bind to their targets, they change their shape and show a bright dye for optical detection or a gene-editing tool for therapeutic delivery. But DNA nanorobots may not be stable and specific in complex biological fluids due to nuclease degradation or off-target binding. So, DNA nanorobots need to be protected by chemical modifications or encapsulation.
- **Shapes**: The shape of the nanorobots can vary, such as round, stick-like, tube-like, spiral, or complex shapes. The shape influences how they move and interact in fluid environments and with biological components, such as cells, proteins, or receptors. For instance, round nanorobots move easily and quickly in fluids because they have low drag force and high diffusion coefficient. But round nanorobots may not enter cells well because they have low aspect ratio and surface area. So round nanorobots may need to have targeting ligands or stimuli-responsive moieties on their surface to improve their cellular uptake. Another example is stick-like or tube-like nanorobots that have high aspect ratio and surface area that can help them enter cells better. But stick-like or tube-like nanorobots move slowly and hard in fluids because they have high drag force and low diffusion coefficient. So, stick-like or tube-like nanorobots may need to be controlled by external forces

or self-propelling mechanisms to overcome the hydrodynamic resistance. A third example is spiral nanorobots that can copy the structure and function of natural spiral microorganisms. Spiral nanorobots can move themselves in fluids by spinning under external magnetic fields. Spiral nanorobots can also catch circulating tumor cells on their surface by applying a gradient magnetic field that pulls them out of the fluid. But spiral nanorobots may have complex fabrication process and low biocompatibility because of their metallic components. So spiral nanorobots need to be covered with biocompatible materials or made by green methods to make them safer.

- **Sizes**: The nanorobots can be different sizes from a few nanometers to a few micrometers. The size influences how they spread, move, and leave in the body. The size also influences how they interact with biological obstacles, such as blood vessels, tumor environment, and immune system. For instance, tiny nanorobots (smaller than 10 nm) move easily and quickly in fluids because they have low drag force and high diffusion coefficient. But tiny nanorobots may not last long and stay in the body because they are quickly removed by kidney filtration or phagocytosis. So tiny nanorobots may need to have stealth agents or targeting ligands on their surface to increase their circulation time and tumor accumulation. Another example is big nanorobots (bigger than 100 nm) move slowly and hard in fluids because they have high drag force and low diffusion coefficient. But big nanorobots may last long and stay in the body because they are less removed by kidney filtration or phagocytosis. So big nanorobots may need to be made to overcome the biological obstacles such as the endothelial gaps, the extracellular matrix, and the interstitial pressure that prevent their tumor penetration and delivery. A third example is medium-sized nanorobots (between 10 and 100 nm) that can balance the trade-off between lasting long and being removed in the body. Medium-sized nanorobots can also use the enhanced permeability and retention (EPR) effect that lets them passively accumulate in tumor tissues due to the leaky blood vessels and impaired lymphatic drainage. But medium-sized nanorobots may have different and unpredictable EPR effect depending on the tumor type, location, and stage. So medium-sized nanorobots may need to be combined with active targeting strategies or stimuli-responsive mechanisms to improve their tumor specificity and efficacy.
- **Biocompatibility:** The nanorobots need to be highly biocompatible to avoid causing harm such as toxicity, inflammation, immunogenicity, or thrombogenicity in the body. Biocompatibility depends on how the nanorobots look and behave, such as their material, shape, size, and surface functionalization. For instance, nanorobots made of materials that can break down or are friendly to the body, such as polymers, lipids, proteins, or DNA, can reduce their toxicity and immunogenicity. But these nanorobots may not last long and work well in the body because they can be degraded or denatured by biological factors. So, these nanorobots need to be adjusted for their physical and chemical properties and biodegradability. Another example is nanorobots made of materials that do not break down or are not friendly to the body, such as metals or synthetic polymers, can last long and work well in the body. But these nanorobots may cause high toxicity and immunogenicity because they can accumulate or be recognized by biological components. So these nanorobots need to be covered with biocompatible materials or made by green methods to make them safer. A third example is nanorobots with different shapes and sizes that can affect their biocompatibility. Nanorobots that are round and tiny can reduce their inflammation and thrombogenicity because they have low interaction with blood components. But these nanorobots may not enter cells well and reach tumor tissues because they have low aspect ratio and surface area. So these nanorobots may need to have targeting ligands or stimuli-responsive moieties on their

surface to improve their cellular uptake and tumor accumulation. On the other hand, nanorobots that are stick-like, tube-like, spiral, or complex shapes and big sizes can enter cells well and reach tumor tissues because they have high aspect ratio and surface area. But these nanorobots may cause high inflammation and thrombogenicity because they have high interaction with blood components. So, these nanorobots may need to have stealth agents or anti-coagulants on their surface to reduce their inflammation and thrombogenicity.

BIONANO ROBOTS FOR CANCER THERAPY: A FOE OR A FRIEND?

Cancer is one of the leading causes of death worldwide, and despite the advances in diagnosis and treatment, it remains a major challenge for medicine (Mavroidis, 2013). Nanotechnology, the manipulation of matter at the nanoscale, offers new possibilities for improving cancer therapy. Among the various nanomaterials, bionano robots, or biohybrid nanomachines, are emerging as a promising and innovative approach. Bionano robots are composed of biological components, such as cells, proteins, or DNA, integrated with synthetic materials, such as metals, polymers, or carbon nanotubes. These hybrid structures can perform specific functions, such as sensing, targeting, drug delivery, or imaging, in response to external stimuli or environmental cues (Webster, 2012). Bionano robots have the potential to overcome some of the limitations of conventional cancer therapies, such as low specificity, systemic toxicity, drug resistance, or poor penetration. However, bionano robots also pose significant challenges and risks for their clinical application. The safety, biocompatibility, stability, and ethical implications of bionano robots need to be carefully evaluated before they can be used in humans. Moreover, the regulation and standardization of bionano robots are still lacking and require further development. Therefore, bionano robots for cancer therapy can be seen as both a foe and a friend, depending on how they are designed, used, and controlled (Elsevier, 2012).

FUTURE DIRECTIONS

Bio-inspired nanorobots are a promising field of research that aims to create nanoscale devices that can perform various tasks in the human body (Fan, 2020), such as cancer diagnosis and treatment. Some of the future directions of bio-inspired nanorobots for cancer diagnosis are:

- Developing nanorobots that can harvest energy from the body or external sources, such as light, magnetic fields, or ultrasound, to power their functions and movements.
- Designing nanorobots that can sense and respond to specific stimuli in the tumor microenvironment (Kong,2023), such as pH, temperature, enzymes, or biomarkers, to activate their diagnostic or therapeutic actions.
- Integrating nanorobots with biosensors, imaging agents, or drug delivery systems to enable simultaneous detection and treatment of cancer cells.
- Creating nanorobots that can self-assemble, self-repair, or self-destruct after completing their tasks to avoid accumulation or toxicity in the body.
- Exploring the use of living cells or biological materials as building blocks for nanorobots that can mimic natural functions and behaviors.

These are some of the possible ways that bio-inspired nanorobots can advance the field of cancer diagnosis and treatment in the future. However, there are also many challenges and limitations that need to be overcome, such as ensuring biocompatibility, safety, specificity, stability, and scalability of nanorobots. Moreover, ethical, social, and regulatory issues also need to be addressed before nanorobots can be widely applied in clinical settings.

CONCLUSION

Bio-inspired nanorobots are a promising approach for cancer diagnosis and therapy, as they can mimic the natural functions of biological systems and interact with the tumor microenvironment. In this paper, we reviewed the recent advances in the design, fabrication, and evaluation of bio-inspired nanorobots for cancer applications(Grumezescu, 2016). We discussed the advantages and challenges of different types of bio-inspired nanorobots, such as DNA origami, bacterial, enzymatic, and cell-based nanorobots. We also highlighted the potential clinical implications and future directions of bio-inspired nanorobotics for cancer management (Jiang, 2010). We concluded that bio-inspired nanorobots have great potential to revolutionize cancer diagnosis and therapy, but they also face significant technical and ethical hurdles that need to be overcome before they can be translated into clinical practice.

REFERENCES

Banerjee, S. S., Andhari, S. S., Wavhale, R. D., Dhobale, K. D., Tawade, B. V., Chate, G. P., & Khandare, J. J. (2020). Self-propelling targeted magneto-nanobots for deep tumor penetration and pH-responsive intracellular drug delivery. *Scientific Reports*, *10*(1), 1–14. PMID:31913322

Cavalcanti, A., Shirinzadeh, B., Freitas, R. A. Jr, & Hogg, T. (2008). Nanorobot architecture for medical target identification. *Nanomedicine; Nanotechnology, Biology, and Medicine*, *4*(2), 134–152. PMID:18455965

Cavalcanti, A., Shirinzadeh, B., Freitas, R. A. Jr, & Hogg, T. (2008). Nanorobot architecture for medical target identification. *Nanomedicine; Nanotechnology, Biology, and Medicine*, *4*(2), 134–152. PMID:18455965

Datta Burton, S. (2022). Not Anytime Soon: The Clinical Translation of Nanorobots and Its Biocompatibility Constraints. In *Interactive Robotics: Legal, Ethical, Social and Economic Aspects* (pp. 123–136). Springer. doi:10.1007/978-3-031-04305-5_35

Jiang, L., & Feng, L. (Eds.). (2010). *Bioinspired Intelligent Nanostructured Interfacial Materials*. Springer. doi:10.1142/7380

Martel, S., Mohammadi, M., Felfoul, O., Lu, Z., & Pouponneau, P. (2009). Flagellated magnetotactic bacteria as controlled MRI-trackable propulsion and steering systems for medical nanorobots operating in the human microvasculature. *The International Journal of Robotics Research*, *28*(4), 571–582. doi:10.1177/0278364908100924 PMID:19890435

Martel, S., Mohammadi, M., Felfoul, O., Lu, Z., & Pouponneau, P. (2009). Flagellated magnetotactic bacteria as controlled MRI-trackable propulsion and steering systems for medical nanorobots operating in the human microvasculature. *The International Journal of Robotics Research*, *28*(4), 571–582. doi:10.1177/0278364908100924 PMID:19890435

Nanobiomaterials in Drug Delivery. (2016). *Applications of Nanobiomaterials* (A. M. Grumezescu, Ed.). Vol. 9). Elsevier.

Wang, J., & Lee, J. (2019). Bioinspired Nanorobots for Cancer Detection and Therapy. In *Bioinspired Nanomaterials and Nanostructures from Nanobiology to Nanomedicine* (pp. 1–31). Springer.

Wavhale, R. D., Andhari, S. S., Dhobale, K. D., Tawade, B. V., Chate, G. P., Patil, Y. N., ... Banerjee, S. S. (2021). Self-propelling magnetic nanorobots for capturing circulating tumor cells in blood samples of cancer patients. *Communications Chemistry*, *4*(1), 1–12. PMID:36697560

Yan, H., & Fan, C. (2019). DNA nanotechnology and its biological applications. In D. N. A. Nanotechnology (Ed.), (pp. 1–22). Springer.

Yan, H., & Fan, C. (2019). DNA nanotechnology and its biological applications. In D. N. A. Nanotechnology (Ed.), (pp. 1–22). Springer.

Chapter 13
Advanced Biomimetic Compound Continuum Robot for Minimally Invasive Surgical Applications

Ranjit Barua
https://orcid.org/0000-0003-2236-3876
Omdayal Group of Institutions, India

ABSTRACT

the bio-inspired compound continuum robot represents a groundbreaking innovation in the realm of minimally invasive surgery (MIS). Drawing inspiration from the flexibility and adaptability observed in nature, this robotic system employs a novel approach to navigating complex anatomical structures with enhanced precision. Mimicking the serpentine motion of snakes, the robot utilizes a compound continuum structure composed of interconnected segments. This design allows for unparalleled maneuverability, enabling the robot to navigate through confined spaces and intricate pathways within the human body. By emulating the biomechanics of natural organisms, the robot can reach anatomical locations that traditional rigid instruments might struggle to access. In this chapter, the authors will discuss the advanced biomimetic compound continuum robot for minimally invasive surgical applications.

INTRODUCTION

The Biomimetic Compound Continuum Robot represents a groundbreaking fusion of robotics and biomimicry, revolutionizing the landscape of minimally invasive surgery (MIS) (Zhang et al., 2022). Drawing inspiration from nature's adaptability, this innovative robotic system is designed to navigate complex anatomical structures with unparalleled precision and flexibility (Barua et al., 2022) (Li et al., 2013). The Advanced Biomimetic Compound Continuum Robot (ABCCR) has emerged as a transformative force in modern minimally invasive surgery, revolutionizing the way surgeons approach and perform intricate procedures. With its soft and flexible structure, inspired by the biomechanics of natural organ-

DOI: 10.4018/979-8-3693-1962-8.ch013

isms, ABCCR addresses key challenges associated with traditional rigid surgical instruments, offering significant advantages in terms of precision, maneuverability, and patient outcomes. One of the primary contributions of ABCCR to minimally invasive surgery lies in its ability to navigate through confined and delicate anatomical structures with unparalleled flexibility (Zhang et al., 2022). Unlike conventional rigid tools, ABCCR's snake-like motion enables it to access hard-to-reach areas, reducing the need for large incisions. This minimizes trauma to surrounding tissues, decreases postoperative pain, and accelerates patient recovery. The biomimetic design of ABCCR is particularly advantageous in procedures where intricate movements are required. In tasks such as suturing, tissue manipulation, and dissection, the robot's adaptability allows surgeons to replicate the nuanced dexterity of their hands with greater precision (Li et al., 2013). This is crucial in surgeries involving complex anatomical structures or procedures requiring meticulous attention to detail. The integration of advanced control systems and machine learning further enhances ABCCR's performance in the surgical arena. Surgeons can manipulate the robot with a high degree of accuracy, aided by real-time feedback from sensors and cameras embedded in the robot (Barua et al., 2022). Additionally, machine learning algorithms enable the robot to learn from each procedure, optimizing its movements and responses over time, ultimately contributing to improved surgical outcomes (Zhang et al., 2022). The impact of ABCCR in modern minimally invasive surgery extends beyond traditional procedures to include novel applications (Barua et al., 2022). For instance, in interventions requiring access to challenging locations, such as brain surgeries or procedures involving intricate vascular structures, ABCCR proves to be an invaluable asset. Its ability to navigate tight spaces and adapt to the contours of the human body makes it a versatile tool for a wide range of surgical specialties. As ABCCR continues to evolve, ongoing research and development promise even greater advancements in its capabilities. The seamless integration of this biomimetic robot into the surgical workflow underscores its potential to redefine the standard of care in minimally invasive surgery, offering patients safer procedures, faster recovery times, and improved overall outcomes. In the modern era of surgical innovation, ABCCR stands as a testament to the successful fusion of nature-inspired design and cutting-edge technology in the service of advancing medical practice.

REVIEWS OF THE LITERATURE

The advanced biomimetic compound continuum robot (ABCCR) represents a revolutionary leap in the field of robotics, drawing inspiration from nature to create a versatile and adaptive robotic system. Its development has been marked by a series of breakthroughs and innovations, transforming the landscape of robotics and opening up new possibilities for applications in various fields (Barua et al., 2022). The roots of ABCCR can be traced back to the early 21st century when researchers began exploring the concept of soft robotics. Traditional rigid robots faced limitations in terms of flexibility, adaptability, and safety. Soft robotics, inspired by the flexibility and resilience of biological organisms, aimed to overcome these challenges (Zhang et al., 2022). This laid the foundation for the development of continuum robots, which emulate the structure and motion of natural organisms. In the early stages of ABCCR development, researchers drew inspiration from the biomechanics of animals like octopuses and snakes. These creatures exhibit exceptional flexibility and dexterity, allowing them to navigate complex environments with ease. The idea was to create a robot that could mimic the continuous, snake-like motion while maintaining the ability to deform and adapt to its surroundings (Li et al., 2013) (Datta et al., 2023). The breakthrough came with the integration of biomimetic materials and advanced control systems. Researchers focused on developing flexible materials that could withstand various environmental conditions and deform without losing structural integrity (Barua et al., 2023) (Anderson et al., 1970). The incorporation of smart materials, such as shape-memory alloys and soft polymers, allowed the robot to bend and twist in ways that were previously unattainable with traditional rigid structures.

The control system of ABCCR was equally crucial to its success. Advanced algorithms enabled precise control of the robot's movements, allowing it to navigate tight spaces and execute complex tasks (Anderson et al., 1970). Machine learning played a significant role, as the robot could adapt and learn from its interactions with the environment, continuously improving its performance over time. As the technology matured, ABCCR found applications in a wide range of fields. In the medical industry, the robot's ability to navigate through delicate and confined spaces made it invaluable for minimally invasive surgeries. Its soft and flexible structure reduced the risk of tissue damage and improved patient recovery times. Surgeons could now perform intricate procedures with unprecedented precision. In search and rescue missions, ABCCR's flexibility allowed it to navigate through rubble and debris, reaching areas inaccessible to traditional robots. Equipped with sensors and cameras, the robot became an essential tool for locating and assisting survivors in disaster-stricken environments. Its adaptability to unpredictable terrains made it a game-changer in emergency response scenarios. The industrial sector also embraced ABCCR for tasks that required precision and flexibility (Zhang et al., 2022). The robot's dexterity made it suitable for intricate assembly processes, and its ability to reach confined spaces increased efficiency in manufacturing. With advancements in automation, ABCCR became an integral part of smart factories, contributing to increased productivity and safety. In the field of exploration, ABCCR demonstrated its capabilities in space missions and underwater exploration. Its adaptability allowed it to traverse uneven surfaces and navigate complex environments, making it an ideal candidate for missions where traditional rigid robots faced limitations. The evolution of ABCCR has been marked by continuous innovation and collaboration across various scientific disciplines. Researchers and engineers worked hand in hand to refine the robot's design, enhance its capabilities, and explore new avenues of application. The synergy of biology, materials science, robotics, and artificial intelligence has propelled ABCCR to the forefront of technological advancements (Li et al., 2013). Looking ahead, the future of ABCCR holds even greater promise. As research continues, the robot is likely to become more sophisticated, with improved sensory capabilities, faster response times, and enhanced adaptability (Barua et al., 2023). The impact of ABCCR on industries, healthcare, and exploration is expected to grow, solidifying its place as a groundbreaking technology that has reshaped the landscape of robotics.Top of Form Biomimicry involves emulating biological systems and processes to inspire technological innovations. In the case of the Biomimetic Compound Continuum Robot, nature serves as the blueprint for its design (Datta et al., 2023) (Barua et al., 2023). The flexible and snake-like movements (Figure 1) observed in various organisms become the basis for creating a robotic system capable of navigating intricate pathways within the human body (Zhang et al., 2022). However, robot manipulators don't always need to be made of rigid linkages. In this study, we address an alternate design possibility: building a robot with a continuous form, or backbone. When compared to the "vertebrate" design of traditional rigid-link robots, these robots, also known as continuum robots, can be thought of as "invertebrate" robots. Continuum robots are able to bend at any point along its structure, as well as frequently stretch or contract and twist (Barua et al., 2023). Their capabilities surpass those of their counterparts with stiff links as a result. In the past, there has been interest in continuous backbone robots since the 1960s. Regarding the initial continuum robot, it seems that the Tensor Arm (Anderson et al., 1970) by Anderson and Horn was the first model documented in the literature. The early version was meant for undersea uses, but it never made it out of the lab.

Initial designs could produce a wide range of shapes, but it soon became apparent that the relationship between the shapes and inputs was extremely complex, considerably more so than for rigid-link robots, and certainly difficult to construct using the computing systems available at the time. Complementary to the aforementioned advancements in hardware, continuum robot modeling, especially with regard to kinematics, saw growth and innovation in the 1990s. "Bottom-up" continuum robot kinematics models were created in a set of papers (Jones et al., 2006) (Gravagne et al., 2000) (Hannan et al., 2003). These works developed backbone kinematics starting from the limitations of physical continuum robot backbones. The first published "top-down" theory of (Chirikjian et al., 1994) can be used to build the models emerging from these "bottom-up" techniques, as is evident. The new models allowed for model-based implementations and consequently real-time computer control of continuum robot shapes because they matched closely to hardware restrictions.

At the heart of the Biomimetic Compound Continuum Robot is its unique compound continuum structure (Russo et al., 2023). Unlike traditional rigid robotic systems, this design mimics the segmented flexibility found in organisms like snakes or invertebrates (Garfjeld Roberts et al., 2023). The robot consists of interconnected segments, allowing it to move with a serpentine motion. This inherent flexibility enables the robot to navigate through confined spaces and complex anatomical structures, providing surgeons with a tool capable of reaching locations that were once challenging with conventional surgical instruments (Barua et al., 2023).Minimally invasive surgery (MIS) has become a preferred approach due to its potential for reduced postoperative pain, shorter recovery times, and smaller incisions (Zhu et al., 2023) (Barua et al., 2023). The Biomimetic Compound Continuum Robot takes these advantages a step further by minimizing tissue trauma during surgery (Das et al., 2023) (Datta et al., 2023). The segmented structure allows the robot to adapt to the contours of the body, reducing the risk of damage to surrounding tissues (Wei et al., 2023). This feature is particularly crucial in delicate procedures where precision is paramount (Waidi et al., 2023). The applications of the Biomimetic Compound Continuum Robot are diverse, spanning various surgical specialties (Zhang et al., 2022) (Dupont et al., 2022). This innovative robotic system, inspired by the flexibility and adaptability of natural organisms, introduces new possibilities for surgical interventions across various medical specialties (Barua et al., 2022). One of the key features of the Biomimetic Compound Continuum Robot (Figure 2) is its ability to achieve

Figure 1. The robotic system attains unparalleled dexterity by leveraging the biological framework inspired by bionic snakes
(Zhang et al., 2022).

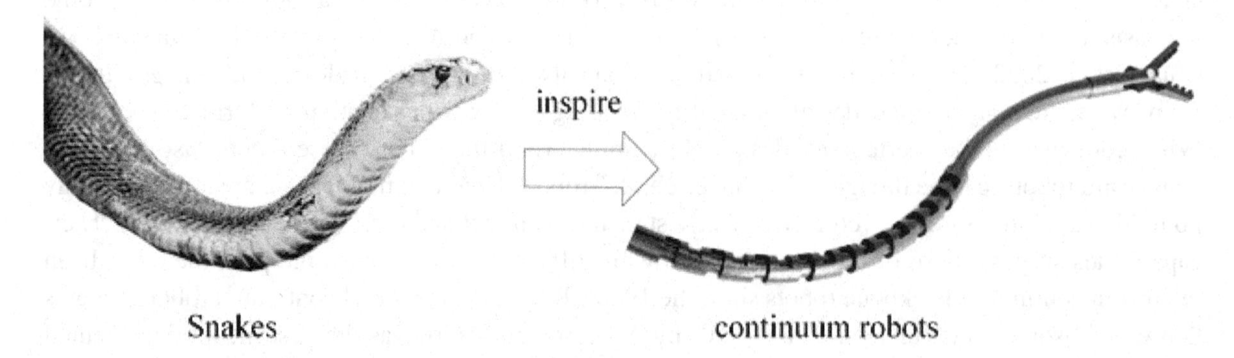

precise navigation within the human body (Wei et al., 2022). Surgeons can remotely control the robot, guiding it with a high degree of accuracy to the targeted surgical site (Morton et al., 2022). The biomimetic design enables the robot to traverse challenging anatomical pathways, making it suitable for a wide range of surgical applications (Barua et al., 2023).

MINIMALLY INVASIVE SURGERY (MIS)

Minimally Invasive Surgery (MIS) has emerged as a transformative approach to medical interventions, revolutionizing the field of surgery over the past few decades (Chen et al., 2022). This essay explores the evolution, techniques, benefits, challenges, and future prospects of minimally invasive procedures, shedding light on the profound impact they have had on patient outcomes and the practice of surgery (Barua et al., 2020) (Li et al., 2021).

Figure 2. The configuration of the compound continuum robot is depicted as follows: (a) The overarching structure, (b) The cable-driven mode and radial dimensions of the robot, and (c) The axial dimensions of the robot joints
(Wei et al., 2022)

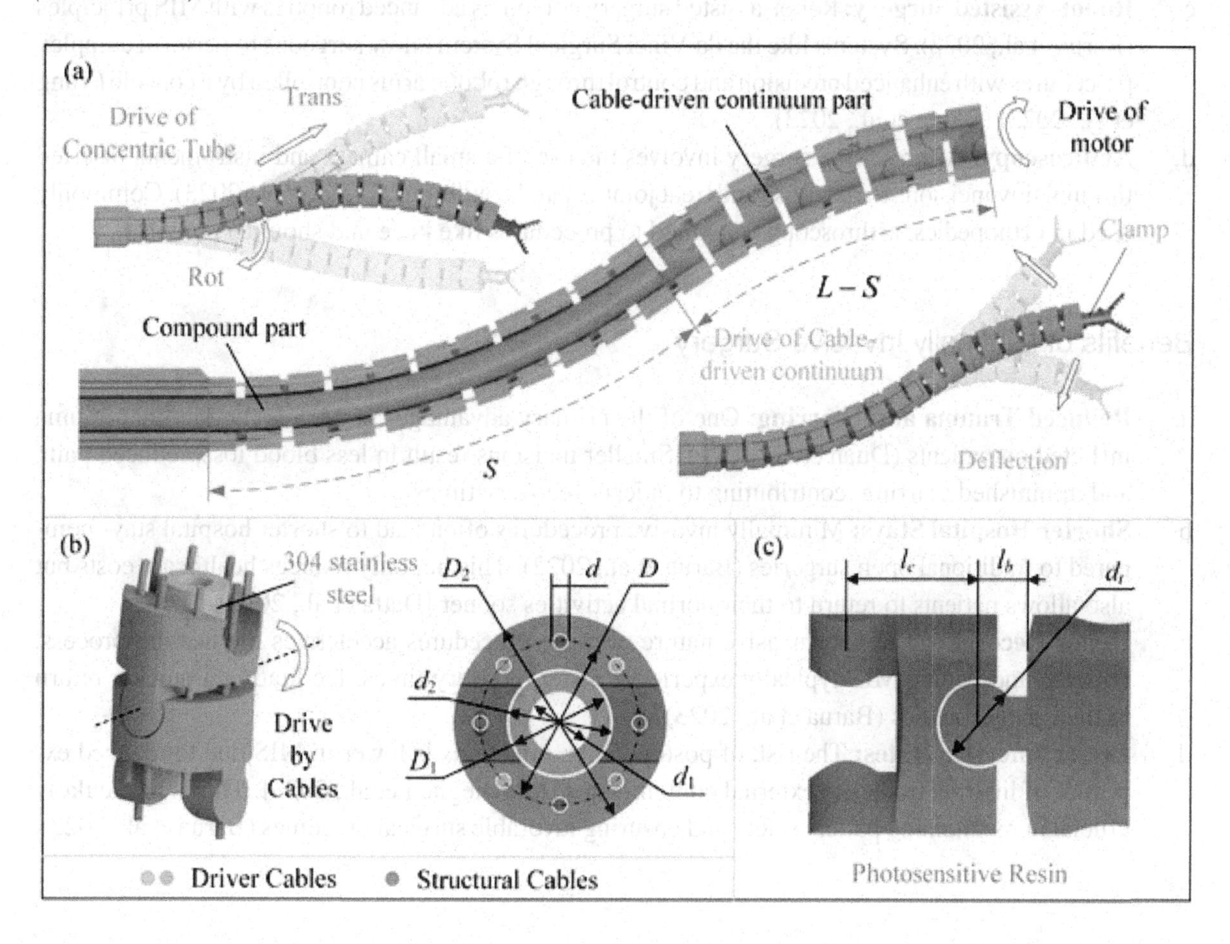

Evolution of Minimally Invasive Surgery

The roots of minimally invasive techniques can be traced back to the early 20th century with the advent of laparoscopy (Morizane et al., 2023) (Shimizu et al., 2022). However, it wasn't until the latter half of the century that technological advancements allowed for the widespread adoption of minimally invasive procedures (Barua et al., 2023) (Azizi et al., 2018). The development of fiber-optic technology, video cameras, and improved instrumentation paved the way for more sophisticated and less invasive surgical approaches.

Techniques in Minimally Invasive Surgery

a. **Laparoscopy:** Laparoscopy, also known as keyhole surgery, involves making small incisions through which a laparoscope and specialized instruments are inserted (Shimizu et al., 2022). It is commonly used in abdominal and pelvic surgeries, such as cholecystectomy, appendectomy, and hysterectomy (Barua et al., 2021).
b. **Endoscopy:** Endoscopic procedures involve the use of a flexible tube with a light and camera (endoscope) to visualize and operate within the body (Siau et al., 2019). Gastrointestinal endoscopy, bronchoscopy, and cystoscopy are examples of endoscopic techniques widely employed in diagnostic and therapeutic interventions.
c. **Robot-Assisted Surgery:** Robot-assisted surgery combines advanced robotics with MIS principles (Barua et al., 2022). Systems like the da Vinci Surgical System allow surgeons to perform complex procedures with enhanced precision and control through robotic arms controlled by a console (Yang et al., 2023) (Barua et al., 2023).
d. **Arthroscopy:** Arthroscopic surgery involves the use of a small camera and instruments inserted through tiny incisions to diagnose and treat joint-related conditions (Barua et al., 2023). Commonly used in orthopedics, arthroscopy is applied to procedures like knee and shoulder surgeries.

Benefits of Minimally Invasive Surgery

a. **Reduced Trauma and Scarring:** One of the primary advantages of MIS is the minimal trauma inflicted on patients (Duan et al., 2023). Smaller incisions result in less blood loss, reduced pain, and diminished scarring, contributing to quicker recovery times.
b. **Shorter Hospital Stays:** Minimally invasive procedures often lead to shorter hospital stays compared to traditional open surgeries (Barua et al., 2022). This not only reduces healthcare costs but also allows patients to return to their normal activities sooner [Datta et al., 2018].
c. **Faster Recovery:** The less invasive nature of these procedures accelerates the healing process. Patients undergoing MIS typically experience faster recovery times, facilitating a quicker return to their daily routines (Barua et al., 2023).
d. **Lower Infection Rates:** The risk of postoperative infections is lower in MIS due to reduced exposure of internal tissues to external contaminants [Marchegiani et al., 2023]. This is particularly crucial in maintaining patient safety and ensuring favorable surgical outcomes (Barua et al., 2022).

Figure 3. Minimally invasive surgery
(Azizi et al., 2018)

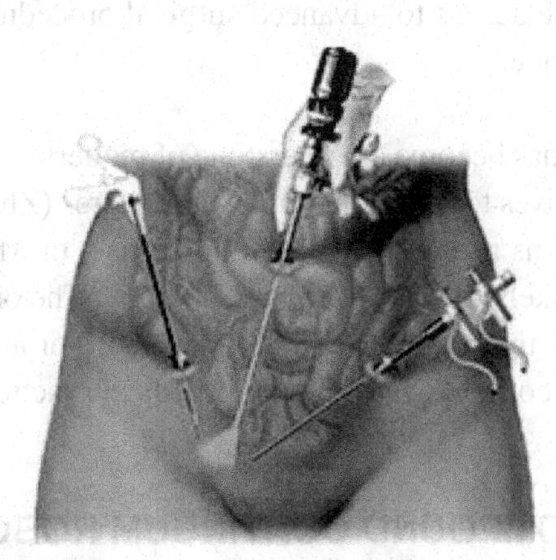

Challenges and Limitations

While minimally invasive surgery has transformed the field of surgery, it is not without challenges:

a. **Learning Curve:** Adopting MIS techniques requires specialized training and a steep learning curve for surgeons. Proficiency in manipulating instruments through small incisions or using robotic interfaces demands dedicated practice (Datta et al., 2023).
b. **Costs and Equipment:** The initial costs associated with acquiring and maintaining advanced equipment for MIS, such as robotic surgical systems, can be substantial. However, the long-term benefits in terms of patient outcomes and reduced postoperative care may offset these costs.
c. **Limited Application in Complex Cases:** In some complex surgeries or cases with anatomical challenges, traditional open procedures may still be preferred (Duan et al., 2023). The adaptability of MIS to all surgical scenarios is an ongoing area of research and development.

Future Prospects of MIS

The future of minimally invasive surgery holds exciting possibilities:

a. **Technological Advancements:** Continued technological innovations, including improved imaging, enhanced robotics, and smarter instrumentation, will further refine MIS techniques, making them more accessible and efficient.
b. **Integration of Artificial Intelligence (AI):** The integration of AI in MIS can contribute to real-time decision-making, automated assistance during surgery, and personalized treatment plans, enhancing the precision and effectiveness of these procedures (Barua et al., 2023).
c. **Expanding Surgical Specialties:** As technology evolves, the application of MIS is expected to expand into various surgical specialties (Yang et al., 2023). Ongoing research explores its feasibility in areas such as cardiac surgery, neurosurgery, and vascular surgery.

d. **Global Access to Advanced Surgery:** The development of cost-effective and portable MIS technologies could increase access to advanced surgical procedures in resource-limited settings, benefiting patients worldwide.

Minimally Invasive Surgery has become a cornerstone of modern surgical practice, offering patients safer and more efficient alternatives to traditional open procedures (Zhang et al., 2022). As technology continues to advance and surgeons refine their skills, the impact of MIS on patient outcomes and the overall healthcare landscape is likely to grow (Yang et al., 2023). The ongoing commitment to research, training, and innovation ensures that the future holds even more promising developments, cementing minimally invasive surgery as a cornerstone of modern medical practice.

WHAT IS BIOMIMETIC COMPOUND CONTINUUM ROBOT?

The term "Biomimetic Compound Continuum Robot" refers to a type of robotic system designed for various applications, particularly in the field of minimally invasive surgery (MIS) (Zhang et al., 2022). Let's break down the key components of this term to understand its meaning:

Biomimetic

"Biomimetic" means imitating or inspired by biological systems and processes found in nature. In the context of robotics, biomimicry involves designing machines that replicate or mimic the form, function, or behavior of living organisms (Zhang et al., 2022).

Compound Continuum

"Compound continuum" refers to a flexible and segmented structure that allows for continuous, snake-like motion (Russo et al., 2023). Unlike traditional rigid robots, a continuum robot is characterized by

Figure 4. Surgical instrumentation criteria for surgical procedures
(Zhang et al., 2022)

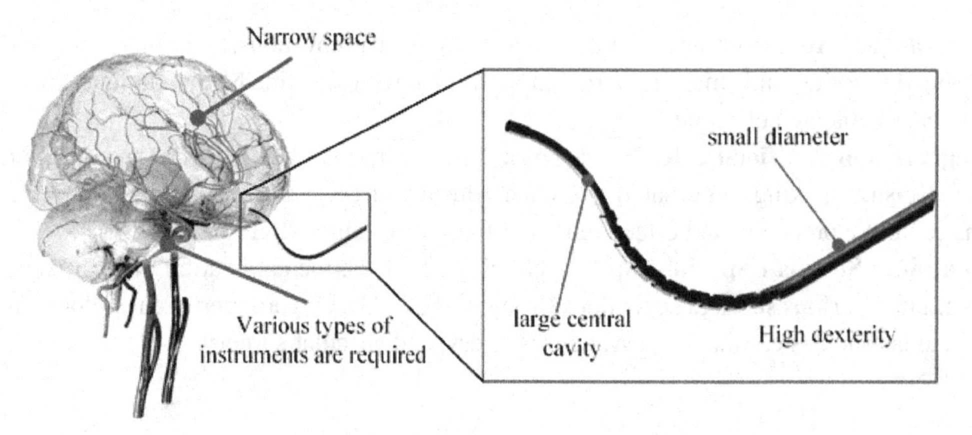

its ability to move in a more flexible and adaptable manner, much like the way some animals, such as snakes or worms, can navigate through confined spaces.

Robot

A "robot" in this context refers to a machine capable of carrying out tasks autonomously or under remote human control (Barua et al., 2023). In the case of the Biomimetic Compound Continuum Robot, the design is inspired by the flexibility and adaptability observed in natural organisms (Datta et al., 2023).

Combining these elements, the Biomimetic Compound Continuum Robot is a robotic system that mimics the flexible and segmented motion seen in certain living organisms [Yang et al., 2023] (Barua et al., 2023). This design is particularly advantageous in applications like minimally invasive surgery, where the robot can navigate through intricate anatomical structures with reduced impact on surrounding tissues (Duan et al., 2023). The term emphasizes the robot's biomimetic inspiration and its compound continuum structure, allowing for more versatile and less invasive movements in various medical and technological scenarios (Li et al., 2021). Here are a few examples of of research prototypes and projects that embody the principles of biomimicry and compound continuum robotics in the context of minimally invasive surgery.

a. **i-Snake:**
    - *Description:* Developed by the European Union-funded project RAS (Robotic Assisted Surgery), i-Snake is a research initiative that explores the use of a flexible robotic system for minimally invasive surgery (Wu et al., 2013). It aims to mimic the dexterity and flexibility of a snake to navigate through complex anatomical structures.

b. **CardioArm:**
    - *Description:* The CardioArm is a robotic catheter system designed for navigating through the cardiovascular system (Wang et al., 2017). Inspired by the agility of snakes, this system allows for precise movements within blood vessels during minimally invasive cardiac procedures.

c. **RoboSnake:**
    - *Description:* RoboSnake is a biomimetic robotic system designed to replicate the undulating motion of snakes (Cuperman et al., 2019). While not exclusively developed for surgery, its bio-inspired design has potential applications in minimally invasive procedures, particularly in navigating confined spaces.

d. **Flex System:**
    - *Description:* The Flex System, developed by FlexDex Surgical, incorporates principles of continuum robotics (Remacle et al., 2015). Although not explicitly biomimetic in its design, it utilizes a flexible robotic arm to enable minimally invasive surgery with enhanced dexterity, providing surgeons with intuitive control.

e. **STIFF-FLOP:**
    - *Description:* STIFF-FLOP (STIFFness controllable Flexible and Learnable manipulator for surgical OPerations) is a research project that focuses on developing a robotic system with a flexible structure inspired by octopus tentacles (Dawood et al., 2021). The aim is to provide surgeons with a tool that can adapt to various surgical scenarios.

Figure 5. Continuum robots: a. Octopus leg); and b. tentacle-inspired
(a), (McMahan et al., 2009); (b), (Sanan et al., 2011)

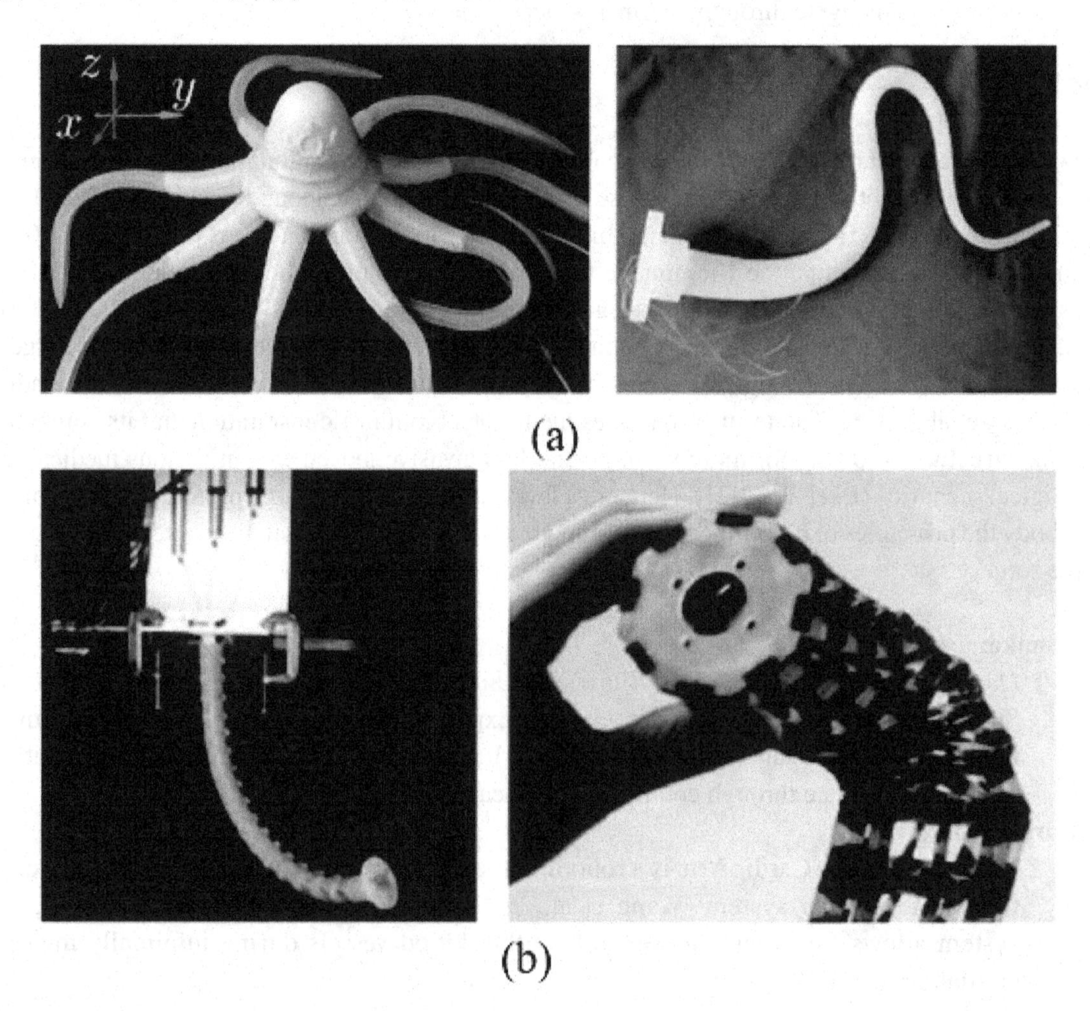

APPLICATIONS IN SURGICAL PROCEDURES

The applications of the Biomimetic Compound Continuum Robot span a wide range of surgical procedures (Figure 5), offering transformative capabilities in the field of minimally invasive surgery (MIS) (Zhang et al., 2022). In neurosurgery, where precision and access to intricate structures are critical, the robot's ability to navigate through tight spaces becomes invaluable. Similarly, in abdominal surgeries, the robot's flexibility allows it to reach and manipulate organs with reduced invasiveness.

Neurosurgery

In neurosurgery, where precision and access to delicate structures are paramount, the Biomimetic Compound Continuum Robot demonstrates significant advantages (Burgner-Kahrs et al., 2015). The snake-like motion of the robot allows it to navigate through intricate pathways in the brain with reduced trauma to surrounding tissues. Surgeons can remotely guide the robot to reach specific regions that may be challenging to access using traditional surgical instruments (Zhang et al., 2022). This capability enhances the accuracy of procedures such as tumor removal, biopsies, and the placement of therapeutic devices.

Abdominal Surgery

The flexibility of the compound continuum structure makes the robot well-suited for abdominal surgeries. It can navigate through the confined spaces of the abdominal cavity, providing surgeons with enhanced dexterity in manipulating organs and tissues (Burgner-Kahrs et al., 2015). The reduced invasiveness of the robot contributes to faster recovery times and minimizes postoperative pain. Applications in abdominal surgery include procedures such as laparoscopic cholecystectomy, gastrointestinal surgeries, and organ resections.

Cardiovascular Interventions

In the realm of cardiovascular surgery, the Biomimetic Compound Continuum Robot holds promise for a range of interventions. The robot's ability to navigate through intricate blood vessels with precision is particularly beneficial for procedures such as angioplasty, stent placement, and the treatment of vascular abnormalities (Zhang et al., 2022). The minimally invasive nature of these procedures, facilitated by the robot, reduces the risk of complications and shortens recovery times for patients undergoing cardiovascular interventions.

Gynecological Procedures

Gynecological surgeries, including hysterectomies and ovarian procedures, can benefit from the Biomimetic Compound Continuum Robot. The robot's flexible structure allows it to navigate through the pelvic region with greater ease, providing surgeons with improved access to target areas (Russo et al., 2023). The minimally invasive approach reduces scarring and enhances patient recovery, making it an attractive option for various gynecological interventions (Barua et al., 2023).

Urological Interventions

In urology, the robot's adaptability is valuable for procedures such as prostate surgeries and kidney interventions. Its snake-like motion enables precise navigation through complex anatomical structures, reducing the risk of damage to adjacent tissues (Zhang et al., 2022). The minimally invasive approach enhances patient outcomes by decreasing postoperative pain and shortening hospital stays (Yang et al., 2023). The robot's remote surgical capabilities also contribute to the surgeon's ability to perform intricate procedures with increased precision.

Orthopedic Surgery

The Biomimetic Compound Continuum Robot holds potential in orthopedic surgery, particularly in procedures involving joints and soft tissues. The robot's flexibility allows for precise navigation around bones and articulations, making it suitable for arthroscopic surgeries (Russo et al., 2023). Surgeons can use the robot to access joint spaces and perform repairs with reduced invasiveness, contributing to quicker recovery and improved functional outcomes for patients undergoing orthopedic interventions (Li et al., 2021).

Head and Neck Surgery

In head and neck surgeries, the Biomimetic Compound Continuum Robot offers advantages in accessing challenging anatomical structures (Zhang et al., 2022). Its flexible design allows for maneuverability in confined spaces, making it suitable for procedures such as throat surgeries, tongue base surgeries, and minimally invasive approaches to tumors in the head and neck region. The robot's ability to navigate through complex anatomies enhances the precision of these interventions (Yang et al., 2023).

Ophthalmic Procedures

The robot's adaptability extends to ophthalmic surgeries, where delicate and precise maneuvers are essential (Li et al., 2021). In procedures such as retinal surgeries and cataract removal, the Biomimetic Compound Continuum Robot's snake-like motion provides surgeons with enhanced control and access to intricate ocular structures (Zhang et al., 2022). The minimally invasive nature of the robot contributes to reduced postoperative complications and faster visual recovery for patients undergoing ophthalmic interventions (Chen et al., 2022).

FUTURE IMPLICATIONS AND ETHICAL AND REGULATORY CONSIDERATIONS

As technology continues to advance, the Biomimetic Compound Continuum Robot holds promise for further developments (Barua et al., 2023). The biomimetic compound continuum robot, inspired by nature's adaptability, revolutionizes minimally invasive surgery (Li et al., 2021). This robotic system emulates the flexible and snake-like movements found in biological organisms. Comprising interconnected segments, it navigates complex anatomical structures with precision, reaching confined spaces traditional tools cannot (Yang et al., 2023). Its biomimicry enhances surgical capabilities, reducing tissue trauma and allowing for intricate procedures. Surgeons remotely guide the robot, offering unparalleled dexterity and access (Datta et al., 2023). This innovative technology holds promise for diverse surgical applications, marking a significant advancement in improving patient outcomes through less invasive and more precise interventions.

Continued research and innovation may lead to even more refined and specialized versions of the robot, catering to specific surgical procedures and anatomical challenges (Li et al., 2021). Integration with advanced imaging technologies could enhance the robot's ability to navigate with even greater precision (Yang et al., 2023). The biomimetic approach of the robot contributes to a reduction in patient trauma. The flexibility of its continuum structure minimizes the impact on surrounding tissues, leading to less postoperative pain and faster recovery times. Patients undergoing surgeries assisted by this robotic system may experience shorter hospital stays and quicker returns to daily activities (Li et al., 2021). The Biomimetic Compound Continuum Robot offers the advantage of remote surgical capabilities. Surgeons can operate the robot from a console, providing them with a more ergonomic and comfortable environment (Zhang et al., 2022). This not only enhances the precision of movements but also allows for improved focus during lengthy procedures. The remote control feature is particularly beneficial in situations where the surgical site may be challenging to access directly (Zhang et al., 2022). The introduction of advanced robotic systems in surgery raises ethical and regulatory considerations. Ensuring patient safety, ethical use of technology, and comprehensive training for surgeons are crucial aspects that need careful attention (Yang et al., 2023). Regulatory bodies play a vital role in establishing guidelines and standards to govern the integration of such innovative technologies into clinical practice.

CONCLUSION

The Biomimetic Compound Continuum Robot presents a transformative paradigm in minimally invasive surgery, with diverse applications across multiple medical specialties. Its biomimetic design, inspired by nature's flexibility, provides surgeons with a tool that enhances precision, reduces tissue trauma, and expands the scope of minimally invasive interventions. As technology continues to advance, the robot's applications are likely to grow, offering new possibilities for improving patient outcomes and redefining the future of surgical care. The application of this bio-inspired technology in MIS holds immense promise. Surgeons can remotely control the robot, guiding it with precision to perform targeted medical interventions. The inherent flexibility of the compound continuum structure reduces the risk of tissue damage and trauma during the surgical procedure, contributing to faster recovery times and improved patient outcomes. While the applications of the Biomimetic Compound Continuum Robot are promising, challenges remain in terms of technological refinement, training, and regulatory considerations. Continued research is necessary to optimize the robot's capabilities, ensuring its safety and efficacy in a variety of surgical scenarios. Training programs for surgeons need to be developed to familiarize them with the unique features and operational aspects of the robot. Additionally, regulatory bodies play a crucial role in establishing guidelines and standards for the integration of such advanced robotic systems into clinical practice. Ethical considerations surrounding patient safety, consent, and the responsible use of technology must be addressed to ensure that the benefits of the robot are maximized while minimizing potential risks. Moreover, the robot's adaptability makes it well-suited for a variety of surgical procedures, ranging from delicate neurosurgery to intricate abdominal surgeries. Its ability to access challenging anatomical sites with minimal invasiveness marks a paradigm shift in surgical techniques, paving the way for more effective and patient-friendly procedures. In brief, the Bio-Inspired Compound Continuum Robot represents a remarkable fusion of engineering and biology, revolutionizing the landscape of minimally invasive surgery by providing surgeons with an advanced tool that combines precision, flexibility, and adaptability for improved patient care.

Conflicts of Interest

The authors declare that there are no conflicts of interest regarding the publication of this chapter.

Funding

No funding has been provided for this work.

Ethical approval

Not required.

ACKNOWLEDGEMENTS

The authors would like to thank IIEST Shibpur, and IISc Bengalore, and thanks to Mrs. Nibedita Bardhan for language proof reading.

REFERENCES

Anderson, V. C., & Horn, R. C. (1970). *U.S. Patent No. 3,497,083*. Washington, DC: U.S. Patent and Trademark Office.

Azizi, A., Hortamani, R., & Zabihollah, A. (2018). Sensing the material by minimally invasive surgery grasper. *International Robotics & Automation Journal*, *4*(3). doi:10.15406/iratj.2018.04.00117

Barua, R., Bhowmik, S., Dey, A., Das, S., & Datta, S. (2022, September). Analysis of Robotically Controlled Percutaneous Needle Insertion into Ex Vivo Kidney Tissue for Minimally Invasive Percutaneous Nephrolithotomy (PCNL) Surgery. *In International Conference on Emergent Converging Technologies and Biomedical Systems* (pp. 249-257). Singapore: Springer Nature Singapore.

Barua, R., Bhowmik, S., Dey, A., & Mondal, J. (2023). Advances of the Robotics Technology in Modern Minimally Invasive Surgery. In M. Mellal (Ed.), *Design and Control Advances in Robotics* (pp. 91–104). IGI Global.

Barua, R., Das, S., Datta, P., & Chowdhury, A. (2022). Computational FEM Application on Percutaneous Nephrolithotomy (PCNL) Minimum Invasive Surgery Through Needle Insertion Process. In P. Pain, S. Banerjee, & G. Bose (Eds.), *Advances in Computational Approaches in Biomechanics* (pp. 210-222). IGI Global. doi:10.4018/978-1-7998-9078-2.ch013

Barua, R., Das, S., Datta, S., Datta, P., & Roy Chowdhury, A. (2021). Analysis of surgical needle insertion modeling and viscoelastic tissue material interaction for minimally invasive surgery (MIS). *Materials Today: Proceedings*, *57*, 259–264. doi:10.1016/j.matpr.2022.02.498

Barua, R., Das, S., Datta, S., Datta, P., & Roy Chowdhury, A. (2023). Study and experimental investigation of insertion force modeling and tissue deformation phenomenon during surgical needle-soft tissue interaction. *Proceedings of the Institution of Mechanical Engineers. Part C, Journal of Mechanical Engineering Science*, *237*(5), 1007–1014. doi:10.1177/09544062221126628

Barua, R., Das, S., Datta, S., Roy Chowdhury, A., & Datta, P. (2022). Experimental study of the robotically controlled surgical needle insertion for analysis of the minimum invasive process. In *Emergent Converging Technologies and Biomedical Systems: Select Proceedings of ETBS 2021* (pp. 473-482). Singapore: Springer Singapore. 10.1007/978-981-16-8774-7_38

Barua, R., Das, S., & Mondal, J. (2023). Emerging Applications of Artificial Intelligence (AI) and Machine Learning (ML) in Modern Urology. In R. Queirós, B. Cunha, & X. Fonseca (Eds.), *Exploring the Convergence of Computer and Medical Science Through Cloud Healthcare* (pp. 117–133). IGI Global. doi:10.4018/978-1-6684-5260-8.ch006

Barua, R., Das, S., Roy Chowdhury, A., & Datta, P. (2023). Experimental and simulation investigation of surgical needle insertion into soft tissue mimic biomaterial for minimally invasive surgery (MIS). *Proceedings of the Institution of Mechanical Engineers. Part H, Journal of Engineering in Medicine*, *237*(2), 254–264. doi:10.1177/09544119221143860 PMID:36527297

Barua, R., Das, S., RoyChowdhury, A., & Datta, P. (2023). Simulation and experimental investigation of the surgical needle deflection model during the rotational and steady insertion process. *The International Journal of Artificial Organs*, *46*(1), 40–51. doi:10.1177/03913988221136154 PMID:36397288

Barua, R., Datta, P., Chowdhury, A. R., & Das, S. (2022). Computational Study of In-Vitro Ureter Urine Flow in DJ Stent. In P. Pain, S. Banerjee, & G. Bose (Eds.), *Advances in Computational Approaches in Biomechanics* (pp. 198–209). IGI Global. doi:10.4018/978-1-7998-9078-2.ch012

Barua, R., & Datta, S. (2022). Study of the surgical needle and biological soft tissue interaction phenomenon during insertion process for medical application: A Survey. *Proceedings of the Institution of Mechanical Engineers, Part H: Journal of Engineering in Medicine, 236*(10), 1465-1477.

Barua, R., & Datta, S. (2023). Emerging Surgical Robotic Applications for Modern Minimally Invasive Surgery (MIS). In M. Habib (Ed.), *Global Perspectives on Robotics and Autonomous Systems: Development and Applications* (pp. 314–332). IGI Global. doi:10.4018/978-1-6684-7791-5.ch014

Barua, R., & Datta, S. (2023). Artificial Intelligence in Modern Medical Science: A Promising Practice. In S. Rajest, B. Singh, A. J. Obaid, R. Regin, & K. Chinnusamy (Eds.), *Recent Developments in Machine and Human Intelligence* (pp. 1–12). IGI Global. doi:10.4018/978-1-6684-9189-8.ch001

Barua, R., Datta, S., & Sarkar, A. (2023). Artificial Intelligence and Robotics-Based Minimally Invasive Surgery: Innovations and Future Perceptions. In G. Karthick & S. Karupusamy (Eds.), *Contemporary Applications of Data Fusion for Advanced Healthcare Informatics* (pp. 350–368). IGI Global. doi:10.4018/978-1-6684-8913-0.ch015

Barua, R., Giria, H., Datta, S., Roy Chowdhury, A., & Datta, P. (2020). Force modeling to develop a novel method for fabrication of hollow channels inside a gel structure. *Proceedings of the Institution of Mechanical Engineers. Part H, Journal of Engineering in Medicine*, *234*(2), 223–231. doi:10.1177/0954411919891654 PMID:31774361

Barua, R., & Mondal, J. (2023). Study of the Current Trends of CAD (Computer-Aided Detection) in Modern Medical Imaging. In L. Panigrahi, S. Biswal, A. Bhoi, A. Kalam, & P. Barsocchi (Eds.), *Machine Learning and AI Techniques in Interactive Medical Image Analysis* (pp. 35–50). IGI Global., doi:10.4018/978-1-6684-4671-3.ch002

Barua, R., Sarkar, A., & Datta, S. (2023). Modern Lab-on-Chip Biosensors Application on Infectious COVID-19 Detection. In R. Singh, R. Phanden, B. Sikarwar, & J. Davim (Eds.), *Advances in MEMS and Microfluidic Systems* (pp. 258–270). IGI Global. doi:10.4018/978-1-6684-6952-1.ch013

Burgner-Kahrs, J., Rucker, D. C., & Choset, H. (2015). Continuum robots for medical applications: A survey. *IEEE Transactions on Robotics*, *31*(6), 1261–1280. doi:10.1109/TRO.2015.2489500

Chen, K., Zhang, J., Beeraka, N. M., Sinelnikov, M. Y., Zhang, X., Cao, Y., & Lu, P. (2022). Robot-Assisted Minimally Invasive Breast Surgery: Recent Evidence with Comparative Clinical Outcomes. *Journal of Clinical Medicine*, *11*(7), 1827. doi:10.3390/jcm11071827 PMID:35407434

Chirikjian, G. S. (1994). Hyper-redundant manipulator dynamics: A continuum approximation. *Advanced Robotics*, *9*(3), 217–243. doi:10.1163/156855395X00175

Cuperman, D., & Verner, I. M. (2019). Fostering analogical reasoning through creating robotic models of biological systems. *Journal of Science Education and Technology*, *28*(2), 90–103. doi:10.1007/s10956-018-9750-4

Das, S., Datta, S., Barman, A., & Barua, R. (2023). Smart Biodegradable and Bio-Based Polymeric Biomaterials for Biomedical Applications. In A. Kumar, P. Kumar, A. Srivastava, & V. Goyat (Eds.), *Modeling, Characterization, and Processing of Smart Materials* (pp. 56–82). IGI Global. doi:10.4018/978-1-6684-9224-6.ch003

Datta, S., & Barua, R. (2023). Fluorescent Nanomaterials and Its Application in Biomedical Engineering. In A. Rakha, A. Munawar, V. Khanna, & S. Bansal (Eds.), *Modeling and Simulation of Functional Nanomaterials for Forensic Investigation* (pp. 164–186). IGI Global. doi:10.4018/978-1-6684-8325-1.ch009

Datta, S., & Barua, R. (2023). Advanced Materials for Surgical Tools and Biomedical Implants. *Advanced Materials and Manufacturing Techniques for Biomedical Applications*, 25-35.

Datta, S., Barua, R., & Das, J. (2020). *A review on electro-rheological fluid (er) and its various technological applications*. Extremophilic Microbes and Metabolites-Diversity, Bioprospecting and Biotechnological Applications.

Datta, S., Barua, R., & Das, S. (2023). Role and Challenges of Bioprinting in Bone Tissue Engineering. In R. Ranjith & J. Davim (Eds.), *Handbook of Research on Advanced Functional Materials for Orthopedic Applications* (pp. 205–218). IGI Global. doi:10.4018/978-1-6684-7412-9.ch012

Datta, S., Barua, R., & Prasad, A. (2023). Additive Manufacturing for the Development of Artificial Organs. *Advanced Materials and Manufacturing Techniques for Biomedical Applications*, 411-427.

Datta, S., Barua, R., Sarkar, R., Barui, A., Chowdhury, A. R., & Datta, P. (2018, September). Design and development of alginate: Poly-l-lysine scaffolds by 3D bio printing and studying their mechanical, structural and cell viability properties. [). IOP Publishing.]. *IOP Conference Series. Materials Science and Engineering*, *402*, 012113. doi:10.1088/1757-899X/402/1/012113

Datta, S., Das, S., & Barua, R. (2023). Self-Sustained Nanobiomaterials: Innovative Materials for Biomedical Applications. *Advanced Materials and Manufacturing Techniques for Biomedical Applications*, 303-323.

Dawood, A. B., Fras, J., Aljaber, F., Mintz, Y., Arezzo, A., Godaba, H., & Althoefer, K. (2021). Fusing dexterity and perception for soft robot-assisted minimally invasive surgery: What we learnt from STIFF-FLOP. *Applied Sciences (Basel, Switzerland)*, *11*(14), 6586. doi:10.3390/app11146586

Duan, W., Akinyemi, T., Du, W., Ma, J., Chen, X., Wang, F., Omisore, O., Luo, J., Wang, H., & Wang, L. (2023). Technical and Clinical Progress on Robot-Assisted Endovascular Interventions: A Review. *Micromachines*, *14*(1), 197. doi:10.3390/mi14010197 PMID:36677258

Dupont, P. E., Simaan, N., Choset, H., & Rucker, C. (2022). Continuum Robots for Medical Interventions. *Proceedings of the IEEE*. Institute of Electrical and Electronics Engineers.

Garfjeld Roberts, P., Glasbey, J. C., Abram, S., Osei-Bordom, D., Bach, S. P., & Beard, D. J. (2020). Research quality and transparency, outcome measurement and evidence for safety and effectiveness in robot-assisted surgery: Systematic review. *BJS Open*, *4*(6), 1084–1099. doi:10.1002/bjs5.50352 PMID:33052029

Gravagne, I., & Walker, I. D. (2000). Kinematics for constrained continuum robots using wavelet decomposition. *Robotics*, *2000*, 292–298.

Hannan, M. W., & Walker, I. D. (2003). Kinematics and the implementation of an elephant's trunk manipulator and other continuum style robots. *Journal of Robotic Systems*, *20*(2), 45–63. doi:10.1002/rob.10070 PMID:14983840

Jones, B. A., & Walker, I. D. (2006). Kinematics for multisection continuum robots. *IEEE Transactions on Robotics*, *22*(1), 43–55. doi:10.1109/TRO.2005.861458

Li, W., Kong, K., Li, P., Wang, G., Cui, B., Zhu, L., & Zhu, S. (2021). Robot-assisted sleeve gastrectomy in patients with obesity with a novel Chinese domestic MicroHand SII surgical system. *BMC Surgery*, *21*(1), 260. doi:10.1186/s12893-021-01259-3 PMID:34034737

Li, Z., & Du, R. (2013). Design and analysis of a bio-inspired wire-driven multi-section flexible robot. *International Journal of Advanced Robotic Systems*, *10*(4), 209. doi:10.5772/56025

Marchegiani, F., Siragusa, L., Zadoroznyj, A., Laterza, V., Mangana, O., Schena, C. A., Ammendola, M., Memeo, R., Bianchi, P. P., Spinoglio, G., Gavriilidis, P., & de'Angelis, N. (2023). New Robotic Platforms in General Surgery: What's the Current Clinical Scenario? *Medicina (Kaunas, Lithuania)*, *59*(7), 1264. doi:10.3390/medicina59071264 PMID:37512075

McMahan, W., & Walker, I. D. (2009, February). Octopus-inspired grasp-synergies for continuum manipulators. In *2008 IEEE International Conference on Robotics and Biomimetics* (pp. 945-950). IEEE. 10.1109/ROBIO.2009.4913126

Morizane, S., Stein, H., Komiya, T., Kaneta, H., & Takenaka, A. (2023). Retroperitoneal robot-assisted laparoscopic nephroureterectomy using the da Vinci Xi and SP systems: Initial experiences in cadaveric models. *Investigative and Clinical Urology*, *64*(4), 380–387. doi:10.4111/icu.20230021 PMID:37417563

Morton, J., Hardwick, R. H., Tilney, H. S., Gudgeon, A. M., Jah, A., Stevens, L., Marecik, S., & Slack, M. (2021). Preclinical evaluation of the versius surgical system, a new robot-assisted surgical device for use in minimal access general and colorectal procedures. *Surgical Endoscopy*, *35*(5), 2169–2177. doi:10.1007/s00464-020-07622-4 PMID:32405893

Remacle, M. M. N., Prasad, V., Lawson, G., Plisson, L., Bachy, V., & Van der Vorst, S. (2015). Transoral robotic surgery (TORS) with the Medrobotics Flex™ System: First surgical application on humans. *European Archives of Oto-Rhino-Laryngology*, *272*, 1451–1455. doi:10.1007/s00405-015-3532-x PMID:25663191

Russo, M., Gautreau, E., Bonnet, X., & Laribi, M. A. (2023). Continuum Robots: From Conventional to Customized Performance Indicators. *Biomimetics*, *8*(2), 147. doi:10.3390/biomimetics8020147 PMID:37092399

Sanan, S., Moidel, J., & Atkeson, C. G. (2011, June). A continuum approach to safe robots for physical human interaction. In *International Symposium on Quality of Life Technology*. IEEE.

Schranz, M., Umlauft, M., Sende, M., & Elmenreich, W. (2020). Swarm Robotic Behaviors and Current Applications. *Frontiers in Robotics and AI*, *7*, 36. doi:10.3389/frobt.2020.00036 PMID:33501204

Shimizu, A., Ito, M., & Lefor, A. K. (2022). Laparoscopic and Robot-Assisted Hepatic Surgery: An Historical Review. *Journal of Clinical Medicine*, *11*(12), 3254. doi:10.3390/jcm11123254 PMID:35743324

Siau, K., Hodson, J., Ingram, R., Baxter, A., Widlak, M. M., Sharratt, C., Baker, G. M., Troth, T., Hicken, B., Tahir, F., Magrabi, M., Yousaf, N., Grant, C., Poon, D., Khalil, H., Lee, H. L., White, J. R., Tan, H., Samani, S., & Major, G. (2019). Time to endoscopy for acute upper gastrointestinal bleeding: Results from a prospective multicentre trainee-led audit. *United European Gastroenterology Journal*, *7*(2), 199–209. doi:10.1177/2050640618811491 PMID:31080604

Waidi, Y. O., Barua, R., & Datta, S. (2023). Metals, Polymers, Ceramics, Composites Biomaterials Used in Additive Manufacturing for Biomedical Applications. In A. Kumar, P. Kumar, A. Srivastava, & V. Goyat (Eds.), *Modeling, Characterization, and Processing of Smart Materials* (pp. 165–184). IGI Global. doi:10.4018/978-1-6684-9224-6.ch008

Wang, H., Zhang, R., Chen, W., Wang, X., & Pfeifer, R. (2017). A cable-driven soft robot surgical system for cardiothoracic endoscopic surgery: Preclinical tests in animals. *Surgical Endoscopy*, *31*(8), 3152–3158. doi:10.1007/s00464-016-5340-9 PMID:27858208

Wei, H., Zhang, G., Wang, S., Zhang, P., Su, J., & Du, F. (2023). Coupling Analysis of Compound Continuum Robots for Surgery: Another Line of Thought. *Sensors (Basel)*, *23*(14), 6407. doi:10.3390/s23146407 PMID:37514701

Wu, X., & Ma, S. (2013). Neurally controlled steering for collision-free behavior of a snake robot. *IEEE Transactions on Control Systems Technology*, *21*(6), 2443–2449. doi:10.1109/TCST.2012.2237519

Yang, Y., Li, D., Sun, Y., Wu, M., Su, J., Li, Y., Yu, X., Li, L., & Yu, J. (2023). Muscle-inspired soft robots based on bilateral dielectric elastomer actuators. *Microsystems & Nanoengineering*, *9*(1), 124. doi:10.1038/s41378-023-00592-2 PMID:37814608

Zhang, G., Du, F., Xue, S., Cheng, H., Zhang, X., Song, R., & Li, Y. (2022). Design and Modeling of a Bio-Inspired Compound Continuum Robot for Minimally Invasive Surgery. *Machines*, *10*(6), 468. doi:10.3390/machines10060468

Zhu, J., Lyu, L., Xu, Y., Liang, H., Zhang, X., Ding, H., & Wu, Z. (2021). Intelligent Soft Surgical Robots for Next-Generation Minimally Invasive Surgery. *Advanced Intelligent Systems*, *3*(5), 2100011. doi:10.1002/aisy.202100011

KEY TERMS AND DEFINITIONS

Laparoscope: A laparoscope is a minimally invasive surgical instrument equipped with a thin, flexible tube and a tiny camera that allows surgeons to visualize the inside of the abdominal or pelvic cavity. It is inserted through small incisions, enabling procedures like gallbladder removal, appendectomy, and exploratory surgery with reduced scarring and shorter recovery times compared to traditional open surgery. The camera transmits real-time images to a monitor, guiding surgeons during the procedure, and offering a detailed view of the organs and tissues.

MIS: A medical technique that uses small incisions and specialized tools to perform procedures within the body, reducing the need for large surgical openings. It offers advantages such as faster recovery, less scarring, and reduced risk of complications compared to traditional open surgery.

Robotics: Basically a interdisciplinary field of designing, building, and operating robots. These autonomous or remote-controlled machines perform tasks, often in manufacturing, healthcare, exploration, and more. Utilizing sensors, actuators, and artificial intelligence, robotics aims to enhance efficiency and safety while advancing automation and technology across various industries.

Soft Tissue: Soft tissue refers to a group of connective tissues in the body that includes muscles, tendons, ligaments, fat, and blood vessels. It provides support and flexibility, enabling bodily movements, and plays a crucial role in various bodily functions. Injuries to soft tissues can lead to pain and mobility issues.

Surgical robot: A sophisticated medical device designed to assist surgeons in performing complex procedures with precision. Controlled by a surgeon from a console, these robots use advanced technology, such as robotic arms and cameras, to provide enhanced dexterity, visualization, and minimally invasive capabilities, improving surgical outcomes and patient recovery.

Chapter 14
An Intelligent Robotic Fogger System for Predicting Dengue Outbreaks

D. Raveena Judie Dolly
https://orcid.org/0000-0001-9837-2213
Karunya Institute of Technology and Sciences, India

D. J Jagannath
Karunya Institute of Technology and Sciences, India

J. Dinesh Peter
https://orcid.org/0000-0002-4357-7163
Karunya Institute of Technology and Sciences, India

ABSTRACT

Humans may encounter an arboviral illness through viruses transmitted by mosquitoes, commonly resulting in a fever known as breakbone fever. This term reflects the severity of muscle spasms and joint pains associated with the illness. While some cases are asymptomatic, others can be fatal. Dengue awareness often arises during seasonal changes. The integration of AI in dengue prediction becomes crucial for early diagnosis and treatment. Utilizing appropriate deep learning classifiers can aid in categorizing cases based on their severity. This article advocates for the implementation of an intelligent robotic fogger system in predicted areas. This approach employs interprofessional strategies to safeguard health workers and residents in regions prone to dengue outbreaks.

INTRODUCTION

Dengue, a viral infection transmitted by mosquitoes, is particularly common during seasonal and monsoonal shifts. The causative agents, Dengue types 1 to 4, and the Chikungunya virus fall under the Flavivirus genus. Without proper diagnosis and treatment, Dengue can lead to severe consequences. The

DOI: 10.4018/979-8-3693-1962-8.ch014

southern regions of India are at increased risk due to climatic variations, with 28,578 cases reported in 2021 and a total of 1.93 lakh cases nationwide during the same year. This emphasizes the urgent need for heightened awareness and proactive measures to address potential outbreaks.

While predicting Dengue outbreaks remains challenging, artificial intelligence emerges as a promising solution. Integrating AI holds the potential for accurate outbreak predictions, potentially saving lives through timely warnings. After identifying an outbreak period, deploying a smart Fogger Robot becomes a valuable strategy to control the virus's spread, ensuring the safety of both the general public and healthcare workers.

DISCUSSION

Two cases of around 14 years of age were reported in southern parts of Tamilnadu where severe fever was observed in both the cases. The first case had high temperature, which didn't ease even after several medications. The patient had several pills before proper diagnosis. The third day of fever instigated the doctors that it could be dengue and the blood test revealed the same. By then the patient felt so weak and was admitted where, the platelet count went on decreasing from 2.55 Lakh to 32,000 and found that the liver was also affected.

Clinical tests were advised for the patient. As per the direction, it was observed that on the onset day of fever, the patient had a test report with 48.0 mg/L of CRP – Turbidometry but the reference range is <6 mg/L. The CBC showed incremental increase in Neutrophils which was about 93%, where the reference range was about 40-75% and the Lymphocytes was 04% where the reference range was around 15-45%. The platelet count was found to be 2.55 Lakhs/cumm. The first day, the platelet count seemed to be normal. As per the literature survey, CRP helps in distinguishing the bacterial and viral infection. The test was repeated on the third day and the CRP – Turbidometry was found to be 46.1 mg/L. The total WBC count was reduced to 2000 cells/cumm from 6800 cells/cumm. The platelet count had a decremental phase of 1.57 Lakhs/cumm. Hence Dengue NS1 Antigen was found to be positive. Dengue IgM Antibody and Dengue IgG Antibody was tested negative. Counselling was provided which boosted the patient mentally. The test was repeated on the fourth day where again the WBC count was noticed to be 1700 cells/cumm. Lymphocytes was observed as 49%. The platelet count was observed to be 1.25 Lakhs/cumm. The next day, the patient was tested with SGOT, Liver function test and was observed to have 201 U/L where, the reference range need to be up to 40U/L. SGPT was observed as 103 U/L where, the reference range need to be up to 41 U/L. Medications started immediately to improve the liver functioning of the patient. The next day, platelet count was 32000 / cumm. IV fluids was suggested and was initiated immediately on the third day. Proper medication was then provided by the doctors for recovering. Again the platelet count was checked on the next day and was found that it was 60000 / cumm. As per the advice, the next day, platelet count was checked and found to be 145000/cumm. The medications for liver continued and the patient was discharged on the next day.

The other case had severe fever but the platelet count was observed to be lingering around 2 Lakh. It was also observed that after the fifth day of the fever, the platelet count started to decrease and on the seventh day the patient could recover. Rashes appeared after the 6th day of the fever.

It is recommended that, A prediction (Brasier et al., 2012; Gomes et al., 2010; Guzman et al., 2010) in this case could warn the mosquito borne areas during the outbreak. Diverse Machine learning algorithms (Chen et al., 2015; Lee et al., 2012; Mushtaque et al., 2020; National Vector Borne Disease

Control Programme (NVBDCP), 2015; Potts et al., 2010; Ranjit & Kissoon, 2011; Rao & Kumar, 2012; Side et al., 2020) have been adopted to predict the outbreak. In many literatures, RNA-Seq data have been widely deployed for evolving predictive model of the different categories of dengue. Several classifiers are also adopted to classify the various types of dengue. Tropical areas are more prone to mosquitoes; hence the outbreak can be suggested by predictive modelling methods as indicated in Husin (2012). The southern Parts of India encounter more dengue cases. As observed in Figure 1, the total reported cases in India has been decreased in 2022 (National Vector Borne Disease Control Programme (NVBDCP), 2015). But when compared to 2017, there is a significant increase in the number of cases. Appropriate prediction and prevention techniques need to be enforced to prevent diseases that spread through mosquitoes.

The number of Dengue cases in India was observed to be 1,88,401 in the year 2017. A decline of 87,209 in the number of cases was observed in the year 2018. Again a substantial increase of around 1,57,315 was observed in the year 2019. But, again a decline of 44,585 was observed in the year 2020. There is a rapid increase in the number of cases around 1,93,245 in the year 2021. Till 30th Sep, 2022, the number of cases was around 63,280. As observed in Fig.2, there is a significant decrease in the states like Andhra Pradesh, Karnataka, Kerala and Tamilnadu. But, comparatively Telangana seems to have a hike in the number of reported cases in the year 2022.

It seems to be a real concern in reducing the number of cases. Artificial intelligence has played a vital role in predicting the dengue outbreak. Learning points are highlighted where, investigation report claims that the Dengue fever has to be treated immediately with greater concern.

Dengue Viral Prediction Using Artificial Neural Networks

Aburas (2010) suggested a neural network model for predicting the Dengue confirmed-cases. The authors have contributed an article to predict dengue using Artificial Neural Networks (ANNs).

Figure 1. Dengue cases in India from 2017 to 2022 (National Vector Borne Disease Control Programme (NVBDCP), 2015)

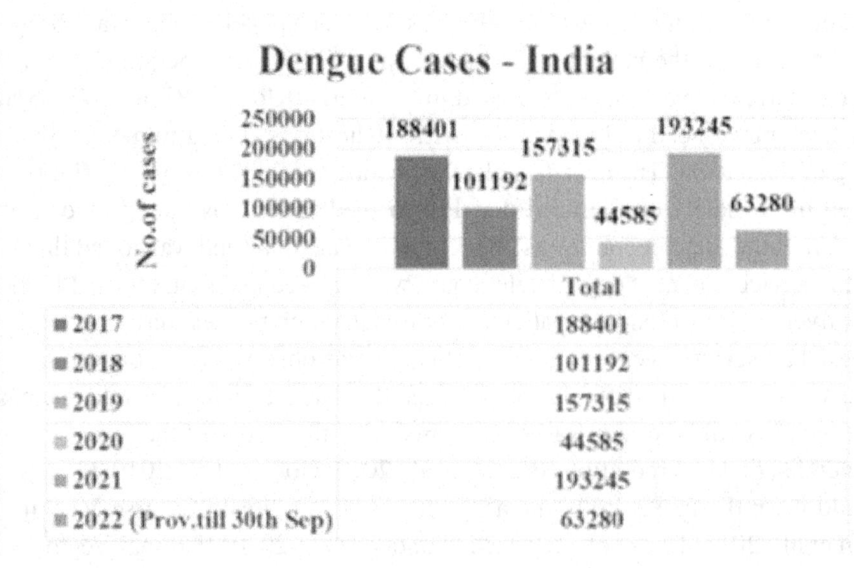

	Total
2017	188401
2018	101192
2019	157315
2020	44585
2021	193245
2022 (Prov.till 30th Sep)	63280

Figure 2. Dengue cases in South India from 2017 to 2022 (National Vector Borne Disease Control Programme (NVBDCP), 2015)

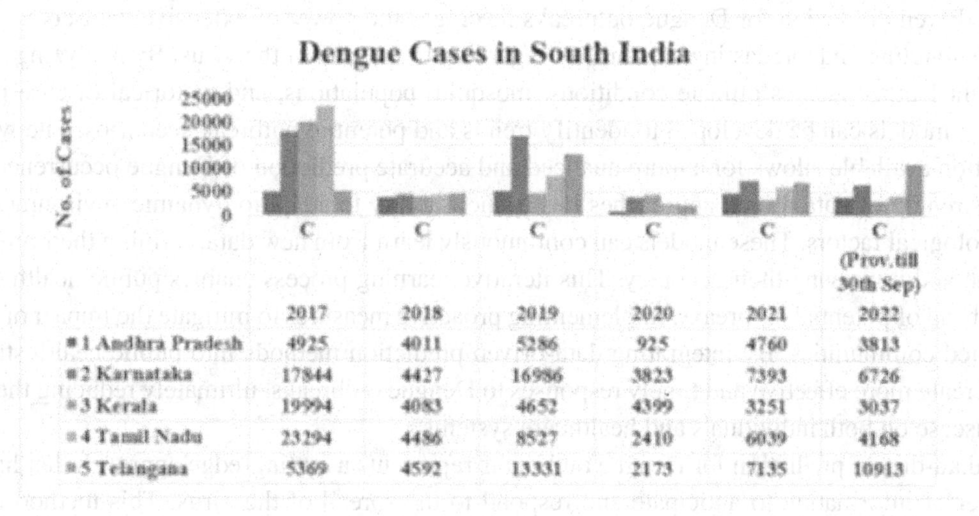

	2017	2018	2019	2020	2021	2022
1 Andhra Pradesh	4925	4011	5286	925	4760	3813
2 Karnataka	17844	4427	16986	3823	7393	6726
3 Kerala	19994	4083	4652	4399	3251	3037
4 Tamil Nadu	23294	4486	8527	2410	6039	4168
5 Telangana	5369	4592	13331	2173	7135	10913

Based on the real data provided by Singaporean National Environment Agency (NEA) the behavior of dengue cases were predicted based on the physical parameters of mean temperature, mean relative humidity and total rainfall. Around 14,209 dengue reported confirmed-cases have been analyzed by using the ANNs.

Forecasting Dengue virus outbreaks poses a considerable challenge, yet Artificial Neural Networks (ANNs) emerge as a promising tool to improve the precision and effectiveness of predictions. Taking inspiration from the neural structure of the human brain, ANNs have the capacity to analyze extensive datasets, uncovering intricate patterns that conventional methods might overlook. In the Dengue context, these networks can process diverse variables, including climate data, mosquito population dynamics, and historical infection rates, to generate forecasts. ANNs' adaptability and capacity to learn from a variety of inputs make them especially well-suited for addressing the dynamic and multifaceted nature of Dengue prediction.

The application of ANNs in predicting Dengue viral activity holds substantial potential for early detection, enabling prompt and targeted public health interventions. Through training on historical data and continuous updates with real-time information, these networks can discern subtle trends preceding outbreaks. This proactive approach empowers authorities to implement preventive measures, allocate resources efficiently, and heighten public awareness during periods of elevated Dengue transmission risk. The integration of Artificial Neural Networks into Dengue prediction thus signifies a pivotal advancement in mitigating the impact of the virus on public health.

Data Driven Prediction for Dengue Outbreak

Buczak (2012) has suggested a epidemiological prediction method for dengue outbreaks using local and remote sensing data. Althouse (2011) suggested prediction using search query. A prediction method employing Fuzzy Association Rule Mining was adopted to extract relationships between clinical, meteorological, climatic, and socio-political data from Peru. The relationships are rule based. The best set of rules are automatically chosen and forms a classifier. HIGH (outbreak) or LOW (no outbreak) is predicted based on the novel methodology

Data-driven prediction for Dengue outbreaks leverages the power of extensive datasets to enhance our understanding and forecasting capabilities regarding the spread of the virus. By analyzing a variety of relevant factors such as climate conditions, mosquito populations, and historical disease patterns, predictive models can be developed to identify trends and potential outbreak scenarios. The wealth of information available allows for a more nuanced and accurate prediction of Dengue occurrences.

The strength of data-driven approaches lies in their ability to adapt to dynamic environmental and epidemiological factors. These models can continuously learn from new data, refining their predictions over time and improving their accuracy. This iterative learning process enables public health officials to stay ahead of potential outbreaks, implementing proactive measures to mitigate the impact of Dengue on affected communities. By integrating data-driven prediction methods into public health strategies, we can create more effective and timely responses to Dengue outbreaks, ultimately reducing the burden of the disease on both individuals and healthcare systems.

The data-driven prediction for Dengue outbreaks represents a cutting-edge approach that harnesses the power of information to anticipate and respond to the spread of the virus. This method not only enhances our ability to forecast potential outbreaks but also empowers public health initiatives to take targeted actions in a timely manner, contributing to the overall control and prevention of Dengue.

Dengue Prediction Based on Linear Regression

Karim (2012), Model for dengue prediction in Dhaka city based on the Climatic factors influencing dengue cases. Githeko (2012) developed a climate-based dengue outbreak model in Dhaka, Bangladesh. Linear regression method is adopted for predicting the outbreak. To normalize data for linear regression, Log transformation is adopted. Average monthly humidity, rainfall, minimum and maximum temperature were used as independent variables and number of dengue cases reported monthly was used as dependent variable. Accuracy of the model for predicting outbreak was assessed through receiver operative characteristics (ROC) curve.

Dengue prediction employing linear regression involves utilizing statistical techniques to model the relationship between various factors and the likelihood of Dengue outbreaks. Linear regression identifies and quantifies the linear associations between independent variables, such as climate parameters, mosquito populations, and historical Dengue incidence, to predict the dependent variable, which is the likelihood of a Dengue outbreak. This method enables a systematic analysis of the impact of different factors on Dengue occurrences, providing a quantitative basis for predictions.

The strength of linear regression lies in its simplicity and interpretability, making it a valuable tool for understanding the potential risk factors contributing to Dengue outbreaks. By establishing a linear relationship between input variables and Dengue incidence, this method facilitates the identification of key drivers and allows for straightforward predictions based on new data. However, it's essential to recognize the limitations of linear regression, particularly in capturing non-linear and complex relationships within the Dengue ecosystem. Despite its simplicity, linear regression can serve as a foundational step in Dengue prediction, offering valuable insights into the factors influencing the likelihood of outbreaks.

Dengue prediction based on linear regression provides a systematic and interpretable approach to forecasting outbreaks by analyzing the relationships between various contributing factors. While it may not capture the full complexity of Dengue dynamics, it serves as a valuable initial step in understanding and predicting the incidence of Dengue, providing a basis for further exploration and refinement of predictive models.

SIRI Model for Dengue

Syafruddin (2020), Analysis and Simulation of SIRI Model for Dengue Fever Transmission was suggested in the article. The re-infection of dengue fever is predicted based on SIRI model. It is mentioned that the dengue cases always occur at a certain time in a region.

The SIRI (Susceptible-Infectious-Recovered-Immune) model offers a structured framework for understanding and predicting the dynamics of Dengue outbreaks. In this model, the population is divided into four compartments: Susceptible, Infectious, Recovered, and Immune. The Susceptible population represents individuals who are at risk of contracting Dengue, the Infectious population includes those currently infected and capable of transmitting the virus, the Recovered group consists of individuals who have overcome the infection, and the Immune group comprises those who have developed immunity. Through a system of differential equations, the SIRI model simulates the flow of individuals between these compartments, allowing for the exploration of how Dengue spreads through a population over time.

The SIRI model for Dengue is particularly valuable for understanding the impact of various intervention strategies and vaccination programs. By adjusting parameters related to the transmission rate and recovery rate, researchers can simulate different scenarios to assess the effectiveness of control measures. This modeling approach aids in optimizing public health strategies, as it provides insights into the potential outcomes of interventions and helps guide decision-making processes. The SIRI model thus serves as a valuable tool in the broader effort to mitigate the impact of Dengue on public health.

The SIRI model offers a comprehensive and dynamic framework for studying Dengue outbreaks. Its compartmental structure and differential equations provide a systematic way to explore the transmission dynamics and assess the effectiveness of interventions, contributing to the development of informed strategies for Dengue prevention and control.

Dengue Forecast Model Using Machine Learning

Guo (2017) suggested a forecast model using machine learning. A case study in China was reported where, it demonstrated the SVR model to track the dynamics of dengue outbreaks in China. The ML technique adopted in the article has provided a superior performance compared to the existing methods.

Creating a Dengue forecast model using machine learning involves leveraging advanced algorithms to analyze historical data and identify patterns that contribute to the spread of the virus. Machine learning models, such as decision trees, random forests, or support vector machines, can process vast datasets containing variables like climate conditions, mosquito populations, and past Dengue incidences. By training the model on this data, it learns to recognize complex relationships and make predictions about future Dengue outbreaks. The predictive power of these models lies in their ability to adapt and improve as more data becomes available, allowing for dynamic and accurate forecasts.

One significant advantage of machine learning in Dengue forecasting is its capability to handle diverse and non-linear relationships within the data. Unlike traditional statistical models, machine learning algorithms can capture intricate patterns and dependencies, providing a more nuanced understanding of the factors influencing Dengue outbreaks. This adaptability is particularly crucial in the context of a dynamic and evolving disease environment. The integration of machine learning into Dengue forecasting not only enhances prediction accuracy but also empowers public health officials to implement timely interventions and allocate resources effectively, ultimately mitigating the impact of Dengue on communities.

Employing machine learning for Dengue forecasting represents a cutting-edge approach that harnesses the power of advanced algorithms to predict outbreaks more accurately. By extracting insights from diverse datasets, machine learning models contribute to a more comprehensive understanding of the complex dynamics of Dengue transmission, enabling proactive measures to be taken for the prevention and control of this infectious disease.

Artificial Intelligence as a Dengue Surveillance and Prediction Tool

Sundram (2019) has proposed the Bayesian network system Using Artificial Intelligence for Dengue Surveillance and Prediction. The authors claim that the proposed system has the ability to predict the next dengue outbreaks in real-time. Data entry and other functionalities are user friendly and the system is able to report advance outbreak predictions.

Artificial Intelligence (AI) is rapidly becoming a pivotal tool in Dengue surveillance and prediction, revolutionizing our approach to understanding and managing the disease. AI applications, such as machine learning algorithms and predictive modeling, can analyze vast and diverse datasets, including climate data, demographic information, and historical Dengue cases, to identify intricate patterns and correlations. These algorithms can then generate accurate predictions and assist in early detection of potential outbreaks, allowing public health officials to implement timely interventions.

The strength of AI in Dengue surveillance lies in its ability to adapt and learn from new data continuously. As it processes real-time information, AI algorithms can dynamically adjust predictions, improving their accuracy over time. This adaptability is crucial in addressing the dynamic nature of Dengue, where factors influencing the spread of the virus can change rapidly. Additionally, AI can provide valuable insights into the risk factors and variables contributing to Dengue outbreaks, aiding in the development of targeted prevention and control strategies.

By harnessing the power of AI, Dengue surveillance and prediction can move beyond traditional methods, offering a more proactive and efficient approach to managing the disease. The integration of AI technologies not only enhances our ability to forecast Dengue outbreaks accurately but also empowers public health systems to take preemptive actions, ultimately contributing to the effective control and reduction of Dengue's impact on communities.

Predictive Models for Dengue using ANN and SVM

Jorge (2019) has proposed that SVM polynomial and ANN-MLP have proved their superiority in objective analysis in terms of Accuracy, Sensitivity and Specificity for prediction of Dengue. ANN and SVM classifiers can be incorporated to forewarn the outbreak of Dengue. The ANN-MLP shows 96% of accuracy.

Predictive models for Dengue utilizing Artificial Neural Networks (ANN) and Support Vector Machines (SVM) represent advanced applications of machine learning in disease forecasting. Artificial Neural Networks, inspired by the human brain's neural structure, can handle complex relationships within Dengue-related data, such as climate variables, mosquito populations, and historical Dengue incidences. ANN excels in recognizing intricate patterns, enabling accurate predictions by learning from diverse inputs. On the other hand, Support Vector Machines, a powerful classification algorithm, are adept at identifying patterns and creating decision boundaries in multidimensional data spaces. SVMs excel in separating different classes within the data, making them effective tools for predicting Dengue outbreaks based on various factors.

The combination of ANN and SVM in Dengue prediction allows for a comprehensive and complementary approach. ANN can capture nonlinear relationships and complex dependencies in the data, while SVM can effectively classify instances and delineate distinct patterns. Integrating these models enhances the overall accuracy and robustness of Dengue predictions. The adaptability of both ANN and SVM to changing conditions, along with their ability to learn from new data, positions them as valuable tools in creating dynamic and responsive predictive models for Dengue. This sophisticated fusion of machine learning methodologies holds great promise for improving our understanding of Dengue dynamics and enhancing our capacity to implement proactive public health measures.

Dengue Predictive Models Using ANN and Discriminant Analysis

Silitonga (2021) has compared Dengue Predictive Models using ANN and Discriminant Analysis. The model developed would aid in predicting the severity level before entering into the critical phase. ANN models based on logistic and hyperbolic tangent activation function with 70% training data yielded the highest accuracy (90.91%), sensitivity (91.11%), and specificity (95.51%). These predictive models can aid in monitoring the severity level based on laboratory test results.

Dengue predictive models, employing Artificial Neural Networks (ANN) and Discriminant Analysis, showcase a synergistic blend of advanced machine learning techniques for accurate disease forecasting. Artificial Neural Networks, inspired by the human brain's neural architecture, excel in capturing intricate patterns and relationships within Dengue-related datasets. They are particularly adept at handling nonlinear and complex dependencies, making them invaluable for discerning the multifaceted factors influencing Dengue outbreaks. In parallel, Discriminant Analysis, a statistical technique, focuses on classifying and distinguishing between different groups within the data. Applied to Dengue prediction, Discriminant Analysis helps identify key variables that contribute significantly to outbreak occurrences.

The combination of ANN and Discriminant Analysis in Dengue predictive modeling brings complementary strengths to the forefront. While ANN excels in recognizing subtle patterns and nonlinear relationships, Discriminant Analysis enhances the interpretability of the model by identifying the most discriminative features for classifying Dengue outbreaks. This integrated approach improves the overall robustness of Dengue predictions, providing a more comprehensive understanding of the complex interplay of variables. The adaptability and learning capabilities of both techniques ensure that the predictive models stay dynamic and responsive to changing conditions, ultimately contributing to more effective and timely public health interventions for Dengue prevention and control.

Deep Learning Classifiers

Employing deep learning classifiers proves instrumental in precisely categorizing the severity of Dengue cases. Specifically, convolutional neural networks (CNNs) excel in image-based analysis, while recurrent neural networks (RNNs) effectively capture temporal dependencies in patient data. These classifiers, leveraging advanced techniques like long short-term memory (LSTM) networks, play a crucial role in distinguishing between Dengue Fever (DF) and the more severe Dengue Hemorrhagic Fever (DHF) by evaluating clinical indicators such as C-reactive protein (CRP). Furthermore, LSTM networks contribute to forecasting the disease's progression, facilitating timely intervention strategies.

In the envisioned robotic fogger system, the success of deployment relies on interprofessional strategies. Collaboration among healthcare professionals, data scientists, and robotics experts proves indispensable for the effective implementation of this technology. Healthcare providers bring crucial domain expertise in comprehending Dengue epidemiology, while data scientists contribute insights into refining predictive models and classifiers. Robotics experts are vital in ensuring the Smart Fogger Robot's efficient operation and navigation. Consistent communication and collaboration among these professionals are paramount to customizing the technology to the specific requirements of Dengue-prone areas.

Despite the immense potential of integrating AI and a Smart Fogger Robot, certain challenges and limitations necessitate careful consideration. Concerns related to data privacy, model interpretability, and the robot's adaptability to changing environmental conditions require focused attention. Additionally, successful implementation requires addressing potential resistance or skepticism within communities and ensuring equitable access to healthcare resources. Tackling these challenges is essential for the sustainable and ethical deployment of advanced technologies in Dengue prevention and control.

Smart Fogger Robot During Dengue Outbreak

Once the prediction is performed using Machine learning or Deep learning based on the varied climatic conditions and the available data, a Mosquito fogger robot can be designed to automatically spray the disinfectant. The engagement of the public health workers need not be physically required thereby ensuring no health hazards among them. The fogger Robot can alarm the surrounding before spraying the disinfectant and based on a threshold level of time, it can move into various places of a home, mall, industry etc. and end spraying. This could help the health workers to stay safe.

The deployment of a Smart Fogger Robot during a Dengue outbreak presents a compelling and forward-thinking approach to disease control. Dengue, transmitted by mosquitoes, poses a significant public health threat, especially in regions susceptible to outbreaks. The Smart Fogger Robot, equipped with intelligent technologies, can play a crucial role in mitigating the spread of the virus. By autonomously navigating through affected areas and dispersing targeted insecticides, the robot can efficiently reduce mosquito populations, interrupting the transmission cycle and minimizing the risk of Dengue infection.

Implementing a Smart Fogger Robot not only enhances the effectiveness of vector control but also addresses challenges faced by human workers in accessing high-risk environments. The robot's precision in delivering interventions can lead to more targeted and resource-efficient strategies, optimizing the impact of disease control efforts. Furthermore, the use of such technology aligns with the broader trend of leveraging automation and artificial intelligence to augment public health initiatives, offering a proactive and innovative solution to combat Dengue outbreaks. Overall, the incorporation of a Smart Fogger Robot represents a promising and scalable tool in the ongoing battle against mosquito-borne diseases.

Intelligent Robotic Fogger System

A Smart Fogger Robot for Dengue is a specialized robotic device designed to combat Dengue fever by efficiently and effectively applying fogging treatments to control and reduce mosquito populations. Dengue is a mosquito-borne viral disease, and controlling mosquito populations is crucial in preventing its spread. Here are key features and functionalities of a Smart Fogger Robot for Dengue:

Autonomous Navigation: The robot is equipped with autonomous navigation capabilities, using sensors, cameras, and mapping technology to move through designated areas, such as neighborhoods, parks, or urban areas.

Mosquito Detection: Some models may incorporate mosquito detection technology, including sensors or cameras that can identify mosquito breeding sites or mosquito activity in real-time.

Fogging Equipment: The robot is equipped with a fogging system that disperses insecticides or larvicides in the form of fine droplets. The fogging equipment should be adjustable to control the droplet size and distribution for maximum effectiveness.

Precision Targeting: The robot uses precision targeting to ensure that the fogging treatment is applied only to areas where mosquito breeding or activity is detected. This reduces the environmental impact and minimizes the exposure of non-target organisms.

Real-Time Monitoring: The robot may have sensors to monitor environmental conditions such as temperature, humidity, and wind speed, which can affect the effectiveness of the fogging treatment. Real-time data analysis allows for adjustments to the fogging process.

Remote Operation: Operators can control the robot remotely, monitoring its progress and making adjustments as needed. Remote operation allows for flexibility and adaptability in response to changing conditions.

Integrated Safety Features: Safety features are crucial to prevent accidents or unintended exposure. This includes emergency stop buttons, obstacle detection sensors, and safety protocols for human interaction.

Data Collection and Reporting: The robot collects data on fogging operations, mosquito activity, and environmental conditions. This data can be used for analysis and reporting to assess the effectiveness of mosquito control efforts.

Sustainability: To promote environmental sustainability, the robot may use environmentally friendly insecticides or larvicides and minimize chemical usage through precision targeting.

Communication: The robot can communicate with local authorities or a central control system to provide real-time updates on its operations and any issues encountered during fogging.

Community Engagement: In some cases, the robot may be equipped with communication features to engage with the local community, providing information about Dengue prevention and mosquito control efforts.

A Smart Fogger Robot for Dengue can significantly enhance mosquito control efforts in areas affected by Dengue outbreaks. By automating and optimizing fogging operations, it can help reduce the spread of the disease, protect public health, and contribute to the overall well-being of communities at risk of Dengue fever.

The illustrated Intelligent Robotic Fogger System depicted in Fig. 3 offers a viable solution for enhancing public safety. Embracing this mechanism can effectively mitigate potential risks and ensure the well-being of the public.

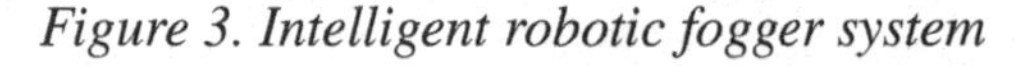

Figure 3. Intelligent robotic fogger system

CONCLUSION

Leveraging Artificial Intelligence in Dengue prediction models proves instrumental in effectively forecasting the occurrence of Dengue fever in regions prone to the disease. By employing sophisticated classifiers, such as those distinguishing between DENV-1, DENV-2, DENV-3, and DENV-4, these models can provide a targeted approach to identify specific strains and tailor appropriate treatments accordingly. Furthermore, the severity of Dengue cases, whether classified as DF or DHF, can be accurately assessed by incorporating classifiers based on factors like CRP (C-reactive protein). Implementing a computer-aided diagnosis approach, driven by AI, offers a proactive strategy to predict Dengue outbreaks, enabling timely warnings and preemptive measures in areas susceptible to the virus.

To enhance control and prevention efforts, a Smart Fogger Robot emerges as a valuable tool deployed at regular intervals in predicted Dengue-prone areas. This autonomous device, guided by AI-driven insights, aids in curtailing the spread of Dengue by precisely dispersing targeted interventions. The integration of AI and the deployment of a smart Fogger Robot represent a comprehensive and forward-thinking approach to Dengue management, utilizing technology to predict, classify, and proactively address the complexities associated with Dengue outbreaks.

Conflict of Interest

The authors declare that the research was conducted in the absence of any commercial or financial relationships that could be construed as a potential conflict of interest.

Author Contributions

DRJD, DJJ and JDP devised the work. DRJD provided the main conceptual ideas, DJJ and JDP verified the proof outline, and worked out almost all of the technical details. DJJ and JDP worked on the manuscript. All authors contributed to the article and approved the submitted version.

ACKNOWLEDGMENT

Authors express their sincere thanks to Karunya Institute of Technology and Sciences for the research environment and infrastructure that has been extended.

REFERENCES

Brasier, A. R., Ju, H., Garcia, J., Spratt, H. M., Victor, S. S., Forshey, B. M., ... Rocha, C. (2012). A three-component biomarker panel for prediction of dengue hemorrhagic fever. *The American Journal of Tropical Medicine and Hygiene*, *86*(2), 341–348. doi:10.4269/ajtmh.2012.11-0469 PMID:22302872

Chen, C. C., Lee, I.-K., Liu, J.-W., Huang, S.-Y., & Wang, L. (2015). Utility of C-Reactive Protein Levels for Early Prediction of Dengue Severity in Adults. *BioMed Research International*, *2015*, 1–6. doi:10.1155/2015/936062 PMID:26247033

Gomes, A. L. V., Wee, L. J., Khan, A. M., Gil, L. H., Marques, E. T. Jr, Calzavara-Silva, C. E., & Tan, T. W. (2010). Classification of dengue fever patients based on gene expression data using support vector machines. *PLoS One*, *5*(6), 11267. doi:10.1371/journal.pone.0011267 PMID:20585645

Guzman, M. G., Halstead, S. B., Artsob, H., Buchy, P., Farrar, J., Gubler, D. J., & Nathan, M. B. (2010). Dengue: A continuing global threat. *Nature Reviews. Microbiology*, *8*(S12), S7–S16. doi:10.1038/nrmicro2460 PMID:21079655

Lee, V. J., Chow, A., Zheng, X., Carrasco, L. R., Cook, A. R., Lye, D. C., Ng, L.-C., & Leo, Y. S. (2012). Simple clinical and laboratory predictors of Chikungunya versus dengue infections in adults. *PLoS Neglected Tropical Diseases*, *6*(9), 1786. doi:10.1371/journal.pntd.0001786 PMID:23029573

Mushtaque, R. S., Ahmad, S. M., Mushtaque, R., & Baloch, S. (2020). Case Report, A Curious Case of Dengue Fever: A Case Report of Unorthodox Manifestations. *Case Reports in Medicine*, *2020*, 1–4. doi:10.1155/2020/1701082 PMID:32774384

National Vector Borne Disease Control Programme (NVBDCP). (2015). Retrieved from https://nvbdcp.gov.in/index4.php?lang=1&level=0&linkid=431&lid=3715

Potts, J. A., Gibbons, R. V., Rothman, A. L., Srikiatkhachorn, A., Thomas, S. J., Supradish, P. O., Lemon, S. C., Libraty, D. H., Green, S., & Kalayanarooj, S. (2010). Prediction of dengue disease severity among pediatric Thai patients using early clinical laboratory indicators. *PLoS Neglected Tropical Diseases*, *4*(8), 769. doi:10.1371/journal.pntd.0000769 PMID:20689812

Ranjit, S., & Kissoon, N. (2011). Dengue hemorrhagic fever and shock syndromes. *Pediatric Critical Care Medicine*, *12*(1), 90–100. doi:10.1097/PCC.0b013e3181e911a7 PMID:20639791

Rao, V. S. H., & Kumar, M. N. (2012). A new intelligence-based approach for computer-aided diagnosis of dengue fever. *IEEE Transactions on Information Technology in Biomedicine*, *16*(1), 112–118. doi:10.1109/TITB.2011.2171978 PMID:22010159

Side, S., Pratama, M. I., Badwi, N., & Sanusi, W. (2020). Analysis and Simulation of SIRI Model for Dengue Fever Transmission. *Indian Journal of Science and Technology*, *13*(3), 340–351. doi:10.17485/ijst/2020/v13i03/147852

Compilation of References

. Ma, Y., Wang, Z., Yang, H., & Yang, L. (2020). Artificial intelligence applications in the development of autonomous vehicles: A survey. *IEEE/CAA Journal of AutomaticaSinica, 7*(2), 315-329.

. Milledge, J. (2020). Hypobaria: high altitude, aviation physiology, and medicine. *Cotes' lung function*, 615-637.

Aggarwal, S., Gupta, D., & Saini, S. (2019, November). A literature survey on robotics in healthcare. In *2019 4th International Conference on Information Systems and Computer Networks (ISCON)* (pp. 55-58). IEEE. 10.1109/ISCON47742.2019.9036253

Agrawal, S., Kumar, V., Anand, N., Agarwal, V. K., & Islam, A. (2016). *Development of Data Acquisition System and Data Analysis Technique for Automotive Applications. 50*(Ic), 3–6.

Aguirre, S., & Rodriguez, A. (2017). *Automation of a Business Process Using Robotic Process Automation (RPA): A Case Study*. Springer. . doi:10.1007/978-3-319-66963-2_7

Akter, S., Michael, K., Uddin, M. R., McCarthy, G., & Rahman, M. (2022). Transforming business using digital innovations: The application of AI, blockchain, cloud and data analytics. *Annals of Operations Research, 308*(1–2), 7–39. doi:10.1007/s10479-020-03620-w

Ala'a, A. M. (2023). Adoption of Roboticsand Robotics in Healthcare: A Systematic Literature Review. [IJCMIT]. *International Journal of Contemporary Management and Information Technology, 3*(6), 1–16.

Al-Ani, R., Zhou, B., Shi, Q., & Sagheer, A. (2018). A survey on secure safety applications in vanet. In *2018 IEEE 20th International Conference on High Performance Computing and Communications; IEEE 16th International Conference on Smart City; IEEE 4th International Conference on Data Science and Systems (HPCC/SmartCity/DSS)* (pp. 1485-1490). IEEE. 10.1109/HPCC/SmartCity/DSS.2018.00245

Alsamhi, S. H., & Lee, B. (2020). Blockchain-empowered multi-robot collaboration to fight COVID-19 and future pandemics. *IEEE Access : Practical Innovations, Open Solutions, 9*, 44173–44197. doi:10.1109/ACCESS.2020.3032450 PMID:34786312

AlShamsi, S., AlSuwaidi, L., & Shaalan, K. (2022). Robotics and AI in Healthcare: A Systematic Review. *Recent Innovations in Roboticsand Smart Applications*, 319-343.

Amestica, O. E., Melin, P. E., Duran-Faundez, C. R., & Lagos, G. R. (2019). An Experimental Comparison of Arduino IDE Compatible Platforms for Digital Control and Data Acquisition Applications. *IEEE CHILEAN Conference on Electrical, Electronics Engineering, Information and Communication Technologies, CHILECON 2019*, (pp. 1–6). IEEE. 10.1109/CHILECON47746.2019.8986865

Amrutkar, C., Satav, A., Sonawwanay, P. D., & Pawar, A. H. (2024). Overview of Autonomous Vehicle and Its Challenges. Techno-Societal 2022. *ICATSA, 2022*, 243–251. doi:10.1007/978-3-031-34648-4_25

Anderson, V. C., & Horn, R. C. (1970). *U.S. Patent No. 3,497,083*. Washington, DC: U.S. Patent and Trademark Office.

Antoniadi, A. M., Du, Y., Guendouz, Y., Wei, L., Mazo, C., Becker, B. A., & Mooney, C. (2021). Current challenges and future opportunities for XAI in machine learning-based clinical decision support systems: A systematic review. *Applied Sciences (Basel, Switzerland)*, *11*(11), 5088. doi:10.3390/app11115088

Antony, V. N., Li, M., Lin, S. H., Li, J., & Huang, C. M. (2024). *Social Robots for Sleep Health: A Scoping Review*. arXiv preprint arXiv:2403.04169.

Aouf, A., Boussaid, L., & Sakly, A. (2019). Same fuzzy logic controller for two-wheeled mobile robot navigation in strange environments. *Journal of Robotics*, *2019*, 2465219. doi:10.1155/2019/2465219

Arrieta, A. B., Díaz-Rodríguez, N., Del Ser, J., Bennetot, A., Tabik, S., Barbado, A., & Herrera, F. (2020). Explainable Artificial Intelligence (XAI): Concepts, taxonomies, opportunities and challenges toward responsible AI. *Information Fusion*, *58*, 82–115. doi:10.1016/j.inffus.2019.12.012

Atakishiyev, S., Salameh, M., Yao, H., & Goebel, R. (2021). Explainable artificial intelligence for autonomous driving: A comprehensive overview and field guide for future research directions. *arXiv preprint arXiv:2112.11561*.

Azizi, A., Hortamani, R., & Zabihollah, A. (2018). Sensing the material by minimally invasive surgery grasper. *International Robotics & Automation Journal*, *4*(3). doi:10.15406/iratj.2018.04.00117

Aziz, L., Salam, M. S. B. H., Sheikh, U. U., & Ayub, S. (2020). Exploring deep learning-based architecture, strategies, applications and current trends in generic object detection: A comprehensive review. *IEEE Access : Practical Innovations, Open Solutions*, *8*, 170461–170495. doi:10.1109/ACCESS.2020.3021508

Bae, I., Moon, J., & Seo, J. (2019). Toward a comfortable driving experience for a self-driving shuttle bus. *Electronics (Basel)*, *8*(9), 943. doi:10.3390/electronics8090943

Bajwa, J., Munir, U., Nori, A., & Williams, B. (2021, July). Artificial intelligence in healthcare: Transforming the practice of medicine. *Future Healthcare Journal*, *8*(2), e188–e194. doi:10.7861/fhj.2021-0095 PMID:34286183

Bakdi, A., Bounoua, W., Guichi, A., & Mekhilef, S. (2021). Real-time fault detection in PV systems under MPPT using PMU and high-frequency multi-sensor data through online PCA-KDE-based multivariate KL divergence. *International Journal of Electrical Power & Energy Systems*, *125*, 106457. doi:10.1016/j.ijepes.2020.106457

Bakshi, G., Kumar, A., & Puranik, A. N. (2021). Adoption of robotics technology in healthcare sector. In Advances in Communication, Devices and Networking [Singapore: Springer Singapore.]. *Proceedings of ICCDN*, *2020*, 405–414.

Banerjee, S. S., Andhari, S. S., Wavhale, R. D., Dhobale, K. D., Tawade, B. V., Chate, G. P., & Khandare, J. J. (2020). Self-propelling targeted magneto-nanobots for deep tumor penetration and pH-responsive intracellular drug delivery. *Scientific Reports*, *10*(1), 1–14. PMID:31913322

Bangotra, D. K., Singh, Y., & Selwal, A. (2018). Machine learning in wireless sensor networks: Challenges and opportunities. In *2018 Fifth International Conference on Parallel, Distributed and Grid Computing (PDGC)* (pp. 534-539). IEEE. 10.1109/PDGC.2018.8745845

Bao, J., Li, D., & Qiao, X. (2020). Integrated navigation for autonomous underwater vehicles in aquaculture: A review. *Information Processing in Agriculture*, *7*(1).

Baoquan, G. (2011). *Time-domain Analysis and Research on blasting vibration signals Based on Fourier I°*. Research Gate.

Bartneck, C., Lütge, C., Wagner, A., & Welsh, S. (2021). *An introduction to ethics in robotics and AI*. Springer Nature. doi:10.1007/978-3-030-51110-4

Barua, R., & Datta, S. (2022). Study of the surgical needle and biological soft tissue interaction phenomenon during insertion process for medical application: A Survey. *Proceedings of the Institution of Mechanical Engineers, Part H: Journal of Engineering in Medicine, 236*(10), 1465-1477.

Barua, R., Bhowmik, S., Dey, A., Das, S., & Datta, S. (2022, September). Analysis of Robotically Controlled Percutaneous Needle Insertion into Ex Vivo Kidney Tissue for Minimally Invasive Percutaneous Nephrolithotomy (PCNL) Surgery. *In International Conference on Emergent Converging Technologies and Biomedical Systems* (pp. 249-257). Singapore: Springer Nature Singapore.

Barua, R., Das, S., Datta, P., & Chowdhury, A. (2022). Computational FEM Application on Percutaneous Nephrolithotomy (PCNL) Minimum Invasive Surgery Through Needle Insertion Process. In P. Pain, S. Banerjee, & G. Bose (Eds.), *Advances in Computational Approaches in Biomechanics* (pp. 210-222). IGI Global. doi:10.4018/978-1-7998-9078-2.ch013

Barua, R., Das, S., Datta, S., Roy Chowdhury, A., & Datta, P. (2022). Experimental study of the robotically controlled surgical needle insertion for analysis of the minimum invasive process. In *Emergent Converging Technologies and Biomedical Systems: Select Proceedings of ETBS 2021* (pp. 473-482). Singapore: Springer Singapore. 10.1007/978-981-16-8774-7_38

Barua, R., Bhowmik, S., Dey, A., & Mondal, J. (2023). Advances of the Robotics Technology in Modern Minimally Invasive Surgery. In M. Mellal (Ed.), *Design and Control Advances in Robotics* (pp. 91–104). IGI Global.

Barua, R., Das, S., Datta, S., Datta, P., & Roy Chowdhury, A. (2021). Analysis of surgical needle insertion modeling and viscoelastic tissue material interaction for minimally invasive surgery (MIS). *Materials Today: Proceedings*, *57*, 259–264. doi:10.1016/j.matpr.2022.02.498

Barua, R., Das, S., Datta, S., Datta, P., & Roy Chowdhury, A. (2023). Study and experimental investigation of insertion force modeling and tissue deformation phenomenon during surgical needle-soft tissue interaction. *Proceedings of the Institution of Mechanical Engineers. Part C, Journal of Mechanical Engineering Science*, *237*(5), 1007–1014. doi:10.1177/09544062221126628

Barua, R., Das, S., & Mondal, J. (2023). Emerging Applications of Artificial Intelligence (AI) and Machine Learning (ML) in Modern Urology. In R. Queirós, B. Cunha, & X. Fonseca (Eds.), *Exploring the Convergence of Computer and Medical Science Through Cloud Healthcare* (pp. 117–133). IGI Global. doi:10.4018/978-1-6684-5260-8.ch006

Barua, R., Das, S., Roy Chowdhury, A., & Datta, P. (2023). Experimental and simulation investigation of surgical needle insertion into soft tissue mimic biomaterial for minimally invasive surgery (MIS). *Proceedings of the Institution of Mechanical Engineers. Part H, Journal of Engineering in Medicine*, *237*(2), 254–264. doi:10.1177/09544119221143860 PMID:36527297

Barua, R., Das, S., RoyChowdhury, A., & Datta, P. (2023). Simulation and experimental investigation of the surgical needle deflection model during the rotational and steady insertion process. *The International Journal of Artificial Organs*, *46*(1), 40–51. doi:10.1177/03913988221136154 PMID:36397288

Barua, R., Datta, P., Chowdhury, A. R., & Das, S. (2022). Computational Study of In-Vitro Ureter Urine Flow in DJ Stent. In P. Pain, S. Banerjee, & G. Bose (Eds.), *Advances in Computational Approaches in Biomechanics* (pp. 198–209). IGI Global. doi:10.4018/978-1-7998-9078-2.ch012

Barua, R., & Datta, S. (2023). Artificial Intelligence in Modern Medical Science: A Promising Practice. In S. Rajest, B. Singh, A. J. Obaid, R. Regin, & K. Chinnusamy (Eds.), *Recent Developments in Machine and Human Intelligence* (pp. 1–12). IGI Global. doi:10.4018/978-1-6684-9189-8.ch001

Barua, R., & Datta, S. (2023). Emerging Surgical Robotic Applications for Modern Minimally Invasive Surgery (MIS). In M. Habib (Ed.), *Global Perspectives on Robotics and Autonomous Systems: Development and Applications* (pp. 314–332). IGI Global. doi:10.4018/978-1-6684-7791-5.ch014

Barua, R., Datta, S., & Sarkar, A. (2023). Artificial Intelligence and Robotics-Based Minimally Invasive Surgery: Innovations and Future Perceptions. In G. Karthick & S. Karupusamy (Eds.), *Contemporary Applications of Data Fusion for Advanced Healthcare Informatics* (pp. 350–368). IGI Global. doi:10.4018/978-1-6684-8913-0.ch015

Barua, R., Giria, H., Datta, S., Roy Chowdhury, A., & Datta, P. (2020). Force modeling to develop a novel method for fabrication of hollow channels inside a gel structure. *Proceedings of the Institution of Mechanical Engineers. Part H, Journal of Engineering in Medicine*, *234*(2), 223–231. doi:10.1177/0954411919891654 PMID:31774361

Barua, R., & Mondal, J. (2023). Study of the Current Trends of CAD (Computer-Aided Detection) in Modern Medical Imaging. In L. Panigrahi, S. Biswal, A. Bhoi, A. Kalam, & P. Barsocchi (Eds.), *Machine Learning and AI Techniques in Interactive Medical Image Analysis* (pp. 35–50). IGI Global., doi:10.4018/978-1-6684-4671-3.ch002

Barua, R., Sarkar, A., & Datta, S. (2023). Modern Lab-on-Chip Biosensors Application on Infectious COVID-19 Detection. In R. Singh, R. Phanden, B. Sikarwar, & J. Davim (Eds.), *Advances in MEMS and Microfluidic Systems* (pp. 258–270). IGI Global. doi:10.4018/978-1-6684-6952-1.ch013

Belk, R. (2021). Ethical issues in service robotics and artificial intelligence. *Service Industries Journal*, *41*(13-14), 860–876. doi:10.1080/02642069.2020.1727892

Bhoi, S. K., & Khilar, P. M. (2013). A secure routing protocol for Vehicular Ad Hoc Network to provide ITS services. In *2013 International Conference on Communication and Signal Processing* (pp. 1170-1174). IEEE. 10.1109/iccsp.2013.6577240

Biswas, A. K., & Dasgupta, M. (2020). A secure hybrid routing protocol for mobile ad-hoc networks (MANETs). In *2020 11th International Conference on Computing, Communication and Networking Technologies (ICCCNT)* (pp. 1-7). IEEE.

Bogue, R. (2011). Robots in healthcare. Industrial Robot. *International Journal (Toronto, Ont.)*, *38*(3), 218–223.

Bourdon, P., Ahmed, O. B., Urruty, T., Djemal, K., & Fernandez-Maloigne, C. (2021). Explainable ai for medical imaging: Knowledge matters. *Multi-faceted Deep Learning: Models and Data.*

Brasier, A. R., Ju, H., Garcia, J., Spratt, H. M., Victor, S. S., Forshey, B. M., ... Rocha, C. (2012). A three-component biomarker panel for prediction of dengue hemorrhagic fever. *The American Journal of Tropical Medicine and Hygiene*, *86*(2), 341–348. doi:10.4269/ajtmh.2012.11-0469 PMID:22302872

Brown, B. (2016). The social life of autonomous cars. *MIT Technology Review, 50*(2).

Burgner-Kahrs, J., Rucker, D. C., & Choset, H. (2015). Continuum robots for medical applications: A survey. *IEEE Transactions on Robotics*, *31*(6), 1261–1280. doi:10.1109/TRO.2015.2489500

Butter, M., Rensma, A., Kalisingh, S., Schoone, M., Leis, M., Gelderblom, G. J., & Korhonen, I. (2008). *Robotics for healthcare.*

Cai, Y., Luan, T., Gao, H., Wang, H., Chen, L., Li, Y., Sotelo, M. A., & Li, Z. (2021). YOLOv4-5D: An effective and efficient object detector for autonomous driving. *IEEE Transactions on Instrumentation and Measurement*, *70*, 1–13. doi:10.1109/TIM.2021.3065438

Cao, K., Chen, Y., Dang, H., Gao, S., & Yan, K. (2022, July). Multi-robot Coverage System Based on Health Optimization Management Algorithm. In *Proceedings of 2021 5th Chinese Conference on Swarm Intelligence and Cooperative Control* (pp. 902-911). Singapore: Springer Nature Singapore.

Cao, C. G., & Rogers, G. (2007). *Robotics in health care: HF issues in surgery. Handbook of Human Factors and Ergonomics in Health Care and Patient Safety*. Lawrence Earlbaum & Associates.

Cao, K., Chen, Y., Gao, S., Zhang, H., & Dang, H. (2022). Multi-Robot Formation Control Based on CVT Algorithm and Health Optimization Management. *Applied Sciences (Basel, Switzerland)*, *12*(2), 755. doi:10.3390/app12020755

Cappiello, A. (2018). Digital Disruption and InsurTech Start-ups: Risks and Challenges. In Technology and the Insurance Industry (pp. 29–50). Springer International Publishing. doi:10.1007/978-3-319-74712-5_3

Carranza-García, M., Torres-Mateo, J., Lara-Benítez, P., & García-Gutiérrez, J. (2021). On the performance of one-stage and two-stage object detectors in autonomous vehicles using camera data. *Remote Sensing (Basel)*, *13*(1), 89. doi:10.3390/rs13010089

Cavalcanti, A., Shirinzadeh, B., Freitas, R. A. Jr, & Hogg, T. (2008). Nanorobot architecture for medical target identification. *Nanomedicine; Nanotechnology, Biology, and Medicine*, *4*(2), 134–152. PMID:18455965

Chaurasia, A., Parashar, B., & Kautish, S. (2024). Artificial Intelligence and Automation for Industry 4.0. In S. Kautish, P. Chatterjee, D. Pamucar, N. Pradeep, & D. Singh (Eds.), *Computational Intelligence for Modern Business Systems. Disruptive Technologies and Digital Transformations for Society 5.0*. doi:10.1007/978-981-99-5354-7_18

Chellaswamy, C., Famitha, H., Anusuya, T., & Amirthavarshini, S. B. (2018). *IoT Based Humps and Pothole Detection on Roads and Information Sharing*. Research Gate.

Chen, E. N. (2009). Remote analysis of mechanical vibration based on client/server architecture. *ICEMI 2009 - Proceedings of 9th International Conference on Electronic Measurement and Instruments*, (pp. 10–13). IEEE. 10.1109/ICEMI.2009.5274399

Chen, S., & Xu, H. (2023). A pneumatic–hydraulic hybrid actuator for underwater soft robot swimming and crawling. *Sensors and Actuators A: Physical, 356*. doi:10.1016/j.sna.2023.114284

Chen, Z., Liu, Q., & Lian, C. (2019, June). Pointlanenet: Efficient end-to-end cnns for accurate real-time lane detection. In 2019 IEEE intelligent vehicles symposium (IV) (pp. 2563-2568). IEEE.

Chen, C. C., Lee, I.-K., Liu, J.-W., Huang, S.-Y., & Wang, L. (2015). Utility of C-Reactive Protein Levels for Early Prediction of Dengue Severity in Adults. *BioMed Research International*, *2015*, 1–6. doi:10.1155/2015/936062 PMID:26247033

Chen, K., Zhang, J., Beeraka, N. M., Sinelnikov, M. Y., Zhang, X., Cao, Y., & Lu, P. (2022). Robot-Assisted Minimally Invasive Breast Surgery: Recent Evidence with Comparative Clinical Outcomes. *Journal of Clinical Medicine*, *11*(7), 1827. doi:10.3390/jcm11071827 PMID:35407434

Chirikjian, G. S. (1994). Hyper-redundant manipulator dynamics: A continuum approximation. *Advanced Robotics*, *9*(3), 217–243. doi:10.1163/156855395X00175

Chiu, Y. C., Tsai, C. Y., Ruan, M. D., Shen, G. Y., & Lee, T. T. (2020, August). Mobilenet-SSDv2: An improved object detection model for embedded systems. In *2020 International conference on system science and engineering (ICSSE)* (pp. 1-5). IEEE. 10.1109/ICSSE50014.2020.9219319

Chopra, H., Baig, A. A., Cavalu, S., Singh, I., & Emran, T. B. (2022, August 17). Robotics in surgery: Current trends. *Annals of Medicine and Surgery (London)*, *81*, 104375. doi:10.1016/j.amsu.2022.104375 PMID:36051814

Cresswell, K., Cunningham-Burley, S., & Sheikh, A. (2018). Health care robotics: Qualitative exploration of key challenges and future directions. *Journal of Medical Internet Research*, *20*(7), e10410. doi:10.2196/10410 PMID:29973336

Cui, Q., Wang, Y., Chen, K.-C., Ni, W., Lin, I.-C., Tao, X., & Zhang, P. (2019). Big data analytics and network calculus enabling intelligent management of autonomous vehicles in a smart city. *IEEE Internet of Things Journal*, *6*(2), 2021–2034. doi:10.1109/JIOT.2018.2872442

Cuperman, D., & Verner, I. M. (2019). Fostering analogical reasoning through creating robotic models of biological systems. *Journal of Science Education and Technology*, *28*(2), 90–103. doi:10.1007/s10956-018-9750-4

Da Silva Assis, L., da Silva Soares, A., Coelho, C. J., & Van Baalen, J. (2016). An evolutionary algorithm for autonomous robot navigation. *Procedia Computer Science*, *80*, 2261–2265. doi:10.1016/j.procs.2016.05.404

Daniel, T. (2023). Autonomous AI Systems in Conflict: Emergent Behavior and Its Impact on Predictability and Reliability. *Journal of Military Ethics*, *22*(1), 2–17. doi:10.1080/15027570.2023.2213985

Das, G. P., McGinnity, T. M., Coleman, S. A., & Behera, L. (2015). A distributed task allocation algorithm for a multi-robot system in healthcare facilities. *Journal of Intelligent & Robotic Systems*, *80*(1), 33–58. doi:10.1007/s10846-014-0154-2

Das, S., Datta, S., Barman, A., & Barua, R. (2023). Smart Biodegradable and Bio-Based Polymeric Biomaterials for Biomedical Applications. In A. Kumar, P. Kumar, A. Srivastava, & V. Goyat (Eds.), *Modeling, Characterization, and Processing of Smart Materials* (pp. 56–82). IGI Global. doi:10.4018/978-1-6684-9224-6.ch003

Datta Burton, S. (2022). Not Anytime Soon: The Clinical Translation of Nanorobots and Its Biocompatibility Constraints. In *Interactive Robotics: Legal, Ethical, Social and Economic Aspects* (pp. 123–136). Springer. doi:10.1007/978-3-031-04305-5_35

Datta, S., & Barua, R. (2023). Advanced Materials for Surgical Tools and Biomedical Implants. *Advanced Materials and Manufacturing Techniques for Biomedical Applications*, 25-35.

Datta, S., Barua, R., & Prasad, A. (2023). Additive Manufacturing for the Development of Artificial Organs. *Advanced Materials and Manufacturing Techniques for Biomedical Applications*, 411-427.

Datta, S., Das, S., & Barua, R. (2023). Self-Sustained Nanobiomaterials: Innovative Materials for Biomedical Applications. *Advanced Materials and Manufacturing Techniques for Biomedical Applications*, 303-323.

Datta, S., & Barua, R. (2023). Fluorescent Nanomaterials and Its Application in Biomedical Engineering. In A. Rakha, A. Munawar, V. Khanna, & S. Bansal (Eds.), *Modeling and Simulation of Functional Nanomaterials for Forensic Investigation* (pp. 164–186). IGI Global. doi:10.4018/978-1-6684-8325-1.ch009

Datta, S., Barua, R., & Das, J. (2020). *A review on electro-rheological fluid (er) and its various technological applications*. Extremophilic Microbes and Metabolites-Diversity, Bioprospecting and Biotechnological Applications.

Datta, S., Barua, R., & Das, S. (2023). Role and Challenges of Bioprinting in Bone Tissue Engineering. In R. Ranjith & J. Davim (Eds.), *Handbook of Research on Advanced Functional Materials for Orthopedic Applications* (pp. 205–218). IGI Global. doi:10.4018/978-1-6684-7412-9.ch012

Datta, S., Barua, R., Sarkar, R., Barui, A., Chowdhury, A. R., & Datta, P. (2018, September). Design and development of alginate: Poly-l-lysine scaffolds by 3D bio printing and studying their mechanical, structural and cell viability properties. [). IOP Publishing.]. *IOP Conference Series. Materials Science and Engineering*, *402*, 012113. doi:10.1088/1757-899X/402/1/012113

Dawood, A. B., Fras, J., Aljaber, F., Mintz, Y., Arezzo, A., Godaba, H., & Althoefer, K. (2021). Fusing dexterity and perception for soft robot-assisted minimally invasive surgery: What we learnt from STIFF-FLOP. *Applied Sciences (Basel, Switzerland)*, *11*(14), 6586. doi:10.3390/app11146586

Deng, W., & Wu, R. (2019). Real-time driver-drowsiness detection system using facial features. *IEEE Access : Practical Innovations, Open Solutions*, *7*, 118727–118738. doi:10.1109/ACCESS.2019.2936663

Dixit, P., & Silakari, S. (2021). Deep learning algorithms for cybersecurity applications: A technological and status review. *Computer Science Review*, *39*, 100317. doi:10.1016/j.cosrev.2020.100317

Dong, J., Chen, S., Miralinaghi, M., Chen, T., Li, P., & Labi, S. (2023). Why did the AI make that decision? Towards an explainable artificial intelligence (XAI) for autonomous driving systems. *Transportation Research Part C, Emerging Technologies*, *156*, 104358. doi:10.1016/j.trc.2023.104358

Dongpo, L., & Xuhui, M. (2011). *Virtual Proving Ground.*

Duan, W., Akinyemi, T., Du, W., Ma, J., Chen, X., Wang, F., Omisore, O., Luo, J., Wang, H., & Wang, L. (2023). Technical and Clinical Progress on Robot-Assisted Endovascular Interventions: A Review. *Micromachines*, *14*(1), 197. doi:10.3390/mi14010197 PMID:36677258

Duggal, A. S., Malik, P. K., Gehlot, A., Singh, R., Gaba, G. S., Masud, M., & Al-Amri, J. F. (2022). A sequential roadmap to Industry 6.0: Exploring future manufacturing trends. *IET Communications*, *16*(5), 521–531. doi:10.1049/cmu2.12284

Dupont, P. E., Simaan, N., Choset, H., & Rucker, C. (2022). Continuum Robots for Medical Interventions. *Proceedings of the IEEE.* Institute of Electrical and Electronics Engineers.

Du, Y., Liu, C., Wu, D., & Li, S. (2016). Application of Vehicle Mounted Accelerometers to Measure Pavement Roughness. *International Journal of Distributed Sensor Networks*, *2016*(6), 8413146. doi:10.1155/2016/8413146

Eckert, C., Neunsinger, C., & Osterrieder, K. (2022). (Repeat J)Managing customer satisfaction: Digital applications for insurance companies. *The Geneva Papers on Risk and Insurance. Issues and Practice*, *47*(3), 569–602. doi:10.1057/s41288-021-00257-z

Elendu, C., Amaechi, D. C., Elendu, T. C., Jingwa, K. A., Okoye, O. K., Okah, M. J., & Alimi, H. A. (2023). Ethical implications of AI and robotics in healthcare. *Revista de Medicina (São Paulo)*, *102*(50), e36671. PMID:38115340

Eling, M., Nuessle, D., & Staubli, J. (2022). The impact of artificial intelligence along the insurance value chain and on the insurability of risks. *The Geneva Papers on Risk and Insurance. Issues and Practice*, *47*(2), 205–241. doi:10.1057/s41288-020-00201-7

Etukudoh, N. S., Esame, N. V., Obeta, U. M., Ejinaka, O. R., & Khang, A. (2024). Automations and Robotics Improves Quality Healthcare in the Era of Digital Medical Laboratory. In Computer Vision and AI-Integrated IoT Technologies in the Medical Ecosystem (pp. 419-434). CRC Press. doi:10.1201/9781003429609-24

Faniadis, E., & Amanatiadis, A. (2020, November). Deep learning inference at the edge for mobile and aerial robotics. In *2020 IEEE International Symposium on Safety, Security, and Rescue Robotics (SSRR)* (pp. 334-340). IEEE. 10.1109/SSRR50563.2020.9292575

Fischell, E., Stanton, T. K., Kukulya, A., & Andone, C. (1806). Lavery, Monitoring of macroalgae (kelp) farms with autonomous underwater vehicle-based split-beam sonar, September 2018. *The Journal of the Acoustical Society of America*, *144*(3).

Flückiger, I., & Duygun, M. (2022). (LR 111)New technologies and data in insurance. In Geneva Papers on Risk and Insurance: Issues and Practice, 47(3). Palgrave Macmillan. doi:10.1057/s41288-022-00274-6

Foroutannia, A., Shoryabi, M., Anaraki, A. A., & Rowhanimanesh, A. (2021, March). SIN: A Programmable Platform for Swarm Robotics. In *2021 26th International Computer Conference, Computer Society of Iran (CSICC)* (pp. 1-5). IEEE. 10.1109/CSICC52343.2021.9420596

Fosch-Villaronga, E., & Drukarch, H. (2021). On Healthcare Robots: Concepts, definitions, and considerations for healthcare robot governance. arXiv preprint arXiv:2106.03468.

Fujiyoshi, H., Hirakawa, T., & Yamashita, T. (2019). Deep learning-based image recognition for autonomous driving. *IATSS Research*, *43*(4), 244–252. doi:10.1016/j.iatssr.2019.11.008

Gadkari, M. Y., & Sambre, N. B. (2012). VANET: Routing protocols, security issues and simulation tools. *IOSR Journal of Computer Engineering*, *3*(3), 28–38. doi:10.9790/0661-0332838

Garfjeld Roberts, P., Glasbey, J. C., Abram, S., Osei-Bordom, D., Bach, S. P., & Beard, D. J. (2020). Research quality and transparency, outcome measurement and evidence for safety and effectiveness in robot-assisted surgery: Systematic review. *BJS Open*, *4*(6), 1084–1099. doi:10.1002/bjs5.50352 PMID:33052029

Gaur, L., Afaq, A., Singh, G., & Dwivedi, Y. K. (2021). Role of artificial intelligence and robotics to foster the touchless travel during a pandemic: A review and research agenda. *International Journal of Contemporary Hospitality Management*, *33*(11), 4079–4098. doi:10.1108/IJCHM-11-2020-1246

Gellweiler, C., & Krishnamurthi, L. (2020). Editorial: How digital innovators achieve customer value. In Journal of Theoretical and Applied Electronic Commerce Research, 15(1). Universidad de Talca. doi:10.4067/S0718-18762020000100101

Ghosh, U., & Datta, R. (2014). SDRP: Secure and dynamic routing protocol for mobile ad-hoc networks. *IET Networks*, *3*(3), 235–243. doi:10.1049/iet-net.2013.0056

Gomes, A. L. V., Wee, L. J., Khan, A. M., Gil, L. H., Marques, E. T. Jr, Calzavara-Silva, C. E., & Tan, T. W. (2010). Classification of dengue fever patients based on gene expression data using support vector machines. *PLoS One*, *5*(6), 11267. doi:10.1371/journal.pone.0011267 PMID:20585645

González, A., Olazagoitia, J. L., & Vinolas, J. (2018). A low-cost data acquisition system for automobile dynamics applications. *Sensors (Basel)*, *18*(2), 366. doi:10.3390/s18020366 PMID:29382039

Gonzalez, L. C., Moreno, R., Escalante, H. J., Martinez, F., & Carlos, M. R. (2017). Learning Roadway Surface Disruption Patterns Using the Bag of Words Representation. *IEEE Transactions on Intelligent Transportation Systems*, *18*(11), 2916–2928. doi:10.1109/TITS.2017.2662483

Gonzalo de Diego, B., González Aguña, A., Fernández Batalla, M., Herrero Jaén, S., Sierra Ortega, A., Barchino Plata, R., & Santamaría García, J. M. (2024, March). Competencies in the Robotics of Care for Nursing Robotics: A Scoping Review. In Healthcare (Vol. 12, No. 6, p. 617). MDPI. doi:10.3390/healthcare12060617

Gorchakova, E. R. (2019). *The impact of digitalization on the health insurance system*. doi:10.18411/lj-03-2019-48

Gravagne, I., & Walker, I. D. (2000). Kinematics for constrained continuum robots using wavelet decomposition. *Robotics*, *2000*, 292–298.

Greener, J. G., Kandathil, S. M., Moffat, L., & Jones, D. T. (2022). A guide to machine learning for biologists. *Nature Reviews. Molecular Cell Biology*, *23*(1), 40–55. doi:10.1038/s41580-021-00407-0 PMID:34518686

Guntur, S. R., Gorrepati, R. R., & Dirisala, V. R. (2019). Robotics in healthcare: an internet of medical robotic things (IoMRT) perspective. In *Machine learning in bio-signal analysis and diagnostic imaging* (pp. 293–318). Academic Press. doi:10.1016/B978-0-12-816086-2.00012-6

Guo, N., Lenzo, B., Zhang, X., Zou, Y., Zhai, R., & Zhang, T. (2020). A real-time nonlinear model predictive controller for yaw motion optimization of distributed drive electric vehicles. *IEEE Transactions on Vehicular Technology*, *69*(5), 4935–4946. doi:10.1109/TVT.2020.2980169

Gupta, A., Anpalagan, A., Guan, L., & Khwaja, A. S. (2021). Deep learning for object detection and scene perception in self-driving cars: Survey, challenges, and open issues. *Array (New York, N.Y.)*, *10*, 100057. doi:10.1016/j.array.2021.100057

Gupta, S., Upadhyay, D., & Dubey, A. K. (2019). Self-Driving Car Using Artificial Intelligence. In M. Kumar, R. Pandey, & V. Kumar (Eds.), *Advances in Interdisciplinary Engineering. Lecture Notes in Mechanical Engineering*. Springer. doi:10.1007/978-981-13-6577-5_49

Gurjeet, S. (2022). *Vijay, Kumar, Banga., Thaweesak, Yingthawornsuk*. Artificial Intelligence and Industrial Robot., doi:10.1109/SITIS57111.2022.00098

Guzman, M. G., Halstead, S. B., Artsob, H., Buchy, P., Farrar, J., Gubler, D. J., & Nathan, M. B. (2010). Dengue: A continuing global threat. *Nature Reviews. Microbiology*, *8*(S12), S7–S16. doi:10.1038/nrmicro2460 PMID:21079655

Guzmán-Ortiz, C. V., Navarro-Acosta, N. G., Florez-Garcia, W., & Vicente-Ramos, W. (2020). (LR 122)Impact of digital transformation on the individual job performance of insurance companies in peru. *International Journal of Data and Network Science*, *4*(4), 337–346. doi:10.5267/j.ijdns.2020.9.005

Haken, H., & Levi, P. (2012). *Synergetic agents: From multi-robot systems to molecular robotics*. John Wiley & Sons. doi:10.1002/9783527659524

Hamdi, M. M., Audah, L., Rashid, S. A., & Alani, S. (2021). VANET-based traffic monitoring and incident detection system: A review. *International Journal of Electrical & Computer Engineering (2088-8708)*, *11*(4).

Hamdi, M. M., Audah, L., Rashid, S. A., Mohammed, A. H., Alani, S., & Mustafa, A. S. (2020). A review of applications, characteristics and challenges in vehicular ad hoc networks (VANETs). In 2020 international congress on human-computer interaction, optimization and robotic applications (HORA) (pp. 1-7). IEEE.

Hannan, M. W., & Walker, I. D. (2003). Kinematics and the implementation of an elephant's trunk manipulator and other continuum style robots. *Journal of Robotic Systems*, *20*(2), 45–63. doi:10.1002/rob.10070 PMID:14983840

Harit, P. (2021). (LR 83)The Rise of Insurtech: The Ups and Downs of New Trend. SSRN *Electronic Journal*. doi:10.2139/ssrn.3799576

Hodge, V. J., Hawkins, R., & Alexander, R. (2021). Deep reinforcement learning for drone navigation using sensor data. *Neural Computing & Applications*, *33*(6), 2015–2033. doi:10.1007/s00521-020-05097-x

Holland, J., Kingston, L., McCarthy, C., Armstrong, E., O'Dwyer, P., Merz, F., & McConnell, M. (2021). Service robots in the healthcare sector. *Robotics (Basel, Switzerland)*, *10*(1), 47. doi:10.3390/robotics10010047

Hong, Y., Pan, H., Sun, W., & Jia, Y. (2021). Deep dual-resolution networks for real-time and accurate semantic segmentation of road scenes. *arXiv preprint arXiv:2101.06085.*

Hu, F., Chen, B., Shi, D., Zhang, X., & Pan, H. Z. (2020). Secure Routing Protocol in Wireless Ad Hoc Networks via Deep Learning. In 2020 IEEE Wireless Communications and Networking Conference (WCNC) (pp. 1-6). IEEE. doi:10.1109/WCNC45663.2020.9120545

Huang, W. (2021). Modeling soft swimming robots using discrete elastic rod method. *Proc. Bioinspired Sensing, Actuation, Control Underwater Soft Robotic Syst.*

Hu, F., He, Q., & Kong, F. (2011)... *Time-Frequency Vibration Representation for Steel Mill Condition Monitoring.*, *1*, 1–5.

Hughes, J., Culha, U., Giardina, F., Guenther, F., Rosendo, A., & Iida, F. (2016). Soft manipulators and grippers: A review. *Frontiers in Robotics and AI*, *3*, 1–12. doi:10.3389/frobt.2016.00069

Humaidi, A. J., Ibraheem, I. K., Azar, A. T., & Sadiq, M. E. (2020). A new adaptive synergetic control design for single link robot arm actuated by pneumatic muscles. *Entropy (Basel, Switzerland)*, *22*(7), 723. doi:10.3390/e22070723 PMID:33286496

Hussain, F., Hussain, R., & Hossain, E. (2021). Explainable artificial intelligence (XAI): An engineering perspective. *arXiv preprint arXiv:2101.03613.*

Ilic, S., Katupitiya, J., & Tordon, M. (n.d.). *In-vehicle data logging system for fatigue analysis of drive shaft.*

Inam, M., Li, Z., Ali, A., & Zahoor, A. (2019). A novel protocol for vehicle cluster formation and vehicle head selection in vehicular ad-hoc networks. *International Journal of Electronics and Information Engineering*, *10*(2), 103–119.

Jabbar, R., Shinoy, M., Kharbeche, M., Al-Khalifa, K., Krichen, M., & Barkaoui, K. (2020, February). Driver drowsiness detection model using convolutional neural networks techniques for android application. In *2020 IEEE International Conference on Informatics, IoT, and Enabling Technologies (ICIoT)* (pp. 237-242). IEEE. 10.1109/ICIoT48696.2020.9089484

Jamaludin, W. A. W. (2021). *Photovoltaic-Thermoelectric Generator Monitoring System using Arduino Based Data Acquisition system Technique.*

Javaid, M., Haleem, A., Singh, R. P., Rab, S., Suman, R., & Kumar, L. (2022). Utilization of Robotics for Healthcare: A Scoping Review. *Journal of Industrial Integration and Management, 2250015.*

Jiang, L., & Feng, L. (Eds.). (2010). *Bioinspired Intelligent Nanostructured Interfacial Materials*. Springer. doi:10.1142/7380

Jonathan, A. (2023). *Development of an AI-Robotics 3D Printed Circle of Command for Enhancing Accessibility and Mobility in Individuals with Mobility Issues*. IEEE. doi:10.5121/csit.2023.130910

Jones, B. A., & Walker, I. D. (2006). Kinematics for multisection continuum robots. *IEEE Transactions on Robotics*, *22*(1), 43–55. doi:10.1109/TRO.2005.861458

Joseph, A., Christian, B., Abiodun, A. A., & Oyawale, F. (2018). A review on humanoid robotics in healthcare. In *MATEC Web of Conferences* (Vol. 153, p. 02004). EDP Sciences. 10.1051/matecconf/201815302004

Joshi, V., Joshi, V. V., Pawar, N., & Acharya, S. (2020). *Digitalisation of Health Care in India: Initiatives and Challenges. 40.* https://www.researchgate.net/publication/354586176

Junnila, S., & Niittylahti, J. (2003). Wireless technologies for data acquisition systems. *Proceedings of the 1st International Symposium on Information and Communication Technologies*, (pp. 132–137). IEEE.

Kamboj, S., & Chawla, S. (2014). Geocast routing in vehicular Ad Hoc networks: A survey. [IJCSIT]. *International Journal of Computer Science and Information Technologies*, *5*(4), 5365.

Kang, Y., Yin, H., & Berger, C. (2019). Test Your Self-Driving Algorithm: An Overview of Publicly Available Driving Datasets and Virtual Testing Environments. [Google Scholar] [CrossRef]. *IEEE Transactions on Intelligent Vehicles*, *4*(2), 171–185. doi:10.1109/TIV.2018.2886678

Kar, S. (2019, October). Robotics in HealthCare. In *2019 2nd International Conference on Power Energy, Environment and Intelligent Control (PEEIC)* (pp. 78-83). IEEE. 10.1109/PEEIC47157.2019.8976668

Karpov, V. E., & Tarassov, V. B. (2017, September). Synergetic artificial intelligence and social robotics. In *International Conference on Intelligent Information Technologies for Industry* (pp. 3-15). Cham: Springer International Publishing.

Katevas, N. (Ed.). (2001). *Mobile robotics in healthcare* (Vol. 7). IOS Press.

Katrolia, J. S., Mirbach, B., El-Sherif, A., Feld, H., Rambach, J., & Stricker, D. (2021). Ticam: A time-of-flight in-car cabin monitoring dataset. *arXiv preprint arXiv:2103.11719*.

Kaur, J. (2024). Fueling Healthcare Transformation: The Nexus of Startups, Venture Capital, and Innovation. In Fostering Innovation in Venture Capital and Startup Ecosystems (pp. 327-351). IGI Global.

Kaur, J. (2024). Green Finance 2.0: Pioneering Pathways for Sustainable Development and Health Through Future Trends and Innovations. In Sustainable Investments in Green Finance (pp. 294-319). IGI Global.

Kaur, J. (2024). Towards a Sustainable Triad: Uniting Energy Management Systems, Smart Cities, and Green Healthcare for a Greener Future. In Emerging Materials, Technologies, and Solutions for Energy Harvesting (pp. 258-285). IGI Global.

Kaur, J., & Arora, R. (2023, June). Exploring the dimensionality of employee silence in healthcare sector. In AIP conference proceedings (Vol. 2782, No. 1). AIP Publishing. doi:10.1063/5.0154178

Kaur, J., & Arora, R. (2023, June). Exploring the impact of employee silence in private hospitals-a structural equation modeling approach. In AIP Conference Proceedings (Vol. 2782, No. 1). AIP Publishing. doi:10.1063/5.0154173

Kavidha, V., Gayathri, N., & Kumar, S. R. (2021). AI, IoT and robotics in the medical and healthcare field. *AI and IoT-Based Intelligent Automation in Robotics*, 165-187.

Kernbach, S. (2011). Robot companions: Technology for humans. arXiv preprint arXiv:1111.5207.

Khamis, A., Patel, D., & Elgazzar, K. (2021). Deep Learning for Unmanned Autonomous Vehicles: A Comprehensive Review. In A. Koubaa & A. T. Azar (Eds.), *Deep Learning for Unmanned Systems. Studies in Computational Intelligence* (Vol. 984). Springer. doi:10.1007/978-3-030-77939-9_1

Khang, A., Rath, K. C., Anh, P. T. N., Rath, S. K., & Bhattacharya, S. (2024). Quantum-Based Robotics in the High-Tech Healthcare Industry: Innovations and Applications. In Medical Robotics and AI-Assisted Diagnostics for a High-Tech Healthcare Industry (pp. 1-27). IGI Global.

Kim, J., Gu, G. M., & Heo, P. (2016). Robotics for healthcare. *Biomedical Engineering: Frontier Research and Converging Technologies*, 489-509.

Kim, H., Kwon, D., Son, J., & Choi, J. (2024). A Novel Robotic Healthcare Device for Treating Chronic Venous Insufficiency in People Who Sit for Prolonged Periods. *IEEE Transactions on Medical Robotics and Bionics*, *6*(2), 618–631. doi:10.1109/TMRB.2024.3373909

Kolpashchikov, D., Gerget, O., & Meshcheryakov, R. (2022). *Robotics in healthcare. Handbook of Roboticsin Healthcare* (Vol. 2). Practicalities and Prospects.

Kumar, V., Mishra, S., & Chand, N. (2013). Applications of VANETs: present & future. *communications and network, 5*(01), 12-15.

Kumar, A. (2017). (LR 117) Impact of Digitalisation on Service Sector in India. *Reviewed Refereed Research Journal, 8*(15). https://www.researchgate.net/publication/360782465

Kumar, A., Kaur, A., & Kumar, M. (2019). Face detection techniques: A review. *Artificial Intelligence Review, 52*(2), 927–948. doi:10.1007/s10462-018-9650-2

Kumar, K. R. (2010). VANET parameters and applications: A review. *Global Journal of Computer Science and Technology, 10*(7), 72–77.

Kyrarini, M., Lygerakis, F., Rajavenkatanarayanan, A., Sevastopoulos, C., Nambiappan, H. R., Chaitanya, K. K., Babu, A. R., Mathew, J., & Makedon, F. (2021). A survey of robots in healthcare. *Technologies*, *9*(1), 8. doi:10.3390/technologies9010008

Lai, W. K., Lin, M. T., & Yang, Y. H. (2015). A machine learning system for routing decision-making in urban vehicular ad hoc networks. *International Journal of Distributed Sensor Networks*, *11*(3), 374391. doi:10.1155/2015/374391

Langarica, S., Rüffelmacher, C., & Núñez, F. (2019). An industrial internet application for real-time fault diagnosis in industrial motors. *IEEE Transactions on Automation Science and Engineering*, *17*(1), 284–295. doi:10.1109/TASE.2019.2913628

Lan, H., Hartonen, K., & Riekkola, M. L. (2020). Miniaturised air sampling techniques for analysis of volatile organic compounds in air. *Trends in Analytical Chemistry*, *126*, 115873. doi:10.1016/j.trac.2020.115873

LastMinuteEngineers.com. (2022). *Interfacing Micro SD Card Module with Arduino*. Last Minute Engineers.

Lazrag, H., Chaibi, H., Saadane, R., & Rahmani, M. D. (2018). An optimal and secure routing protocol for wireless sensor networks. In *2018 6th International Conference on Multimedia Computing and Systems (ICMCS)* (pp. 1-5). IEEE. 10.1109/ICMCS.2018.8525911

Lee, V. J., Chow, A., Zheng, X., Carrasco, L. R., Cook, A. R., Lye, D. C., Ng, L.-C., & Leo, Y. S. (2012). Simple clinical and laboratory predictors of Chikungunya versus dengue infections in adults. *PLoS Neglected Tropical Diseases*, *6*(9), 1786. doi:10.1371/journal.pntd.0001786 PMID:23029573

Lestingi, L., Sbrolli, C., Scarmozzino, P., Romeo, G., Bersani, M. M., & Rossi, M. (2022, May). Formal modeling and verification of multi-robot interactive scenarios in service settings. In *Proceedings of the IEEE/ACM 10th International Conference on Formal Methods in Software Engineering* (pp. 80-90). IEEE. 10.1145/3524482.3527653

Liang, W., Li, Z., Zhang, H., Wang, S., & Bie, R. (2015). Vehicular ad hoc networks: Architectures, research issues, methodologies, challenges, and trends. *International Journal of Distributed Sensor Networks*, *11*(8), 745303. doi:10.1155/2015/745303

Li, D., Zhang, Z., Liu, P., Wang, Z., & Zhang, L. (2020). Battery fault diagnosis for electric vehicles based on voltage abnormality by combining the long short-term memory neural network and the equivalent circuit model. *IEEE Transactions on Power Electronics*, *36*(2), 1303–1315. doi:10.1109/TPEL.2020.3008194

Lin, L., & Chen, C. (2020). (LR 82)The promise and perils of insurtech. *SSRN*, *2020*, 115–142. doi:10.2139/ssrn.3463533

Liu, L., Lu, S., Zhong, R., Wu, B., Yao, Y., Zhang, Q., & Shi, W. (2020). Computing systems for autonomous driving: State of the art and challenges. *IEEE Internet of Things Journal*, *8*(8), 6469–6486. doi:10.1109/JIOT.2020.3043716

Liu, Z., Cai, Y., Wang, H., Chen, L., Gao, H., Jia, Y., & Li, Y. (2021). Robust target recognition and tracking of self-driving cars with radar and camera information fusion under severe weather conditions. *IEEE Transactions on Intelligent Transportation Systems*, *23*(7), 6640–6653. doi:10.1109/TITS.2021.3059674

Li, W., Kong, K., Li, P., Wang, G., Cui, B., Zhu, L., & Zhu, S. (2021). Robot-assisted sleeve gastrectomy in patients with obesity with a novel Chinese domestic MicroHand SII surgical system. *BMC Surgery*, *21*(1), 260. doi:10.1186/s12893-021-01259-3 PMID:34034737

Li, Y., Jiao, X. Y., Sun, B. Q., Zhang, Q. H., & Yang, J. Y. (2021, March). Multi-welfare-robot cooperation framework for multi-task assignment in healthcare facilities based on multi-agent system. In *2021 IEEE International Conference on Intelligence and Safety for Robotics (ISR)* (pp. 413-416). IEEE. 10.1109/ISR50024.2021.9419496

Li, Z., & Du, R. (2013). Design and analysis of a bio-inspired wire-driven multi-section flexible robot. *International Journal of Advanced Robotic Systems*, *10*(4), 209. doi:10.5772/56025

Lloyd, P., Dall'Armellina, E., Schneider, J. E., & Valdastri, P. (2024). *Future cardiovascular healthcare via magnetic resonance imaging-driven robotics*.

Maalouf, N., Sidaoui, A., Elhajj, I. H., & Asmar, D. (2018). Robotics in nursing: A scoping review. *Journal of Nursing Scholarship*, *50*(6), 590–600. doi:10.1111/jnu.12424 PMID:30260093

Malika Shuxratovna, U., & Nuriddin Rustam Ugli, A. (2021). (LR 114)DIGITALIZATION OF THE INSURANCE MARKET. In Multidisciplinary Peer Reviewed Journal ISSN, 7(2).

Mallaboyev, N. M., Sharifjanovna, Q. M., Muxammadjon, Q., & Shukurullo, C. (2022, May). INFORMATION SECURITY ISSUES. In Conference Zone (pp. 241-245). IEEE.

Mankodiya, H., Jadav, D., Gupta, R., Tanwar, S., Hong, W. C., & Sharma, R. (2022). Od-xai: Explainable ai-based semantic object detection for autonomous vehicles. *Applied Sciences (Basel, Switzerland)*, *12*(11), 5310. doi:10.3390/app12115310

Manoharan, D. S. (2019). An improved safety algorithm for artificial intelligence enabled processors in self driving cars. *Journal of Artificial Intelligence and Capsule Networks*, *1*(2), 95–104. doi:10.36548/jaicn.2019.2.005

Marano, P. (2019). (LR 112)Navigating insurtech: The digital intermediaries of insurance products and customer protection in the EU. *Maastricht Journal of European and Comparative Law*, *26*(2), 294–315. doi:10.1177/1023263X19830345

Marchegiani, F., Siragusa, L., Zadoroznyj, A., Laterza, V., Mangana, O., Schena, C. A., Ammendola, M., Memeo, R., Bianchi, P. P., Spinoglio, G., Gavriilidis, P., & de'Angelis, N. (2023). New Robotic Platforms in General Surgery: What's the Current Clinical Scenario? *Medicina (Kaunas, Lithuania)*, *59*(7), 1264. doi:10.3390/medicina59071264 PMID:37512075

Martel, S., Mohammadi, M., Felfoul, O., Lu, Z., & Pouponneau, P. (2009). Flagellated magnetotactic bacteria as controlled MRI-trackable propulsion and steering systems for medical nanorobots operating in the human microvasculature. *The International Journal of Robotics Research*, *28*(4), 571–582. doi:10.1177/0278364908100924 PMID:19890435

Ma, Y., Zhang, Y., Wan, J., Zhang, D., & Pan, N. (2015). Robot and cloud-assisted multi-modal healthcare system. *Cluster Computing*, *18*(3), 1295–1306. doi:10.1007/s10586-015-0453-9

McMahan, W., & Walker, I. D. (2009, February). Octopus-inspired grasp-synergies for continuum manipulators. In *2008 IEEE International Conference on Robotics and Biomimetics* (pp. 945-950). IEEE. 10.1109/ROBIO.2009.4913126

Mehta, P. L., Kalra, R., & Prasad, R. (2021). A Backdrop Case Study of AI-Drones in Indian Demographic Characteristics Emphasizing the Role of AI in Global Cities Digitalization. *Wireless Personal Communications*, *118*(1), 301–321. doi:10.1007/s11277-020-08014-6 PMID:33424130

Mejri, M. N., & Ben-Othman, J. (2016). GDVAN: A new greedy behavior attack detection algorithm for VANETs. *IEEE Transactions on Mobile Computing*, *16*(3), 759–771. doi:10.1109/TMC.2016.2577035

Men, X., Liu, H., Chen, N., & Li, F. (2018). *A new time domain filtering method for calculating the RMS value of vibration signals. 51605191*, 2016–2019.

Mir, U. B., Sharma, S., Kar, A. K., & Gupta, M. P. (2020). Critical success factors for integrating artificial intelligence and robotics. *Digital Policy. Regulation & Governance*, *22*(4), 307–331. doi:10.1108/DPRG-03-2020-0032

Mishra, A., Cha, J., & Kim, S. (2022). Privacy-preserved in-cabin monitoring system for autonomous vehicles. *Computational Intelligence and Neuroscience*, *2022*, 2022. doi:10.1155/2022/5389359 PMID:35498178

Mohamed, A., Fouad, M. M. M., & Elhariri, E. (2014). *RoadMonitor: An Intelligent Road Surface Condition Monitoring System RoadMonitor: An Intelligent Road Surface Condition Monitoring System.* Springer. doi:10.1007/978-3-319-11310-4

Mohan, P., Sabarwal, T., & Preethiya, T. (2023). Indian Sign Language Character Recognition System. *2023 4th International Conference on Electronics and Sustainable Communication Systems (ICESC).* IEEE. 10.1109/ICESC57686.2023.10193309

Mois, G., & Beer, J. M. (2020). The role of healthcare robotics in providing support to older adults: A socio-ecological perspective. *Current Geriatrics Reports*, *9*(2), 82–89. doi:10.1007/s13670-020-00314-w PMID:32435576

Mo, K., Suh, K., & Hong, S. (2000). New Approach in Vehicle Durability Evaluation, Virtual Proving Ground. *Test*, (March), 1–5.

Monterrey, C. (2018). *Road Load Data Acquisition system with SAE-J1939 Communications Network: Integration and Laboratory Test.* [Thesis, Instituto Tecnológico y de Estudios Superiores de Monterrey].

Moradi, M., Yan, K., Colwell, D., Samwald, M., & Asgari, R. (2023). Model-agnostic explainable artificial intelligence for object detection in image data. *arXiv preprint arXiv:2303.17249.* doi:10.2139/ssrn.4429462

Moreno, C., González, A., Olazagoitia, J. L., & Vinolas, J. (2020). The acquisition rate and soundness of a low-cost data acquisition system (LC-DAQ) for high frequency applications. *Sensors (Basel)*, *20*(2), 524. doi:10.3390/s20020524 PMID:31963552

Morgan, A. A., Abdi, J., Syed, M. A., Kohen, G. E., Barlow, P., & Vizcaychipi, M. P. (2022). Robots in healthcare: A scoping review. *Current Robotics Reports*, *3*(4), 271–280. doi:10.1007/s43154-022-00095-4 PMID:36311256

Morizane, S., Stein, H., Komiya, T., Kaneta, H., & Takenaka, A. (2023). Retroperitoneal robot-assisted laparoscopic nephroureterectomy using the da Vinci Xi and SP systems: Initial experiences in cadaveric models. *Investigative and Clinical Urology*, *64*(4), 380–387. doi:10.4111/icu.20230021 PMID:37417563

Morton, J., Hardwick, R. H., Tilney, H. S., Gudgeon, A. M., Jah, A., Stevens, L., Marecik, S., & Slack, M. (2021). Preclinical evaluation of the versius surgical system, a new robot-assisted surgical device for use in minimal access general and colorectal procedures. *Surgical Endoscopy*, *35*(5), 2169–2177. doi:10.1007/s00464-020-07622-4 PMID:32405893

Moura, V. (2012). Magnetically Actuated Multiscale Medical Robots. In *IROS 2012 Full-day Workshop* (p. 48). IEEE.

Mueller, J. (n.d.). *InsurTech Rising: A Profile of the InsurTech Landscape DECEMBER 2018.*

Muhammad, K., Hussain, T., Ullah, H., Del Ser, J., Rezaei, M., Kumar, N., & de Albuquerque, V. H. C. (2022). Vision-based semantic segmentation in scene understanding for autonomous driving: Recent achievements, challenges, and outlooks. *IEEE Transactions on Intelligent Transportation Systems*, *23*(12), 22694–22715. doi:10.1109/TITS.2022.3207665

Muhammad, K., Ullah, A., Lloret, J., Del Ser, J., & de Albuquerque, V. H. C. (2020). Deep learning for safe autonomous driving: Current challenges and future directions. *IEEE Transactions on Intelligent Transportation Systems*, *22*(7), 4316–4336. doi:10.1109/TITS.2020.3032227

Mushtaque, R. S., Ahmad, S. M., Mushtaque, R., & Baloch, S. (2020). Case Report, A Curious Case of Dengue Fever: A Case Report of Unorthodox Manifestations. *Case Reports in Medicine*, *2020*, 1–4. doi:10.1155/2020/1701082 PMID:32774384

Nagadevi, D. (2021). *A Comprehensive Study on Artificial Intelligence and Robotics for Machine Intelligence.* Springer. doi:10.4018/978-1-7998-7701-1.ch011

Nam, K., Dutt, C. S., Chathoth, P., Daghfous, A., & Khan, M. S. (2021). The adoption of artificial intelligence and robotics in the hotel industry: Prospects and challenges. *Electronic Markets*, *31*(3), 553–574. doi:10.1007/s12525-020-00442-3

Nanobiomaterials in Drug Delivery. (2016). *Applications of Nanobiomaterials* (A. M. Grumezescu, Ed.). Vol. 9). Elsevier.

Narula, A., Narula, N. K., Khanna, S., Narula, R., Narula, J., & Narula, A. (2014). Future prospects of Roboticsin robotics software, a healthcare perspective. *International Journal of Applied Engineering Research: IJAER*, *9*(22), 10271–10280.

National Vector Borne Disease Control Programme (NVBDCP). (2015). Retrieved from https://nvbdcp.gov.in/index4.php?lang=1&level=0&linkid=431&lid=3715

Nayak, B., Bhattacharyya, S. S., & Krishnamoorthy, B. (2019). LR 77: Integrating wearable technology products and big data analytics in business strategy: A study of health insurance firms. *Journal of Systems and Information Technology*, *21*(2), 255–275. doi:10.1108/JSIT-08-2018-0109

Nehal, R. (2023). AI in Robotics: Advancements, Applications and Challenges. *Journal of Information Technology and Digital World.* doi:10.36548/jitdw.2023.2.009

Nordin, I. N. A. M., Razif, M. R. M., & Natarajan, E. (2013). 3-D finite-element analysis of fiber-reinforced soft bending actuator for finger flexion. IEEE/ASME international conference on advanced intelligent mechatronic, Wollongong, Australia.

Of, A., & Signal, V. (2013). *A LabVIEW BASED DATA ACQUISITION SYSTEM FOR MONITORING Sunita Mohanta Department of Electronics and Communication Engineering National Institute of Technology Rourkela, Odisha-769008 Sunita Mohanta Dr. Umesh Chandra Pati Department of Electronics and Co.*

Ogunoiki, A. O. (2015). *University of Birmingham Research Archive.* University of Birmingham.

Oksanen, A., Savela, N., Latikka, R., & Koivula, A. (2020). Trust toward robots and artificial intelligence: An experimental approach to human–technology interactions online. *Frontiers in Psychology*, *11*, 568256. doi:10.3389/fpsyg.2020.568256 PMID:33343447

Olszewska, J. I. (2022, February). Snakes in Trees: An Explainable Artificial Intelligence Approach for Automatic Object Detection and Recognition. In ICAART (3) (pp. 996-1002).

Omeiza, D., Webb, H., Jirotka, M., & Kunze, L. (2021). Explanations in autonomous driving: A survey. *IEEE Transactions on Intelligent Transportation Systems*, *23*(8), 10142–10162. doi:10.1109/TITS.2021.3122865

Oña, E. D., Garcia-Haro, J. M., Jardón, A., & Balaguer, C. (2019). Robotics in health care: Perspectives of robot-aided interventions in clinical practice for rehabilitation of upper limbs. *Applied Sciences (Basel, Switzerland)*, *9*(13), 2586. doi:10.3390/app9132586

Ostrowska, M. (2021). (LR 110)Does new technology put an end to policyholder risk declaration? The impact of digitalisation on insurance relationships. *The Geneva Papers on Risk and Insurance. Issues and Practice*, *46*(4), 573–592. doi:10.1057/s41288-020-00191-6

Ouyang, Z., Niu, J., Liu, Y., & Guizani, M. (2019). Deep CNN-based real-time traffic light detector for self-driving vehicles. *IEEE Transactions on Mobile Computing*, *19*(2), 300–313. doi:10.1109/TMC.2019.2892451

Pal Singh, H., & Singh, R. (2021). Low Cost Data Acquisition System for Road-Vehicle Interaction Using Arduino Board. *Journal of Physics: Conference Series*, *1831*(1), 012031. doi:10.1088/1742-6596/1831/1/012031

Pandit, C. M., & Ladhe, S. A. (2014). Secure routing protocol in MANET using TAC. In *2014 First International Conference on Networks & Soft Computing (ICNSC2014)* (pp. 107-112). IEEE. 10.1109/CNSC.2014.6906693

Paramasivan, B., Prakash, M. J. V., & Kaliappan, M. (2015). Development of a secure routing protocol using game theory model in mobile ad hoc networks. *Journal of Communications and Networks (Seoul)*, *17*(1), 75–83. doi:10.1109/JCN.2015.000012

Patel, A. R., Patel, R. S., Singh, N. M., & Kazi, F. S. (2017). Vitality of robotics in healthcare industry: Robotics perspective. *Robotics and big data technologies for next generation healthcare*, 91-109.

Patil, C., & Nataraj, K. R. (2014). A Secure Routing Protocol for VANET. *International Journal of Engine Research*, *3*(5).

Pee, L. G., Pan, S. L., & Cui, L. (2019). Roboticsin healthcare robots: A social informatics study of knowledge embodiment. *Journal of the Association for Information Science and Technology*, *70*(4), 351–369. doi:10.1002/asi.24145

Perttunen, M., Mazhelis, O., Cong, F., Ristaniemi, T., & Riekki, J. (2011). Distributed Road Surface Condition Monitoring. Lecture Notes in Computer Science. Springer. doi:10.1007/978-3-642-23641-9

PiMyLife. (2022). *Arduino Accelerometer using the ADXL345.*

Polat, F. (2017). *Journal of Engineering Research and Applied Science. Dece.*

Ponn, T., Kröger, T., & Diermeyer, F. (2020). Identification and explanation of challenging conditions for camera-based object detection of automated vehicles. *Sensors (Basel)*, *20*(13), 3699. doi:10.3390/s20133699 PMID:32630350

Poon, Y. S., Lin, C. C., Liu, Y. H., & Fan, C. P. (2022, January). YOLO-based deep learning design for in-cabin monitoring system with fisheye-lens camera. In *2022 IEEE International Conference on Consumer Electronics (ICCE)* (pp. 1-4). IEEE. 10.1109/ICCE53296.2022.9730235

Potts, J. A., Gibbons, R. V., Rothman, A. L., Srikiatkhachorn, A., Thomas, S. J., Supradish, P. O., Lemon, S. C., Libraty, D. H., Green, S., & Kalayanarooj, S. (2010). Prediction of dengue disease severity among pediatric Thai patients using early clinical laboratory indicators. *PLoS Neglected Tropical Diseases*, *4*(8), 769. doi:10.1371/journal.pntd.0000769 PMID:20689812

Preethiya, T., Muthukumar, A., & Durairaj, S. (2018) Providing Secured Data Aggregation in Mobile Wireless Sensor Network. *Proceedings of 4th IEEE International Symposium on Robotics and Manufacturing Automation*. IEEE. 10.1109/ROMA46407.2018.8986735

Preethiya, T., Muthukumar, A., & Durairaj, S. (2019a). Double Cluster Head Heterogeneous Clustering for Optimization in Hybrid Wireless Sensor Network. *Wireless Personal Communications. Wireless Personal Communications*, *110*(4), 1751–1768. doi:10.1007/s11277-019-06810-3

Preethiya, T., Muthukumar, A., & Durairaj, S. (2019b). Mobility Handling in Cluster based Mobile Wireless Sensor Network. *Proceedings of 2019 IEEE International Conference on Clean Energy and Energy Efficient Electronics Circuit for Sustainable Development (INCCES)*. IEEE. 10.1109/INCCES47820.2019.9167692

Preethiya, T., Muthukumar, A., & Durairaj, S. (2020). An energy efficient clustering and multipath routing for mobile wireless sensor network using game theory. *International Journal of Communication Systems*, *33*(7), 1–18.

Priyadarshi, R., Gupta, B., & Anurag, A. (2020). Deployment techniques in wireless sensor networks: A survey, classification, challenges, and future research issues. *The Journal of Supercomputing*, *76*(9), 7333–7373. doi:10.1007/s11227-020-03166-5

RadhaKrishna Karne, D. T. (2021). Review on vanet architecture and applications. [TURCOMAT]. *Turkish Journal of Computer and Mathematics Education*, *12*(4), 1745–1749.

Ragno, L., Borboni, A., Vannetti, F., Amici, C., & Cusano, N. (2023). Application of Social Robots in Healthcare: Review on Characteristics, Requirements, Technical Solutions. *Sensors (Basel), 23*(15), 6820. doi:10.3390/s23156820 PMID:37571603

Rajkumar, M. N., Nithya, M., & HemaLatha, P. (2016). Overview of VANETs with its features and security attacks. *International Research Journal of Engineering and Technology, 3*(1).

Ramos-Sorroche, E., Rubio-Aparicio, J., Santa, J., Guardiola, C., & Egea-Lopez, E. (2023). In-cabin and outdoor environmental monitoring in vehicular scenarios with distributed computing. *Internet of Things : Engineering Cyber Physical Human Systems*, 101009.

Ranjit, S., & Kissoon, N. (2011). Dengue hemorrhagic fever and shock syndromes. *Pediatric Critical Care Medicine, 12*(1), 90–100. doi:10.1097/PCC.0b013e3181e911a7 PMID:20639791

Rao, V. S. H., & Kumar, M. N. (2012). A new intelligence-based approach for computer-aided diagnosis of dengue fever. *IEEE Transactions on Information Technology in Biomedicine, 16*(1), 112–118. doi:10.1109/TITB.2011.2171978 PMID:22010159

Rathi, B. S., Kumar, P. S., & Vo, D. V. N. (2021). Critical review on hazardous pollutants in water environment: Occurrence, monitoring, fate, removal technologies and risk assessment. *The Science of the Total Environment, 797*, 149134. doi:10.1016/j.scitotenv.2021.149134 PMID:34346357

Rehman, S. U., Khan, M., Zia, T., & Zheng, L. (2013). Vehicular ad-hoc networks (VANETs): an overview and challenges. *Journal of Wireless Networking and communications, 3*(3), 29-38.

Remacle, M. M. N., Prasad, V., Lawson, G., Plisson, L., Bachy, V., & Van der Vorst, S. (2015). Transoral robotic surgery (TORS) with the Medrobotics Flex™ System: First surgical application on humans. *European Archives of Oto-Rhino-Laryngology, 272*, 1451–1455. doi:10.1007/s00405-015-3532-x PMID:25663191

Rizk, Y., Awad, M., & Tunstel, E. W. (2019). Cooperative heterogeneous multi-robot systems: A survey. *ACM Computing Surveys, 52*(2), 1–31. doi:10.1145/3303848

Rohan, A., Rabah, M., & Kim, S. H. (2019). Convolutional neural network-based real-time object detection and tracking for parrot AR drone 2. *IEEE Access : Practical Innovations, Open Solutions, 7*, 69575–69584. doi:10.1109/ACCESS.2019.2919332

Russo, M., Gautreau, E., Bonnet, X., & Laribi, M. A. (2023). Continuum Robots: From Conventional to Customized Performance Indicators. *Biomimetics, 8*(2), 147. doi:10.3390/biomimetics8020147 PMID:37092399

Saggi, M. K., & Sandhu, R. K. (2014). A survey of vehicular ad hoc network on attacks and security threats in VANETs. In *International Conference on Research and Innovations in Engineering and Technology (ICRIET 2014)* (pp. 19-20). IEEE.

Sahu, B., Das, P. K., Kabat, M. R., & Kumar, R. (2022). Prevention of Covid-19 affected patient using multi robot cooperation and Q-learning approach: A solution. *Quality & Quantity, 56*(2), 793–821. doi:10.1007/s11135-021-01155-1 PMID:33972809

Saini, M., & Singh, H. (2016). VANET its characteristics attacks and routing techniques: A survey. *International Journal of Scientific Research, 5*(5), 1595–1599.

Salokhe, N., Thakre, P., Awale, R. N., & Kambale, S. (2016). *Vibration Based Damage Detection using Overall Frequency Response and Time Domain.*

Sampoornam, K. P., Saranya, S., Vigneshwaran, S., Sofiarani, P., Sarmitha, S., & Sarumathi, N. (2020). A comparative study on reactive routing protocols in VANET. In *2020 4th International Conference on Electronics, Communication and Aerospace Technology (ICECA)* (pp. 726-731). IEEE. 10.1109/ICECA49313.2020.9297550

Sanan, S., Moidel, J., & Atkeson, C. G. (2011, June). A continuum approach to safe robots for physical human interaction. In *International Symposium on Quality of Life Technology*. IEEE.

Sarkar, S. (2021). LR 81: The Evolving Role of Insurtech in India: Trends, Challenges and The Road Ahead. *The Management Accountant Journal*, *56*(12), 30–37. doi:10.33516/maj.v56i12.30-37p

Sarker, S., Jamal, L., Ahmed, S. F., & Irtisam, N. (2021). Robotics and Roboticsin healthcare during COVID-19 pandemic: A systematic review. *Robotics and Autonomous Systems*, *146*, 103902. doi:10.1016/j.robot.2021.103902 PMID:34629751

Sarma, A. H. K. D., Kar, B. A., & Mall, C. R. (2011). Secure routing protocol for mobile wireless sensor network. In *2011 IEEE Sensors Applications Symposium* (pp. 93-99). IEEE. 10.1109/SAS.2011.5739778

Saxena, A., Khanna, A., & Gupta, D. (2020). Emotion recognition and detection methods: A comprehensive survey. *Journal of Artificial Intelligence and Systems*, *2*(1), 53–79. doi:10.33969/AIS.2020.21005

Schranz, M., Umlauft, M., Sende, M., & Elmenreich, W. (2020). Swarm Robotic Behaviors and Current Applications. *Frontiers in Robotics and AI*, *7*, 36. doi:10.3389/frobt.2020.00036 PMID:33501204

Sener, A. S. (n.d.). *Determination Of Vehicle Components Fatigue Life Based On Fea Method And Experimental Analysis Determination Of Vehicle Components Fatigue Life Based On Fea Method And.* *2*(Lcv), 133–146.

Şen, M. O., Okumuş, F., & Kocamaz, F.ŞEN. (2022). Application of blockchain powered mobile robots in healthcare: Use cases, research challenges and future trends. *Türk Doğa ve Fen Dergisi*, *11*(2), 27–35. doi:10.46810/tdfd.1017499

Shafiullah, A. K. M., & Wu, C. Q. (2013). Generation and validation of loading profiles for highly accelerated durability tests of ground vehicle components. *Engineering Failure Analysis*, *33*, 1–16. doi:10.1016/j.engfailanal.2013.04.008

Shah, A., Mehrotra, P., Sinha, S., & Sha, J. (2021). Bcg Insurtech Report India Insurtech Landscape And Trends. In Boston Consulting Group.

Shaktawat, R. S., Singh, D., & Choudhary, N. (2014). An efficient secure routing protocol in MANET Security-Enhanced AODV (SE-AODV). *International Journal of Computer Applications*, *97*(8). Advance online publication. doi:10.5120/17030-7329

Shallal, A. H., Ucan, O. N., Humaidi, A. J., & Bayat, O. (2020). Multi-robot systems formation control with maneuvring target in system applicable in the hospitality and care-health industry of medical internet of things. *Journal of Medical Imaging and Health Informatics*, *10*(1), 268–278. doi:10.1166/jmihi.2020.2840

Shaukat, K., Iqbal, F., Alam, T. M., Aujla, G. K., Devnath, L., Khan, A. G., & Rubab, A. (2020). The impact of artificial intelligence and robotics on the future employment opportunities. *Trends in Computer Science and Information Technology*, *5*(1), 50-54.

Shen, J., Wang, N., Wan, Z., Luo, Y., Sato, T., Hu, Z., & Chen, Q. A. (2022). Sok: On the semantic ai security in autonomous driving. *arXiv preprint arXiv:2203.05314*.

Shevchuk, O., Kondrat, I., & Stanienda, J. (2020). (LR 115)Pandemic as an accelerator of digital transformation in the insurance industry: Evidence from Ukraine. *Insurance Markets and Companies*, *11*(1), 30–41. doi:10.21511/ins.11(1).2020.04

Shimizu, A., Ito, M., & Lefor, A. K. (2022). Laparoscopic and Robot-Assisted Hepatic Surgery: An Historical Review. *Journal of Clinical Medicine*, *11*(12), 3254. doi:10.3390/jcm11123254 PMID:35743324

Shuai, J., Heng, S., Jing, Y., Feng, Y., Han, N., & Mengqin, Y. (2021). *Load Spectrum Acquisition, Analysis and Application of The Electric Bus in Road Simulation Test Based on*. 479–483. toptechboy.com. (n.d.). *ARDUINO LESSON 21: LOG SENSOR DATA TO AN SD CARD*.

Siau, K., Hodson, J., Ingram, R., Baxter, A., Widlak, M. M., Sharratt, C., Baker, G. M., Troth, T., Hicken, B., Tahir, F., Magrabi, M., Yousaf, N., Grant, C., Poon, D., Khalil, H., Lee, H. L., White, J. R., Tan, H., Samani, S., & Major, G. (2019). Time to endoscopy for acute upper gastrointestinal bleeding: Results from a prospective multicentre trainee-led audit. *United European Gastroenterology Journal*, *7*(2), 199–209. doi:10.1177/2050640618811491 PMID:31080604

Side, S., Pratama, M. I., Badwi, N., & Sanusi, W. (2020). Analysis and Simulation of SIRI Model for Dengue Fever Transmission. *Indian Journal of Science and Technology*, *13*(3), 340–351. doi:10.17485/ijst/2020/v13i03/147852

Silvera-Tawil, D. (2024). Robotics in Healthcare: A Survey. *SN Computer Science*, *5*(1), 189. doi:10.1007/s42979-023-02551-0

Simon, M., Amende, K., Kraus, A., Honer, J., Samann, T., Kaulbersch, H., & Michael Gross, H. (2019). Complexer-yolo: Real-time 3d object detection and tracking on semantic point clouds. In *Proceedings of the IEEE/CVF Conference on Computer Vision and Pattern Recognition Workshops* (pp. 0-0). IEEE. 10.1109/CVPRW.2019.00158

Singh, R., Kathuria, K., & Sagar, A. K. (2018). Secure routing protocols for wireless sensor networks. In *2018 4th international conference on computing communication and automation (ICCCA)* (pp. 1-5). IEEE. 10.1109/CCAA.2018.8777557

Singh, G., Rohil, H., Rishi, R., & Ranga, V. (2019). LETSRP: A secure routing protocol for MANETs. *Int J Eng Adv Technol (IJEAT). ISSN*, *9*(1), 2249–8958.

Soft Robotics Toolkit. (n.d.). Modeling and Design Tool for Soft Pneumatic Actuators. Soft Robotics Toolkit. https://softroboticstoolkit.com/book/modeling-soft-pneumatic-actuators

Sonker, A., & Gupta, R. K. (2020). A new combination of machine learning algorithms using stacking approach for misbehavior detection in VANETs. *International Journal of Computer Science and Network Security*, *20*(10), 94–100.

Soriano, G. P., Yasuhara, Y., Ito, H., Matsumoto, K., Osaka, K., Kai, Y., & Tanioka, T. (2022, August). Robots and robotics in nursing. In Healthcare (Vol. 10, No. 8, p. 1571). MDPI. doi:10.3390/healthcare10081571

Stahl, B. C., & Coeckelbergh, M. (2016). Ethics of healthcare robotics: Towards responsible research and innovation. *Robotics and Autonomous Systems*, *86*, 152–161. doi:10.1016/j.robot.2016.08.018

Stenius, I., Folkesson, J., Bhat, S., Sprague, C. I., Ling, L., Özkahraman, Ö., Bore, N., Cong, Z., Severholt, J., Ljung, C., Arnwald, A., Torroba, I., Gröndahl, F., & Thomas, J.-B. (2022). A System for Autonomous Seaweed Farm Inspection with an Underwater Robot. *Sensors (Basel)*, *22*(13), 5064. doi:10.3390/s22135064 PMID:35808560

Stoeckli, E., Dremel, C., & Uebernickel, F. (2018). LR 63; Exploring characteristics and transformational capabilities of InsurTech innovations to understand insurance value creation in a digital world. *Electronic Markets*, *28*(3), 287–305. Advance online publication. doi:10.1007/s12525-018-0304-7

Sugiyama, O., Shinozawa, K., Akimoto, T., & Hagita, N. (2010). Case study of a multi-robot healthcare system: Effects of docking and metaphor on persuasion. In *Social Robotics: Second International Conference on Social Robotics*. Singapore.

Sun, L., Yang, K., Hu, X., Hu, W., & Wang, K. (2020). Real-time fusion network for RGB-D semantic segmentation incorporating unexpected obstacle detection for road-driving images. *IEEE Robotics and Automation Letters*, *5*(4), 5558–5565. doi:10.1109/LRA.2020.3007457

Tarnawski, J. M., Phanishayee, A., Devanur, N., Mahajan, D., & Nina Paravecino, F. (2020). Efficient algorithms for device placement of dnn graph operators. *Advances in Neural Information Processing Systems*, *33*, 15451–15463.

Thakker, D., Mishra, B. K., Abdullatif, A., Mazumdar, S., & Simpson, S. (2020). Explainable artificial intelligence for developing smart cities solutions. *Smart Cities*, *3*(4), 1353–1382. doi:10.3390/smartcities3040065

Tian, Y., Ma, S., Wen, M., Liu, Y., Cheung, S. C., & Zhang, X. (2021). To what extent do DNN-based image classification models make unreliable inferences? *Empirical Software Engineering*, *26*(5), 84. doi:10.1007/s10664-021-09985-1

Tong, W., Hussain, A., Bo, W. X., & Maharjan, S. (2019). Artificial intelligence for vehicle-to-everything: A survey. *IEEE Access : Practical Innovations, Open Solutions*, *7*, 10823–10843. doi:10.1109/ACCESS.2019.2891073

Trewartha, A., Walker, N., Huo, H., Lee, S., Cruse, K., Dagdelen, J., Dunn, A., Persson, K. A., Ceder, G., & Jain, A. (2022). Quantifying the advantage of domain-specific pre-training on named entity recognition tasks in materials science. *Patterns (New York, N.Y.)*, *3*(4), 100488. doi:10.1016/j.patter.2022.100488 PMID:35465225

Troccaz, J., Dagnino, G., & Yang, G. Z. (2019). Frontiers of medical robotics: From concept to systems to clinical translation. *Annual Review of Biomedical Engineering*, *21*(1), 193–218. doi:10.1146/annurev-bioeng-060418-052502 PMID:30822100

Tyagi, A. K., & Aswathy, S. U. (2021). Autonomous Intelligent Vehicles (AIV): Research statements, open issues, challenges and road for future. *International Journal of Intelligent Networks*, *2*, 83–102. doi:10.1016/j.ijin.2021.07.002

Ur Rehman, S., Mustafa, H., & Larik, A. R. (2021). IoT Based Substation Monitoring Control System Using Arduino with Data Logging. *Proceedings - 2021 IEEE 4th International Conference on Computing and Information Sciences, ICCIS 2021*. IEEE. 10.1109/ICCIS54243.2021.9676384

Vallès-Peris, N., Barat-Auleda, O., & Domènech, M. (2021). Robots in healthcare? What patients say. *International Journal of Environmental Research and Public Health*, *18*(18), 9933. doi:10.3390/ijerph18189933 PMID:34574861

Vallès-Peris, N., & Domènech, M. (2023). Caring in the in-between: A proposal to introduce responsible AI and robotics to healthcare. *AI & Society*, *38*(4), 1685–1695. doi:10.1007/s00146-021-01330-w

Van Wyk, F., Wang, Y., Khojandi, A., & Masoud, N. (2019). Real-time sensor anomaly detection and identification in automated vehicles. *IEEE Transactions on Intelligent Transportation Systems*, *21*(3), 1264–1276. doi:10.1109/TITS.2019.2906038

Venkat, Y., Chand, K. P., & Preethiya, T. (2023). An intrusion detection system for the Internet of Things based on machine learning. *2023 International Conference on Recent Advances in Electrical, Electronics, Ubiquitous Communication, and Computational Intelligence (RAEEUCCI)*. IEEE. 10.1109/RAEEUCCI57140.2023.10134432

Verma, A. K., Nagpal, S., Desai, A., & Sudha, R. (2021). An efficient neural-network model for real-time fault detection in industrial machine. *Neural Computing & Applications*, *33*(4), 1297–1310. doi:10.1007/s00521-020-05033-z

Vesnic-Alujevic, L., Nascimento, S., & Pólvora, A. (2020). Societal and ethical impacts of artificial intelligence: Critical notes on European policy frameworks. *Telecommunications Policy*, *44*(6), 2020. doi:10.1016/j.telpol.2020.101961

Waidi, Y. O., Barua, R., & Datta, S. (2023). Metals, Polymers, Ceramics, Composites Biomaterials Used in Additive Manufacturing for Biomedical Applications. In A. Kumar, P. Kumar, A. Srivastava, & V. Goyat (Eds.), *Modeling, Characterization, and Processing of Smart Materials* (pp. 165–184). IGI Global. doi:10.4018/978-1-6684-9224-6.ch008

Wali, S., & Areeb, M. (2018). Development of Low-Cost DAQ for Power System Signals Using Arduino. *2018 IEEE 21st International Multi-Topic Conference (INMIC)*, (pp. 1–5). IEEE. 10.1109/INMIC.2018.8595519

Wang, J., Zhao, Y., Yang, Y., & Yang, J. (2019). *Fatigue Analysis for Bogie Frame of Urban Transit Rail Vehicle under Overload Situation. Qr2mse*. Research Gate.

Wang, Q., Yan, K., & Li, H. (2009). *Motor Noise Source Identification Based on Frequency Domain Analysis*. Research Gate.

Wang, H., Zhang, R., Chen, W., Wang, X., & Pfeifer, R. (2017). A cable-driven soft robot surgical system for cardiothoracic endoscopic surgery: Preclinical tests in animals. *Surgical Endoscopy*, *31*(8), 3152–3158. doi:10.1007/s00464-016-5340-9 PMID:27858208

Wang, J., & Lee, J. (2019). Bioinspired Nanorobots for Cancer Detection and Therapy. In *Bioinspired Nanomaterials and Nanostructures from Nanobiology to Nanomedicine* (pp. 1–31). Springer.

Wang, J., Li, Y., Zhou, Z., Wang, C., Hou, Y., Zhang, L., & Chen, S. (2022). When, where and how does it fail? A spatial-temporal visual analytics approach for interpretable object detection in autonomous driving. *IEEE Transactions on Visualization and Computer Graphics*. PMID:36040948

Wavhale, R. D., Andhari, S. S., Dhobale, K. D., Tawade, B. V., Chate, G. P., Patil, Y. N., ... Banerjee, S. S. (2021). Self-propelling magnetic nanorobots for capturing circulating tumor cells in blood samples of cancer patients. *Communications Chemistry*, *4*(1), 1–12. PMID:36697560

Wei, H., Zhang, G., Wang, S., Zhang, P., Su, J., & Du, F. (2023). Coupling Analysis of Compound Continuum Robots for Surgery: Another Line of Thought. *Sensors (Basel)*, *23*(14), 6407. doi:10.3390/s23146407 PMID:37514701

Wei, Y., Chen, Y., Ren, T., Chen, Q., Yan, C., Yang, Y., & Li, Y. (2016). A novel, variable stiffness robotic gripper based on integrated soft actuating and particle jamming. *Soft Robotics*, *3*(3), 134143. doi:10.1089/soro.2016.0027

Wu, X., & Ma, S. (2013). Neurally controlled steering for collision-free behavior of a snake robot. *IEEE Transactions on Control Systems Technology*, *21*(6), 2443–2449. doi:10.1109/TCST.2012.2237519

Xavier, M. S., Fleming, A. J., & Yong, Y. K. (2021, February). Finite element modeling of soft uidic actuators: Overview and recent developments. *Advanced Intelligent Systems*, *3*(2), 2000187. doi:10.1002/aisy.202000187

Yadav, S., Rajput, N. K., Sagar, A. K., & Maheshwari, D. (2018). Secure and reliable routing protocols for VANETs. In *2018 4th International Conference on Computing Communication and Automation (ICCCA)* (pp. 1-5). IEEE. 10.1109/CCAA.2018.8777690

Yadav, N., & Chug, U. (2019). Secure Routing in MANET: A Review. In *2019 International Conference on Machine Learning, Big Data, Cloud and Parallel Computing (COMITCon)* (pp. 375-379). IEEE.

Yaneva, T. (n.d.). *Digital Transformation of Insurance Sector*.

Yang, T., Xiangyang, X., Peng, L., Tonghui, L., & Leina, P. (2018). A secure routing of wireless sensor networks based on trust evaluation model. *Procedia Computer Science*, *131*, 1156–1163. doi:10.1016/j.procs.2018.04.289

Yang, Y., Li, D., Sun, Y., Wu, M., Su, J., Li, Y., Yu, X., Li, L., & Yu, J. (2023). Muscle-inspired soft robots based on bilateral dielectric elastomer actuators. *Microsystems & Nanoengineering*, *9*(1), 124. doi:10.1038/s41378-023-00592-2 PMID:37814608

Yan, H., & Fan, C. (2019). DNA nanotechnology and its biological applications. In D. N. A. Nanotechnology (Ed.), (pp. 1–22). Springer.

Yao, L., Fang, Z., Xiao, Y., Hou, J., & Fu, Z. (2021). An intelligent fault diagnosis method for lithium battery systems based on grid search support vector machine. *Energy*, *214*, 118866. doi:10.1016/j.energy.2020.118866

Yao, L., Xiao, Y., Gong, X., Hou, J., & Chen, X. (2020). A novel intelligent method for fault diagnosis of electric vehicle battery system based on wavelet neural network. *Journal of Power Sources*, *453*, 227870. doi:10.1016/j.jpowsour.2020.227870

Yaqoob, I., Khan, L. U., Kazmi, S. A., Imran, M., Guizani, N., & Hong, C. S. (2019). Autonomous driving cars in smart cities: Recent advances, requirements, and challenges. *IEEE Network*, *34*(1), 174–181. doi:10.1109/MNET.2019.1900120

Yinong, C. (2021). Technologies Supporting Artificial Intelligence and Robotics Application Development. doi:10.37965/jait.2020.0065

Yin, S., Rodriguez-Andina, J. J., & Jiang, Y. (2019). Real-time monitoring and control of industrial cyberphysical systems: With integrated plant-wide monitoring and control framework. *IEEE Industrial Electronics Magazine*, *13*(4), 38–47. doi:10.1109/MIE.2019.2938025

Yoon, S. N., & Lee, D. (2018). Roboticsand robots in healthcare: What are the success factors for technology-based service encounters? *International Journal of Healthcare Management*.

Zablocki, É., Ben-Younes, H., Pérez, P., & Cord, M. (2022). Explainability of deep vision-based autonomous driving systems: Review and challenges. *International Journal of Computer Vision*, *130*(10), 2425–2452. doi:10.1007/s11263-022-01657-x

Zhang, G., Du, F., Xue, S., Cheng, H., Zhang, X., Song, R., & Li, Y. (2022). Design and Modeling of a Bio-Inspired Compound Continuum Robot for Minimally Invasive Surgery. *Machines*, *10*(6), 468. doi:10.3390/machines10060468

Zhao, L., Li, Y., Meng, C., Gong, C., & Tang, X. (2016). A SVM based routing scheme in VANETs. In *2016 16th International Symposium on Communications and Information Technologies (ISCIT)* (pp. 380-383). IEEE. 10.1109/ISCIT.2016.7751655

Zhao, J., Zhao, W., Deng, B., Wang, Z., Zhang, F., Zheng, W., & Burke, A. F. (2023). Autonomous driving system: A comprehensive survey. *Expert Systems with Applications*, 122836.

Zhu, J., Lyu, L., Xu, Y., Liang, H., Zhang, X., Ding, H., & Wu, Z. (2021). Intelligent Soft Surgical Robots for Next-Generation Minimally Invasive Surgery. *Advanced Intelligent Systems*, *3*(5), 2100011. doi:10.1002/aisy.202100011

Related References

To continue our tradition of advancing academic research, we have compiled a list of recommended IGI Global readings. These references will provide additional information and guidance to further enrich your knowledge and assist you with your own research and future publications.

Abbasnejad, B., Moeinzadeh, S., Ahankoob, A., & Wong, P. S. (2021). The Role of Collaboration in the Implementation of BIM-Enabled Projects. In J. Underwood & M. Shelbourn (Eds.), *Handbook of Research on Driving Transformational Change in the Digital Built Environment* (pp. 27–62). IGI Global. https://doi.org/10.4018/978-1-7998-6600-8.ch002

Abdulrahman, K. O., Mahamood, R. M., & Akinlabi, E. T. (2022). Additive Manufacturing (AM): Processing Technique for Lightweight Alloys and Composite Material. In K. Kumar, B. Babu, & J. Davim (Ed.), *Handbook of Research on Advancements in the Processing, Characterization, and Application of Lightweight Materials* (pp. 27-48). IGI Global. https://doi.org/10.4018/978-1-7998-7864-3.ch002

Agrawal, R., Sharma, P., & Saxena, A. (2021). A Diamond Cut Leather Substrate Antenna for BAN (Body Area Network) Application. In V. Singh, V. Dubey, A. Saxena, R. Tiwari, & H. Sharma (Eds.), *Emerging Materials and Advanced Designs for Wearable Antennas* (pp. 54–59). IGI Global. https://doi.org/10.4018/978-1-7998-7611-3.ch004

Ahmad, F., Al-Ammar, E. A., & Alsaidan, I. (2022). Battery Swapping Station: A Potential Solution to Address the Limitations of EV Charging Infrastructure. In M. Alam, R. Pillai, & N. Murugesan (Eds.), *Developing Charging Infrastructure and Technologies for Electric Vehicles* (pp. 195–207). IGI Global. doi:10.4018/978-1-7998-6858-3.ch010

Aikhuele, D. (2018). A Study of Product Development Engineering and Design Reliability Concerns. *International Journal of Applied Industrial Engineering*, *5*(1), 79–89. doi:10.4018/IJAIE.2018010105

Al-Khatri, H., & Al-Atrash, F. (2021). Occupants' Habits and Natural Ventilation in a Hot Arid Climate. In R. González-Lezcano (Ed.), *Advancements in Sustainable Architecture and Energy Efficiency* (pp. 146–168). IGI Global. https://doi.org/10.4018/978-1-7998-7023-4.ch007

Al-Shebeeb, O. A., Rangaswamy, S., Gopalakrishan, B., & Devaru, D. G. (2017). Evaluation and Indexing of Process Plans Based on Electrical Demand and Energy Consumption. *International Journal of Manufacturing, Materials, and Mechanical Engineering*, *7*(3), 1–19. doi:10.4018/IJMMME.2017070101

Amuda, M. O., Lawal, T. F., & Akinlabi, E. T. (2017). Research Progress on Rheological Behavior of AA7075 Aluminum Alloy During Hot Deformation. *International Journal of Materials Forming and Machining Processes*, *4*(1), 53–96. doi:10.4018/IJMFMP.2017010104

Amuda, M. O., Lawal, T. F., & Mridha, S. (2021). Microstructure and Mechanical Properties of Silicon Carbide-Treated Ferritic Stainless Steel Welds. In L. Burstein (Ed.), *Handbook of Research on Advancements in Manufacturing, Materials, and Mechanical Engineering* (pp. 395–411). IGI Global. https://doi.org/10.4018/978-1-7998-4939-1.ch019

Anikeev, V., Gasem, K. A., & Fan, M. (2021). Application of Supercritical Technologies in Clean Energy Production: A Review. In L. Chen (Ed.), *Handbook of Research on Advancements in Supercritical Fluids Applications for Sustainable Energy Systems* (pp. 792–821). IGI Global. https://doi.org/10.4018/978-1-7998-5796-9.ch022

Arafat, M. Y., Saleem, I., & Devi, T. P. (2022). Drivers of EV Charging Infrastructure Entrepreneurship in India. In M. Alam, R. Pillai, & N. Murugesan (Eds.), *Developing Charging Infrastructure and Technologies for Electric Vehicles* (pp. 208–219). IGI Global. https://doi.org/10.4018/978-1-7998-6858-3.ch011

Araujo, A., & Manninen, H. (2022). Contribution of Project-Based Learning on Social Skills Development: An Industrial Engineer Perspective. In A. Alves & N. van Hattum-Janssen (Eds.), *Training Engineering Students for Modern Technological Advancement* (pp. 119–145). IGI Global. https://doi.org/10.4018/978-1-7998-8816-1.ch006

Armutlu, H. (2018). Intelligent Biomedical Engineering Operations by Cloud Computing Technologies. In U. Kose, G. Guraksin, & O. Deperlioglu (Eds.), *Nature-Inspired Intelligent Techniques for Solving Biomedical Engineering Problems* (pp. 297–317). Hershey, PA: IGI Global. doi:10.4018/978-1-5225-4769-3.ch015

Atik, M., Sadek, M., & Shahrour, I. (2017). Single-Run Adaptive Pushover Procedure for Shear Wall Structures. In V. Plevris, G. Kremmyda, & Y. Fahjan (Eds.), *Performance-Based Seismic Design of Concrete Structures and Infrastructures* (pp. 59–83). Hershey, PA: IGI Global. doi:10.4018/978-1-5225-2089-4.ch003

Attia, H. (2021). Smart Power Microgrid Impact on Sustainable Building. In R. González-Lezcano (Ed.), *Advancements in Sustainable Architecture and Energy Efficiency* (pp. 169–194). IGI Global. https://doi.org/10.4018/978-1-7998-7023-4.ch008

Aydin, A., Akyol, E., Gungor, M., Kaya, A., & Tasdelen, S. (2018). Geophysical Surveys in Engineering Geology Investigations With Field Examples. In N. Ceryan (Ed.), *Handbook of Research on Trends and Digital Advances in Engineering Geology* (pp. 257–280). Hershey, PA: IGI Global. doi:10.4018/978-1-5225-2709-1.ch007

Ayoobkhan, M. U. D., Y., A., J., Easwaran, B., & R., T. (2021). Smart Connected Digital Products and IoT Platform With the Digital Twin. In P. Vasant, G. Weber, & W. Punurai (Ed.), Research Advancements in Smart Technology, Optimization, and Renewable Energy (pp. 330-350). IGI Global. https://doi.org/ doi:10.4018/978-1-7998-3970-5.ch016

Baeza Moyano, D., & González Lezcano, R. A. (2021). The Importance of Light in Our Lives: Towards New Lighting in Schools. In R. González-Lezcano (Ed.), *Advancements in Sustainable Architecture and Energy Efficiency* (pp. 239–256). IGI Global. https://doi.org/10.4018/978-1-7998-7023-4.ch011

Bagdadee, A. H. (2021). A Brief Assessment of the Energy Sector of Bangladesh. *International Journal of Energy Optimization and Engineering*, *10*(1), 36–55. doi:10.4018/IJEOE.2021010103

Baklezos, A. T., & Hadjigeorgiou, N. G. (2021). Magnetic Sensors for Space Applications and Magnetic Cleanliness Considerations. In C. Nikolopoulos (Ed.), *Recent Trends on Electromagnetic Environmental Effects for Aeronautics and Space Applications* (pp. 147–185). IGI Global. https://doi.org/10.4018/978-1-7998-4879-0.ch006

Bas, T. G. (2017). Nutraceutical Industry with the Collaboration of Biotechnology and Nutrigenomics Engineering: The Significance of Intellectual Property in the Entrepreneurship and Scientific Research Ecosystems. In T. Bas & J. Zhao (Eds.), *Comparative Approaches to Biotechnology Development and Use in Developed and Emerging Nations* (pp. 1–17). Hershey, PA: IGI Global. doi:10.4018/978-1-5225-1040-6.ch001

Bazeer Ahamed, B., & Periakaruppan, S. (2021). Taxonomy of Influence Maximization Techniques in Unknown Social Networks. In P. Vasant, G. Weber, & W. Punurai (Eds.), *Research Advancements in Smart Technology, Optimization, and Renewable Energy* (pp. 351-363). IGI Global. https://doi.org/10.4018/978-1-7998-3970-5.ch017

Beale, R., & André, J. (2017). *Design Solutions and Innovations in Temporary Structures*. Hershey, PA: IGI Global. doi:10.4018/978-1-5225-2199-0

Behnam, B. (2017). Simulating Post-Earthquake Fire Loading in Conventional RC Structures. In P. Samui, S. Chakraborty, & D. Kim (Eds.), *Modeling and Simulation Techniques in Structural Engineering* (pp. 425–444). Hershey, PA: IGI Global. doi:10.4018/978-1-5225-0588-4.ch015

Ben Hamida, I., Salah, S. B., Msahli, F., & Mimouni, M. F. (2018). Distribution Network Reconfiguration Using SPEA2 for Power Loss Minimization and Reliability Improvement. *International Journal of Energy Optimization and Engineering*, *7*(1), 50–65. doi:10.4018/IJEOE.2018010103

Bentarzi, H. (2021). Fault Tree-Based Root Cause Analysis Used to Study Mal-Operation of a Protective Relay in a Smart Grid. In A. Recioui & H. Bentarzi (Eds.), *Optimizing and Measuring Smart Grid Operation and Control* (pp. 289–308). IGI Global. https://doi.org/10.4018/978-1-7998-4027-5.ch012

Beysens, D. A., Garrabos, Y., & Zappoli, B. (2021). Thermal Effects in Near-Critical Fluids: Piston Effect and Related Phenomena. In L. Chen (Ed.), *Handbook of Research on Advancements in Supercritical Fluids Applications for Sustainable Energy Systems* (pp. 1–31). IGI Global. https://doi.org/10.4018/978-1-7998-5796-9.ch001

Bhaskar, S. V., & Kudal, H. N. (2017). Effect of TiCN and AlCrN Coating on Tribological Behaviour of Plasma-nitrided AISI 4140 Steel. *International Journal of Surface Engineering and Interdisciplinary Materials Science*, *5*(2), 1–17. doi:10.4018/IJSEIMS.2017070101

Bhuyan, D. (2018). Designing of a Twin Tube Shock Absorber: A Study in Reverse Engineering. In K. Kumar & J. Davim (Eds.), *Design and Optimization of Mechanical Engineering Products* (pp. 83–104). Hershey, PA: IGI Global. doi:10.4018/978-1-5225-3401-3.ch005

Blumberg, G. (2021). Blockchains for Use in Construction and Engineering Projects. In J. Underwood & M. Shelbourn (Eds.), *Handbook of Research on Driving Transformational Change in the Digital Built Environment* (pp. 179–208). IGI Global. https://doi.org/10.4018/978-1-7998-6600-8.ch008

Bolboaca, A. M. (2021). Considerations Regarding the Use of Fuel Cells in Combined Heat and Power for Stationary Applications. In G. Badea, R. Felseghi, & I. Aşchilean (Eds.), *Hydrogen Fuel Cell Technology for Stationary Applications* (pp. 239–275). IGI Global. https://doi.org/10.4018/978-1-7998-4945-2.ch010

Burstein, L. (2021). Simulation Tool for Cable Design. In L. Burstein (Ed.), *Handbook of Research on Advancements in Manufacturing, Materials, and Mechanical Engineering* (pp. 54–74). IGI Global. https://doi.org/10.4018/978-1-7998-4939-1.ch003

Calderon, F. A., Giolo, E. G., Frau, C. D., Rengel, M. G., Rodriguez, H., Tornello, M., ... Gallucci, R. (2018). Seismic Microzonation and Site Effects Detection Through Microtremors Measures: A Review. In N. Ceryan (Ed.), *Handbook of Research on Trends and Digital Advances in Engineering Geology* (pp. 326–349). Hershey, PA: IGI Global. doi:10.4018/978-1-5225-2709-1.ch009

Ceryan, N., & Can, N. K. (2018). Prediction of The Uniaxial Compressive Strength of Rocks Materials. In N. Ceryan (Ed.), *Handbook of Research on Trends and Digital Advances in Engineering Geology* (pp. 31–96). Hershey, PA: IGI Global. doi:10.4018/978-1-5225-2709-1.ch002

Ceryan, S. (2018). Weathering Indices Used in Evaluation of the Weathering State of Rock Material. In N. Ceryan (Ed.), *Handbook of Research on Trends and Digital Advances in Engineering Geology* (pp. 132–186). Hershey, PA: IGI Global. doi:10.4018/978-1-5225-2709-1.ch004

Chen, H., Padilla, R. V., & Besarati, S. (2017). Supercritical Fluids and Their Applications in Power Generation. In L. Chen & Y. Iwamoto (Eds.), *Advanced Applications of Supercritical Fluids in Energy Systems* (pp. 369–402). Hershey, PA: IGI Global. doi:10.4018/978-1-5225-2047-4.ch012

Chen, H., Padilla, R. V., & Besarati, S. (2021). Supercritical Fluids and Their Applications in Power Generation. In L. Chen (Ed.), *Handbook of Research on Advancements in Supercritical Fluids Applications for Sustainable Energy Systems* (pp. 566–599). IGI Global. https://doi.org/10.4018/978-1-7998-5796-9.ch016

Chen, L. (2017). Principles, Experiments, and Numerical Studies of Supercritical Fluid Natural Circulation System. In L. Chen & Y. Iwamoto (Eds.), *Advanced Applications of Supercritical Fluids in Energy Systems* (pp. 136–187). Hershey, PA: IGI Global. doi:10.4018/978-1-5225-2047-4.ch005

Chen, L. (2021). Principles, Experiments, and Numerical Studies of Supercritical Fluid Natural Circulation System. In L. Chen (Ed.), *Handbook of Research on Advancements in Supercritical Fluids Applications for Sustainable Energy Systems* (pp. 219–269). IGI Global. https://doi.org/10.4018/978-1-7998-5796-9.ch007

Chiba, Y., Marif, Y., Henini, N., & Tlemcani, A. (2021). Modeling of Magnetic Refrigeration Device by Using Artificial Neural Networks Approach. *International Journal of Energy Optimization and Engineering, 10*(4), 68–76. https://doi.org/10.4018/IJEOE.2021100105

Clementi, F., Di Sciascio, G., Di Sciascio, S., & Lenci, S. (2017). Influence of the Shear-Bending Interaction on the Global Capacity of Reinforced Concrete Frames: A Brief Overview of the New Perspectives. In V. Plevris, G. Kremmyda, & Y. Fahjan (Eds.), *Performance-Based Seismic Design of Concrete Structures and Infrastructures* (pp. 84–111). Hershey, PA: IGI Global. doi:10.4018/978-1-5225-2089-4.ch004

Codinhoto, R., Fialho, B. C., Pinti, L., & Fabricio, M. M. (2021). BIM and IoT for Facilities Management: Understanding Key Maintenance Issues. In J. Underwood & M. Shelbourn (Eds.), *Handbook of Research on Driving Transformational Change in the Digital Built Environment* (pp. 209–231). IGI Global. doi:10.4018/978-1-7998-6600-8.ch009

Cortés-Polo, D., Calle-Cancho, J., Carmona-Murillo, J., & González-Sánchez, J. (2017). Future Trends in Mobile-Fixed Integration for Next Generation Networks: Classification and Analysis. *International Journal of Vehicular Telematics and Infotainment Systems, 1*(1), 33–53. doi:10.4018/IJVTIS.2017010103

Costa, H. G., Sheremetieff, F. H., & Araújo, E. A. (2022). Influence of Game-Based Methods in Developing Engineering Competences. In A. Alves & N. van Hattum-Janssen (Eds.), *Training Engineering Students for Modern Technological Advancement* (pp. 69–88). IGI Global. https://doi.org/10.4018/978-1-7998-8816-1.ch004

Cui, X., Zeng, S., Li, Z., Zheng, Q., Yu, X., & Han, B. (2018). Advanced Composites for Civil Engineering Infrastructures. In K. Kumar & J. Davim (Eds.), *Composites and Advanced Materials for Industrial Applications* (pp. 212–248). Hershey, PA: IGI Global. doi:10.4018/978-1-5225-5216-1.ch010

Dalgıç, S., & Kuşku, İ. (2018). Geological and Geotechnical Investigations in Tunneling. In N. Ceryan (Ed.), *Handbook of Research on Trends and Digital Advances in Engineering Geology* (pp. 482–529). Hershey, PA: IGI Global. doi:10.4018/978-1-5225-2709-1.ch014

Dang, C., & Hihara, E. (2021). Study on Cooling Heat Transfer of Supercritical Carbon Dioxide Applied to Transcritical Carbon Dioxide Heat Pump. In L. Chen (Ed.), *Handbook of Research on Advancements in Supercritical Fluids Applications for Sustainable Energy Systems* (pp. 451–493). IGI Global. https://doi.org/10.4018/978-1-7998-5796-9.ch013

Daus, Y., Kharchenko, V., & Yudaev, I. (2021). Research of Solar Energy Potential of Photovoltaic Installations on Enclosing Structures of Buildings. *International Journal of Energy Optimization and Engineering, 10*(4), 18–34. https://doi.org/10.4018/IJEOE.2021100102

Daus, Y., Kharchenko, V., & Yudaev, I. (2021). Optimizing Layout of Distributed Generation Sources of Power Supply System of Agricultural Object. *International Journal of Energy Optimization and Engineering, 10*(3), 70–84. https://doi.org/10.4018/IJEOE.2021070104

de la Varga, D., Soto, M., Arias, C. A., van Oirschot, D., Kilian, R., Pascual, A., & Álvarez, J. A. (2017). Constructed Wetlands for Industrial Wastewater Treatment and Removal of Nutrients. In Á. Val del Río, J. Campos Gómez, & A. Mosquera Corral (Eds.), *Technologies for the Treatment and Recovery of Nutrients from Industrial Wastewater* (pp. 202–230). Hershey, PA: IGI Global. doi:10.4018/978-1-5225-1037-6.ch008

Deb, S., Ammar, E. A., AlRajhi, H., Alsaidan, I., & Shariff, S. M. (2022). V2G Pilot Projects: Review and Lessons Learnt. In M. Alam, R. Pillai, & N. Murugesan (Eds.), *Developing Charging Infrastructure and Technologies for Electric Vehicles* (pp. 252–267). IGI Global. https://doi.org/10.4018/978-1-7998-6858-3.ch014

Dekhandji, F. Z., & Rais, M. C. (2021). A Comparative Study of Power Quality Monitoring Using Various Techniques. In A. Recioui & H. Bentarzi (Eds.), *Optimizing and Measuring Smart Grid Operation and Control* (pp. 259–288). IGI Global. https://doi.org/10.4018/978-1-7998-4027-5.ch011

Deperlioglu, O. (2018). Intelligent Techniques Inspired by Nature and Used in Biomedical Engineering. In U. Kose, G. Guraksin, & O. Deperlioglu (Eds.), *Nature-Inspired Intelligent Techniques for Solving Biomedical Engineering Problems* (pp. 51–77). Hershey, PA: IGI Global. doi:10.4018/978-1-5225-4769-3.ch003

Dhurpate, P. R., & Tang, H. (2021). Quantitative Analysis of the Impact of Inter-Line Conveyor Capacity for Throughput of Manufacturing Systems. *International Journal of Manufacturing, Materials, and Mechanical Engineering, 11*(1), 1–17. https://doi.org/10.4018/IJMMME.2021010101

Dinkar, S., & Deep, K. (2021). A Survey of Recent Variants and Applications of Antlion Optimizer. *International Journal of Energy Optimization and Engineering, 10*(2), 48–73. doi:10.4018/IJEOE.2021040103

Dixit, A. (2018). Application of Silica-Gel-Reinforced Aluminium Composite on the Piston of Internal Combustion Engine: Comparative Study of Silica-Gel-Reinforced Aluminium Composite Piston With Aluminium Alloy Piston. In K. Kumar & J. Davim (Eds.), *Composites and Advanced Materials for Industrial Applications* (pp. 63–98). Hershey, PA: IGI Global. doi:10.4018/978-1-5225-5216-1.ch004

Drabecki, M. P., & Kułak, K. B. (2021). Global Pandemics on European Electrical Energy Markets: Lessons Learned From the COVID-19 Outbreak. *International Journal of Energy Optimization and Engineering, 10*(3), 24–46. https://doi.org/10.4018/IJEOE.2021070102

Dutta, M. M. (2021). Nanomaterials for Food and Agriculture. In M. Bhat, I. Wani, & S. Ashraf (Eds.), *Applications of Nanomaterials in Agriculture, Food Science, and Medicine* (pp. 75–97). IGI Global. doi:10.4018/978-1-7998-5563-7.ch004

Dutta, M. M., & Goswami, M. (2021). Coating Materials: Nano-Materials. In S. Roy & G. Bose (Eds.), *Advanced Surface Coating Techniques for Modern Industrial Applications* (pp. 1–30). IGI Global. doi:10.4018/978-1-7998-4870-7.ch001

Elsayed, A. M., Dakkama, H. J., Mahmoud, S., Al-Dadah, R., & Kaialy, W. (2017). Sustainable Cooling Research Using Activated Carbon Adsorbents and Their Environmental Impact. In T. Kobayashi (Ed.), *Applied Environmental Materials Science for Sustainability* (pp. 186–221). Hershey, PA: IGI Global. doi:10.4018/978-1-5225-1971-3.ch009

Ercanoglu, M., & Sonmez, H. (2018). General Trends and New Perspectives on Landslide Mapping and Assessment Methods. In N. Ceryan (Ed.), *Handbook of Research on Trends and Digital Advances in Engineering Geology* (pp. 350–379). Hershey, PA: IGI Global. doi:10.4018/978-1-5225-2709-1.ch010

Faroz, S. A., Pujari, N. N., Rastogi, R., & Ghosh, S. (2017). Risk Analysis of Structural Engineering Systems Using Bayesian Inference. In P. Samui, S. Chakraborty, & D. Kim (Eds.), *Modeling and Simulation Techniques in Structural Engineering* (pp. 390–424). Hershey, PA: IGI Global. doi:10.4018/978-1-5225-0588-4.ch014

Fekik, A., Hamida, M. L., Denoun, H., Azar, A. T., Kamal, N. A., Vaidyanathan, S., Bousbaine, A., & Benamrouche, N. (2022). Multilevel Inverter for Hybrid Fuel Cell/PV Energy Conversion System. In A. Fekik & N. Benamrouche (Eds.), *Modeling and Control of Static Converters for Hybrid Storage Systems* (pp. 233–270). IGI Global. https://doi.org/10.4018/978-1-7998-7447-8.ch009

Fekik, A., Hamida, M. L., Houassine, H., Azar, A. T., Kamal, N. A., Denoun, H., Vaidyanathan, S., & Sambas, A. (2022). Power Quality Improvement for Grid-Connected Photovoltaic Panels Using Direct Power Control. In A. Fekik & N. Benamrouche (Eds.), *Modeling and Control of Static Converters for Hybrid Storage Systems* (pp. 107–142). IGI Global. https://doi.org/10.4018/978-1-7998-7447-8.ch005

Fernando, P. R., Hamigah, T., Disne, S., Wickramasingha, G. G., & Sutharshan, A. (2018). The Evaluation of Engineering Properties of Low Cost Concrete Blocks by Partial Doping of Sand with Sawdust: Low Cost Sawdust Concrete Block. *International Journal of Strategic Engineering*, *1*(2), 26–42. doi:10.4018/IJoSE.2018070103

Ferro, G., Minciardi, R., Parodi, L., & Robba, M. (2022). Optimal Charging Management of Microgrid-Integrated Electric Vehicles. In M. Alam, R. Pillai, & N. Murugesan (Eds.), *Developing Charging Infrastructure and Technologies for Electric Vehicles* (pp. 133–155). IGI Global. https://doi.org/10.4018/978-1-7998-6858-3.ch007

Flumerfelt, S., & Green, C. (2022). Graduate Lean Leadership Education: A Case Study of a Program. In A. Alves & N. van Hattum-Janssen (Eds.), *Training Engineering Students for Modern Technological Advancement* (pp. 202–224). IGI Global. https://doi.org/10.4018/978-1-7998-8816-1.ch010

Galli, B. J. (2021). Implications of Economic Decision Making to the Project Manager. *International Journal of Strategic Engineering*, *4*(1), 19–32. https://doi.org/10.4018/IJoSE.2021010102

Gento, A. M., Pimentel, C., & Pascual, J. A. (2022). Teaching Circular Economy and Lean Management in a Learning Factory. In A. Alves & N. van Hattum-Janssen (Eds.), *Training Engineering Students for Modern Technological Advancement* (pp. 183–201). IGI Global. https://doi.org/10.4018/978-1-7998-8816-1.ch009

Ghosh, S., Mitra, S., Ghosh, S., & Chakraborty, S. (2017). Seismic Reliability Analysis in the Framework of Metamodelling Based Monte Carlo Simulation. In P. Samui, S. Chakraborty, & D. Kim (Eds.), *Modeling and Simulation Techniques in Structural Engineering* (pp. 192–208). Hershey, PA: IGI Global. doi:10.4018/978-1-5225-0588-4.ch006

Gil, M., & Otero, B. (2017). Learning Engineering Skills through Creativity and Collaboration: A Game-Based Proposal. In R. Alexandre Peixoto de Queirós & M. Pinto (Eds.), *Gamification-Based E-Learning Strategies for Computer Programming Education* (pp. 14–29). Hershey, PA: IGI Global. doi:10.4018/978-1-5225-1034-5.ch002

Gill, J., Ayre, M., & Mills, J. (2017). Revisioning the Engineering Profession: How to Make It Happen! In M. Gray & K. Thomas (Eds.), *Strategies for Increasing Diversity in Engineering Majors and Careers* (pp. 156–175). Hershey, PA: IGI Global. doi:10.4018/978-1-5225-2212-6.ch008

Godzhaev, Z., Senkevich, S., Kuzmin, V., & Melikov, I. (2021). Use of the Neural Network Controller of Sprung Mass to Reduce Vibrations From Road Irregularities. In P. Vasant, G. Weber, & W. Punurai (Ed.), *Research Advancements in Smart Technology, Optimization, and Renewable Energy* (pp. 69-87). IGI Global. https://doi.org/10.4018/978-1-7998-3970-5.ch005

Gomes de Gusmão, C. M. (2022). Digital Competencies and Transformation in Higher Education: Upskilling With Extension Actions. In A. Alves & N. van Hattum-Janssen (Eds.), *Training Engineering Students for Modern Technological Advancement* (pp. 313–328). IGI Global. https://doi.org/10.4018/978-1-7998-8816-1.ch015A

Goyal, N., Ram, M., & Kumar, P. (2017). Welding Process under Fault Coverage Approach for Reliability and MTTF. In M. Ram & J. Davim (Eds.), *Mathematical Concepts and Applications in Mechanical Engineering and Mechatronics* (pp. 222–245). Hershey, PA: IGI Global. doi:10.4018/978-1-5225-1639-2.ch011

Gray, M., & Lundy, C. (2017). Engineering Study Abroad: High Impact Strategy for Increasing Access. In M. Gray & K. Thomas (Eds.), *Strategies for Increasing Diversity in Engineering Majors and Careers* (pp. 42–59). Hershey, PA: IGI Global. doi:10.4018/978-1-5225-2212-6.ch003

Güler, O., & Varol, T. (2021). Fabrication of Functionally Graded Metal and Ceramic Powders Synthesized by Electroless Deposition. In S. Roy & G. Bose (Eds.), *Advanced Surface Coating Techniques for Modern Industrial Applications* (pp. 150–187). IGI Global. https://doi.org/10.4018/978-1-7998-4870-7.ch007

Guraksin, G. E. (2018). Internet of Things and Nature-Inspired Intelligent Techniques for the Future of Biomedical Engineering. In U. Kose, G. Guraksin, & O. Deperlioglu (Eds.), *Nature-Inspired Intelligent Techniques for Solving Biomedical Engineering Problems* (pp. 263–282). Hershey, PA: IGI Global. doi:10.4018/978-1-5225-4769-3.ch013

Hamida, M. L., Fekik, A., Denoun, H., Ardjal, A., & Bokhtache, A. A. (2022). Flying Capacitor Inverter Integration in a Renewable Energy System. In A. Fekik & N. Benamrouche (Eds.), *Modeling and Control of Static Converters for Hybrid Storage Systems* (pp. 287–306). IGI Global. https://doi.org/10.4018/978-1-7998-7447-8.ch011

Hasegawa, N., & Takahashi, Y. (2021). Control of Soap Bubble Ejection Robot Using Facial Expressions. *International Journal of Manufacturing, Materials, and Mechanical Engineering*, *11*(2), 1–16. https://doi.org/10.4018/IJMMME.2021040101

Hejazi, T., & Akbari, L. (2017). A Multiresponse Optimization Model for Statistical Design of Processes with Discrete Variables. In M. Ram & J. Davim (Eds.), *Mathematical Concepts and Applications in Mechanical Engineering and Mechatronics* (pp. 17–37). Hershey, PA: IGI Global. doi:10.4018/978-1-5225-1639-2.ch002

Hejazi, T., & Hejazi, A. (2017). Monte Carlo Simulation for Reliability-Based Design of Automotive Complex Subsystems. In M. Ram & J. Davim (Eds.), *Mathematical Concepts and Applications in Mechanical Engineering and Mechatronics* (pp. 177–200). Hershey, PA: IGI Global. doi:10.4018/978-1-5225-1639-2.ch009

Hejazi, T., & Poursabbagh, H. (2017). Reliability Analysis of Engineering Systems: An Accelerated Life Testing for Boiler Tubes. In M. Ram & J. Davim (Eds.), *Mathematical Concepts and Applications in Mechanical Engineering and Mechatronics* (pp. 154–176). Hershey, PA: IGI Global. doi:10.4018/978-1-5225-1639-2.ch008

Henao, J., Poblano-Salas, C. A., Vargas, F., Giraldo-Betancur, A. L., Corona-Castuera, J., & Sotelo-Mazón, O. (2021). Principles and Applications of Thermal Spray Coatings. In S. Roy & G. Bose (Eds.), *Advanced Surface Coating Techniques for Modern Industrial Applications* (pp. 31–70). IGI Global. https://doi.org/10.4018/978-1-7998-4870-7.ch002

Henao, J., & Sotelo, O. (2018). Surface Engineering at High Temperature: Thermal Cycling and Corrosion Resistance. In A. Pakseresht (Ed.), *Production, Properties, and Applications of High Temperature Coatings* (pp. 131–159). Hershey, PA: IGI Global. doi:10.4018/978-1-5225-4194-3.ch006

Hrnčič, M. K., Cör, D., & Knez, Ž. (2021). Supercritical Fluids as a Tool for Green Energy and Chemicals. In L. Chen (Ed.), *Handbook of Research on Advancements in Supercritical Fluids Applications for Sustainable Energy Systems* (pp. 761–791). IGI Global. doi:10.4018/978-1-7998-5796-9.ch021

Ibrahim, O., Erdem, S., & Gurbuz, E. (2021). Studying Physical and Chemical Properties of Graphene Oxide and Reduced Graphene Oxide and Their Applications in Sustainable Building Materials. In R. González-Lezcano (Ed.), *Advancements in Sustainable Architecture and Energy Efficiency* (pp. 221–238). IGI Global. https://doi.org/10.4018/978-1-7998-7023-4.ch010

Ihianle, I. K., Islam, S., Naeem, U., & Ebenuwa, S. H. (2021). Exploiting Patterns of Object Use for Human Activity Recognition. In A. Nwajana & I. Ihianle (Eds.), *Handbook of Research on 5G Networks and Advancements in Computing, Electronics, and Electrical Engineering* (pp. 382–401). IGI Global. https://doi.org/10.4018/978-1-7998-6992-4.ch015

Ijemaru, G. K., Ngharamike, E. T., Oleka, E. U., & Nwajana, A. O. (2021). An Energy-Efficient Model for Opportunistic Data Collection in IoV-Enabled SC Waste Management. In A. Nwajana & I. Ihianle (Eds.), *Handbook of Research on 5G Networks and Advancements in Computing, Electronics, and Electrical Engineering* (pp. 1–19). IGI Global. https://doi.org/10.4018/978-1-7998-6992-4.ch001

Ilori, O. O., Adetan, D. A., & Umoru, L. E. (2017). Effect of Cutting Parameters on the Surface Residual Stress of Face-Milled Pearlitic Ductile Iron. *International Journal of Materials Forming and Machining Processes*, *4*(1), 38–52. doi:10.4018/IJMFMP.2017010103

Imam, M. H., Tasadduq, I. A., Ahmad, A., Aldosari, F., & Khan, H. (2017). Automated Generation of Course Improvement Plans Using Expert System. *International Journal of Quality Assurance in Engineering and Technology Education*, *6*(1), 1–12. doi:10.4018/IJQAETE.2017010101

Injeti, S. K., & Kumar, T. V. (2018). A WDO Framework for Optimal Deployment of DGs and DSCs in a Radial Distribution System Under Daily Load Pattern to Improve Techno-Economic Benefits. *International Journal of Energy Optimization and Engineering*, *7*(2), 1–38. doi:10.4018/IJEOE.2018040101

Ishii, N., Anami, K., & Knisely, C. W. (2018). *Dynamic Stability of Hydraulic Gates and Engineering for Flood Prevention*. Hershey, PA: IGI Global. doi:10.4018/978-1-5225-3079-4

Iwamoto, Y., & Yamaguchi, H. (2021). Application of Supercritical Carbon Dioxide for Solar Water Heater. In L. Chen (Ed.), *Handbook of Research on Advancements in Supercritical Fluids Applications for Sustainable Energy Systems* (pp. 370–387). IGI Global. https://doi.org/10.4018/978-1-7998-5796-9.ch010

Jayapalan, S. (2018). A Review of Chemical Treatments on Natural Fibers-Based Hybrid Composites for Engineering Applications. In K. Kumar & J. Davim (Eds.), *Composites and Advanced Materials for Industrial Applications* (pp. 16–37). Hershey, PA: IGI Global. doi:10.4018/978-1-5225-5216-1.ch002

Kapetanakis, T. N., Vardiambasis, I. O., Ioannidou, M. P., & Konstantaras, A. I. (2021). Modeling Antenna Radiation Using Artificial Intelligence Techniques: The Case of a Circular Loop Antenna. In C. Nikolopoulos (Ed.), *Recent Trends on Electromagnetic Environmental Effects for Aeronautics and Space Applications* (pp. 186–225). IGI Global. https://doi.org/10.4018/978-1-7998-4879-0.ch007

Karkalos, N. E., Markopoulos, A. P., & Dossis, M. F. (2017). Optimal Model Parameters of Inverse Kinematics Solution of a 3R Robotic Manipulator Using ANN Models. *International Journal of Manufacturing, Materials, and Mechanical Engineering*, *7*(3), 20–40. doi:10.4018/IJMMME.2017070102

Kelly, M., Costello, M., Nicholson, G., & O'Connor, J. (2021). The Evolving Integration of BIM Into Built Environment Programmes in a Higher Education Institute. In J. Underwood & M. Shelbourn (Eds.), *Handbook of Research on Driving Transformational Change in the Digital Built Environment* (pp. 294–326). IGI Global. https://doi.org/10.4018/978-1-7998-6600-8.ch012

Kesimal, A., Karaman, K., Cihangir, F., & Ercikdi, B. (2018). Excavatability Assessment of Rock Masses for Geotechnical Studies. In N. Ceryan (Ed.), *Handbook of Research on Trends and Digital Advances in Engineering Geology* (pp. 231–256). Hershey, PA: IGI Global. doi:10.4018/978-1-5225-2709-1.ch006

Knoflacher, H. (2017). The Role of Engineers and Their Tools in the Transport Sector after Paradigm Change: From Assumptions and Extrapolations to Science. In H. Knoflacher & E. Ocalir-Akunal (Eds.), *Engineering Tools and Solutions for Sustainable Transportation Planning* (pp. 1–29). Hershey, PA: IGI Global. doi:10.4018/978-1-5225-2116-7.ch001

Kose, U. (2018). Towards an Intelligent Biomedical Engineering With Nature-Inspired Artificial Intelligence Techniques. In U. Kose, G. Guraksin, & O. Deperlioglu (Eds.), *Nature-Inspired Intelligent Techniques for Solving Biomedical Engineering Problems* (pp. 1–26). Hershey, PA: IGI Global. doi:10.4018/978-1-5225-4769-3.ch001

Kostić, S. (2018). A Review on Enhanced Stability Analyses of Soil Slopes Using Statistical Design. In N. Ceryan (Ed.), *Handbook of Research on Trends and Digital Advances in Engineering Geology* (pp. 446–481). Hershey, PA: IGI Global. doi:10.4018/978-1-5225-2709-1.ch013

Kumar, A., Patil, P. P., & Prajapati, Y. K. (2018). *Advanced Numerical Simulations in Mechanical Engineering*. Hershey, PA: IGI Global. doi:10.4018/978-1-5225-3722-9

Kumar, G. R., Rajyalakshmi, G., & Manupati, V. K. (2017). Surface Micro Patterning of Aluminium Reinforced Composite through Laser Peening. *International Journal of Manufacturing, Materials, and Mechanical Engineering*, *7*(4), 15–27. doi:10.4018/IJMMME.2017100102

Kumar, N., Basu, D. N., & Chen, L. (2021). Effect of Flow Acceleration and Buoyancy on Thermalhydraulics of sCO2 in Mini/Micro-Channel. In L. Chen (Ed.), *Handbook of Research on Advancements in Supercritical Fluids Applications for Sustainable Energy Systems* (pp. 161–182). IGI Global. doi:10.4018/978-1-7998-5796-9.ch005

Kumari, N., & Kumar, K. (2018). Fabrication of Orthotic Calipers With Epoxy-Based Green Composite. In K. Kumar & J. Davim (Eds.), *Composites and Advanced Materials for Industrial Applications* (pp. 157–176). Hershey, PA: IGI Global. doi:10.4018/978-1-5225-5216-1.ch008

Kuppusamy, R. R. (2018). Development of Aerospace Composite Structures Through Vacuum-Enhanced Resin Transfer Moulding Technology (VERTMTy): Vacuum-Enhanced Resin Transfer Moulding. In K. Kumar & J. Davim (Eds.), *Composites and Advanced Materials for Industrial Applications* (pp. 99–111). Hershey, PA: IGI Global. doi:10.4018/978-1-5225-5216-1.ch005

Kurganov, V. A., Zeigarnik, Y. A., & Maslakova, I. V. (2021). Normal and Deteriorated Heat Transfer Under Heating Turbulent Supercritical Pressure Coolants Flows in Round Tubes. In L. Chen (Ed.), *Handbook of Research on Advancements in Supercritical Fluids Applications for Sustainable Energy Systems* (pp. 494–532). IGI Global. https://doi.org/10.4018/978-1-7998-5796-9.ch014

Li, H., & Zhang, Y. (2021). Heat Transfer and Fluid Flow Modeling for Supercritical Fluids in Advanced Energy Systems. In L. Chen (Ed.), *Handbook of Research on Advancements in Supercritical Fluids Applications for Sustainable Energy Systems* (pp. 388–422). IGI Global. https://doi.org/10.4018/978-1-7998-5796-9.ch011

Loy, J., Howell, S., & Cooper, R. (2017). Engineering Teams: Supporting Diversity in Engineering Education. In M. Gray & K. Thomas (Eds.), *Strategies for Increasing Diversity in Engineering Majors and Careers* (pp. 106–129). Hershey, PA: IGI Global. doi:10.4018/978-1-5225-2212-6.ch006

Macher, G., Armengaud, E., Kreiner, C., Brenner, E., Schmittner, C., Ma, Z., ... Krammer, M. (2018). Integration of Security in the Development Lifecycle of Dependable Automotive CPS. In N. Druml, A. Genser, A. Krieg, M. Menghin, & A. Hoeller (Eds.), *Solutions for Cyber-Physical Systems Ubiquity* (pp. 383–423). Hershey, PA: IGI Global. doi:10.4018/978-1-5225-2845-6.ch015

Madhu, M. N., Singh, J. G., Mohan, V., & Ongsakul, W. (2021). Transmission Risk Optimization in Interconnected Systems: Risk-Adjusted Available Transfer Capability. In P. Vasant, G. Weber, & W. Punurai (Ed.), *Research Advancements in Smart Technology, Optimization, and Renewable Energy* (pp. 183-199). IGI Global. https://doi.org/10.4018/978-1-7998-3970-5.ch010

Mahendramani, G., & Lakshmana Swamy, N. (2018). Effect of Weld Groove Area on Distortion of Butt Welded Joints in Submerged Arc Welding. *International Journal of Manufacturing, Materials, and Mechanical Engineering*, *8*(2), 33–44. doi:10.4018/IJMMME.2018040103

Makropoulos, G., Koumaras, H., Setaki, F., Filis, K., Lutz, T., Montowtt, P., Tomaszewski, L., Dybiec, P., & Järvet, T. (2021). 5G and Unmanned Aerial Vehicles (UAVs) Use Cases: Analysis of the Ecosystem, Architecture, and Applications. In A. Nwajana & I. Ihianle (Eds.), *Handbook of Research on 5G Networks and Advancements in Computing, Electronics, and Electrical Engineering* (pp. 36–69). IGI Global. https://doi.org/10.4018/978-1-7998-6992-4.ch003

Meric, E. M., Erdem, S., & Gurbuz, E. (2021). Application of Phase Change Materials in Construction Materials for Thermal Energy Storage Systems in Buildings. In R. González-Lezcano (Ed.), *Advancements in Sustainable Architecture and Energy Efficiency* (pp. 1–20). IGI Global. https://doi.org/10.4018/978-1-7998-7023-4.ch001

Mihret, E. T., & Yitayih, K. A. (2021). Operation of VANET Communications: The Convergence of UAV System With LTE/4G and WAVE Technologies. *International Journal of Smart Vehicles and Smart Transportation*, *4*(1), 29–51. https://doi.org/10.4018/IJSVST.2021010103

Mir, M. A., Bhat, B. A., Sheikh, B. A., Rather, G. A., Mehraj, S., & Mir, W. R. (2021). Nanomedicine in Human Health Therapeutics and Drug Delivery: Nanobiotechnology and Nanobiomedicine. In M. Bhat, I. Wani, & S. Ashraf (Eds.), *Applications of Nanomaterials in Agriculture, Food Science, and Medicine* (pp. 229–251). IGI Global. doi:10.4018/978-1-7998-5563-7.ch013

Mohammadzadeh, S., & Kim, Y. (2017). Nonlinear System Identification of Smart Buildings. In P. Samui, S. Chakraborty, & D. Kim (Eds.), *Modeling and Simulation Techniques in Structural Engineering* (pp. 328–347). Hershey, PA: IGI Global. doi:10.4018/978-1-5225-0588-4.ch011

Molina, G. J., Aktaruzzaman, F., Soloiu, V., & Rahman, M. (2017). Design and Testing of a Jet-Impingement Instrument to Study Surface-Modification Effects by Nanofluids. *International Journal of Surface Engineering and Interdisciplinary Materials Science*, *5*(2), 43–61. doi:10.4018/IJSEIMS.2017070104

Moreno-Rangel, A., & Carrillo, G. (2021). Energy-Efficient Homes: A Heaven for Respiratory Illnesses. In R. González-Lezcano (Ed.), *Advancements in Sustainable Architecture and Energy Efficiency* (pp. 49–71). IGI Global. https://doi.org/10.4018/978-1-7998-7023-4.ch003

Msomi, V., & Jantjies, B. T. (2021). Correlative Analysis Between Tensile Properties and Tool Rotational Speeds of Friction Stir Welded Similar Aluminium Alloy Joints. *International Journal of Surface Engineering and Interdisciplinary Materials Science*, *9*(2), 58–78. https://doi.org/10.4018/IJSEIMS.2021070104

Muigai, M. N., Mwema, F. M., Akinlabi, E. T., & Obiko, J. O. (2021). Surface Engineering of Materials Through Weld-Based Technologies: An Overview. In S. Roy & G. Bose (Eds.), *Advanced Surface Coating Techniques for Modern Industrial Applications* (pp. 247–260). IGI Global. doi:10.4018/978-1-7998-4870-7.ch011

Mukherjee, A., Saeed, R. A., Dutta, S., & Naskar, M. K. (2017). Fault Tracking Framework for Software-Defined Networking (SDN). In C. Singhal & S. De (Eds.), *Resource Allocation in Next-Generation Broadband Wireless Access Networks* (pp. 247–272). Hershey, PA: IGI Global. doi:10.4018/978-1-5225-2023-8.ch011

Mukhopadhyay, A., Barman, T. K., & Sahoo, P. (2018). Electroless Nickel Coatings for High Temperature Applications. In K. Kumar & J. Davim (Eds.), *Composites and Advanced Materials for Industrial Applications* (pp. 297–331). Hershey, PA: IGI Global. doi:10.4018/978-1-5225-5216-1.ch013

Mwema, F. M., & Wambua, J. M. (2022). Machining of Poly Methyl Methacrylate (PMMA) and Other Olymeric Materials: A Review. In K. Kumar, B. Babu, & J. Davim (Eds.), *Handbook of Research on Advancements in the Processing, Characterization, and Application of Lightweight Materials* (pp. 363–379). IGI Global. https://doi.org/10.4018/978-1-7998-7864-3.ch016

Mykhailyshyn, R., Savkiv, V., Boyko, I., Prada, E., & Virgala, I. (2021). Substantiation of Parameters of Friction Elements of Bernoulli Grippers With a Cylindrical Nozzle. *International Journal of Manufacturing, Materials, and Mechanical Engineering*, *11*(2), 17–39. https://doi.org/10.4018/IJMMME.2021040102

Náprstek, J., & Fischer, C. (2017). Dynamic Stability and Post-Critical Processes of Slender Auto-Parametric Systems. In V. Plevris, G. Kremmyda, & Y. Fahjan (Eds.), *Performance-Based Seismic Design of Concrete Structures and Infrastructures* (pp. 128–171). Hershey, PA: IGI Global. doi:10.4018/978-1-5225-2089-4.ch006

Nautiyal, L., Shivach, P., & Ram, M. (2018). Optimal Designs by Means of Genetic Algorithms. In M. Ram & J. Davim (Eds.), *Soft Computing Techniques and Applications in Mechanical Engineering* (pp. 151–161). Hershey, PA: IGI Global. doi:10.4018/978-1-5225-3035-0.ch007

Nazir, R. (2017). Advanced Nanomaterials for Water Engineering and Treatment: Nano-Metal Oxides and Their Nanocomposites. In T. Saleh (Ed.), *Advanced Nanomaterials for Water Engineering, Treatment, and Hydraulics* (pp. 84–126). Hershey, PA: IGI Global. doi:10.4018/978-1-5225-2136-5.ch005

Nikolopoulos, C. D. (2021). Recent Advances on Measuring and Modeling ELF-Radiated Emissions for Space Applications. In C. Nikolopoulos (Ed.), *Recent Trends on Electromagnetic Environmental Effects for Aeronautics and Space Applications* (pp. 1–38). IGI Global. https://doi.org/10.4018/978-1-7998-4879-0.ch001

Nogueira, A. F., Ribeiro, J. C., Fernández de Vega, F., & Zenha-Rela, M. A. (2018). Evolutionary Approaches to Test Data Generation for Object-Oriented Software: Overview of Techniques and Tools. In M. Khosrow-Pour, D.B.A. (Ed.), Incorporating Nature-Inspired Paradigms in Computational Applications (pp. 162-194). Hershey, PA: IGI Global. https://doi.org/ doi:10.4018/978-1-5225-5020-4.ch006

Nwajana, A. O., Obi, E. R., Ijemaru, G. K., Oleka, E. U., & Anthony, D. C. (2021). Fundamentals of RF/Microwave Bandpass Filter Design. In A. Nwajana & I. Ihianle (Eds.), *Handbook of Research on 5G Networks and Advancements in Computing, Electronics, and Electrical Engineering* (pp. 149–164). IGI Global. https://doi.org/10.4018/978-1-7998-6992-4.ch005

Ogbodo, E. A. (2021). Comparative Study of Transmission Line Junction vs. Asynchronously Coupled Junction Diplexers. In A. Nwajana & I. Ihianle (Eds.), *Handbook of Research on 5G Networks and Advancements in Computing, Electronics, and Electrical Engineering* (pp. 326–336). IGI Global. https://doi.org/10.4018/978-1-7998-6992-4.ch013

Orosa, J. A., Vergara, D., Fraguela, F., & Masdías-Bonome, A. (2021). Statistical Understanding and Optimization of Building Energy Consumption and Climate Change Consequences. In R. González-Lezcano (Ed.), *Advancements in Sustainable Architecture and Energy Efficiency* (pp. 195–220). IGI Global. https://doi.org/10.4018/978-1-7998-7023-4.ch009

Osho, M. B. (2018). Industrial Enzyme Technology: Potential Applications. In S. Bharati & P. Chaurasia (Eds.), *Research Advancements in Pharmaceutical, Nutritional, and Industrial Enzymology* (pp. 375–394). Hershey, PA: IGI Global. doi:10.4018/978-1-5225-5237-6.ch017

Ouadi, A., & Zitouni, A. (2021). Phasor Measurement Improvement Using Digital Filter in a Smart Grid. In A. Recioui & H. Bentarzi (Eds.), *Optimizing and Measuring Smart Grid Operation and Control* (pp. 100–117). IGI Global. https://doi.org/10.4018/978-1-7998-4027-5.ch005

Padmaja, P., & Marutheswar, G. (2017). Certain Investigation on Secured Data Transmission in Wireless Sensor Networks. *International Journal of Mobile Computing and Multimedia Communications*, *8*(1), 48–61. doi:10.4018/IJMCMC.2017010104

Palmer, S., & Hall, W. (2017). An Evaluation of Group Work in First-Year Engineering Design Education. In R. Tucker (Ed.), *Collaboration and Student Engagement in Design Education* (pp. 145–168). Hershey, PA: IGI Global. doi:10.4018/978-1-5225-0726-0.ch007

Panchenko, V. (2021). Prospects for Energy Supply of the Arctic Zone Objects of Russia Using Frost-Resistant Solar Modules. In P. Vasant, G. Weber, & W. Punurai (Eds.), *Research Advancements in Smart Technology, Optimization, and Renewable Energy* (pp. 149-169). IGI Global. https://doi.org/10.4018/978-1-7998-3970-5.ch008

Panchenko, V. (2021). Photovoltaic Thermal Module With Paraboloid Type Solar Concentrators. *International Journal of Energy Optimization and Engineering*, *10*(2), 1–23. https://doi.org/10.4018/IJEOE.2021040101

Pandey, K., & Datta, S. (2021). Dry Machining of Inconel 825 Superalloys: Performance of Tool Inserts (Carbide, Cermet, and SiAlON). *International Journal of Manufacturing, Materials, and Mechanical Engineering*, *11*(4), 26–39. doi:10.4018/IJMMME.2021100102

Panneer, R. (2017). Effect of Composition of Fibers on Properties of Hybrid Composites. *International Journal of Manufacturing, Materials, and Mechanical Engineering*, *7*(4), 28–43. doi:10.4018/IJMMME.2017100103

Pany, C. (2021). Estimation of Correct Long-Seam Mismatch Using FEA to Compare the Measured Strain in a Non-Destructive Testing of a Pressurant Tank: A Reverse Problem. *International Journal of Smart Vehicles and Smart Transportation*, *4*(1), 16–28. doi:10.4018/IJSVST.2021010102

Paul, S., & Roy, P. (2018). Optimal Design of Power System Stabilizer Using a Novel Evolutionary Algorithm. *International Journal of Energy Optimization and Engineering*, *7*(3), 24–46. doi:10.4018/IJEOE.2018070102

Paul, S., & Roy, P. K. (2021). Oppositional Differential Search Algorithm for the Optimal Tuning of Both Single Input and Dual Input Power System Stabilizer. In P. Vasant, G. Weber, & W. Punurai (Eds.), *Research Advancements in Smart Technology, Optimization, and Renewable Energy* (pp. 256-282). IGI Global. https://doi.org/10.4018/978-1-7998-3970-5.ch013

Pavaloiu, A. (2018). Artificial Intelligence Ethics in Biomedical-Engineering-Oriented Problems. In U. Kose, G. Guraksin, & O. Deperlioglu (Eds.), *Nature-Inspired Intelligent Techniques for Solving Biomedical Engineering Problems* (pp. 219–231). Hershey, PA: IGI Global. doi:10.4018/978-1-5225-4769-3.ch010

Pioro, I., Mahdi, M., & Popov, R. (2017). Application of Supercritical Pressures in Power Engineering. In L. Chen & Y. Iwamoto (Eds.), *Advanced Applications of Supercritical Fluids in Energy Systems* (pp. 404–457). Hershey, PA: IGI Global. doi:10.4018/978-1-5225-2047-4.ch013

Plaksina, T., & Gildin, E. (2017). Rigorous Integrated Evolutionary Workflow for Optimal Exploitation of Unconventional Gas Assets. *International Journal of Energy Optimization and Engineering*, *6*(1), 101–122. doi:10.4018/IJEOE.2017010106

Popat, J., Kakadiya, H., Tak, L., Singh, N. K., Majeed, M. A., & Mahajan, V. (2021). Reliability of Smart Grid Including Cyber Impact: A Case Study. In R. Singh, A. Singh, A. Dwivedi, & P. Nagabhushan (Eds.), *Computational Methodologies for Electrical and Electronics Engineers* (pp. 163–174). IGI Global. https://doi.org/10.4018/978-1-7998-3327-7.ch013

Quiza, R., La Fé-Perdomo, I., Rivas, M., & Ramtahalsing, V. (2021). Triple Bottom Line-Focused Optimization of Oblique Turning Processes Based on Hybrid Modeling: A Study Case on AISI 1045 Steel Turning. In L. Burstein (Ed.), *Handbook of Research on Advancements in Manufacturing, Materials, and Mechanical Engineering* (pp. 215–241). IGI Global. https://doi.org/10.4018/978-1-7998-4939-1.ch010

Rahmani, M. K. (2022). Blockchain Technology: Principles and Algorithms. In S. Khan, M. Syed, R. Hammad, & A. Bushager (Eds.), *Blockchain Technology and Computational Excellence for Society 5.0* (pp. 16–27). IGI Global. https://doi.org/10.4018/978-1-7998-8382-1.ch002

Ramdani, N., & Azibi, M. (2018). Polymer Composite Materials for Microelectronics Packaging Applications: Composites for Microelectronics Packaging. In K. Kumar & J. Davim (Eds.), *Composites and Advanced Materials for Industrial Applications* (pp. 177–211). Hershey, PA: IGI Global. doi:10.4018/978-1-5225-5216-1.ch009

Ramesh, M., Garg, R., & Subrahmanyam, G. V. (2017). Investigation of Influence of Quenching and Annealing on the Plane Fracture Toughness and Brittle to Ductile Transition Temperature of the Zinc Coated Structural Steel Materials. *International Journal of Surface Engineering and Interdisciplinary Materials Science*, *5*(2), 33–42. doi:10.4018/IJSEIMS.2017070103

Robinson, J., & Beneroso, D. (2022). Project-Based Learning in Chemical Engineering: Curriculum and Assessment, Culture and Learning Spaces. In A. Alves & N. van Hattum-Janssen (Eds.), *Training Engineering Students for Modern Technological Advancement* (pp. 1–19). IGI Global. https://doi.org/10.4018/978-1-7998-8816-1.ch001

Rondon, B. (2021). Experimental Characterization of Admittance Meter With Crude Oil Emulsions. *International Journal of Electronics, Communications, and Measurement Engineering*, *10*(2), 51–59. https://doi.org/10.4018/IJECME.2021070104

Rudolf, S., Biryuk, V. V., & Volov, V. (2018). Vortex Effect, Vortex Power: Technology of Vortex Power Engineering. In V. Kharchenko & P. Vasant (Eds.), *Handbook of Research on Renewable Energy and Electric Resources for Sustainable Rural Development* (pp. 500–533). Hershey, PA: IGI Global. doi:10.4018/978-1-5225-3867-7.ch021

Sah, A., Bhadula, S. J., Dumka, A., & Rawat, S. (2018). A Software Engineering Perspective for Development of Enterprise Applications. In A. Elçi (Ed.), *Handbook of Research on Contemporary Perspectives on Web-Based Systems* (pp. 1–23). Hershey, PA: IGI Global. doi:10.4018/978-1-5225-5384-7.ch001

Sahli, Y., Zitouni, B., & Hocine, B. M. (2021). Three-Dimensional Numerical Study of Overheating of Two Intermediate Temperature P-AS-SOFC Geometrical Configurations. In G. Badea, R. Felseghi, & I. Aşchilean (Eds.), *Hydrogen Fuel Cell Technology for Stationary Applications* (pp. 186–222). IGI Global. https://doi.org/10.4018/978-1-7998-4945-2.ch008

Sahoo, P., & Roy, S. (2017). Tribological Behavior of Electroless Ni-P, Ni-P-W and Ni-P-Cu Coatings: A Comparison. *International Journal of Surface Engineering and Interdisciplinary Materials Science*, *5*(1), 1–15. doi:10.4018/IJSEIMS.2017010101

Sahoo, S. (2018). Laminated Composite Hypar Shells as Roofing Units: Static and Dynamic Behavior. In K. Kumar & J. Davim (Eds.), *Composites and Advanced Materials for Industrial Applications* (pp. 249–269). Hershey, PA: IGI Global. doi:10.4018/978-1-5225-5216-1.ch011

Sahu, H., & Hungyo, M. (2018). Introduction to SDN and NFV. In A. Dumka (Ed.), *Innovations in Software-Defined Networking and Network Functions Virtualization* (pp. 1–25). Hershey, PA: IGI Global. doi:10.4018/978-1-5225-3640-6.ch001

Salem, A. M., & Shmelova, T. (2018). Intelligent Expert Decision Support Systems: Methodologies, Applications, and Challenges. In T. Shmelova, Y. Sikirda, N. Rizun, A. Salem, & Y. Kovalyov (Eds.), *Socio-Technical Decision Support in Air Navigation Systems: Emerging Research and Opportunities* (pp. 215–242). Hershey, PA: IGI Global. doi:10.4018/978-1-5225-3108-1.ch007

Samal, M. (2017). FE Analysis and Experimental Investigation of Cracked and Un-Cracked Thin-Walled Tubular Components to Evaluate Mechanical and Fracture Properties. In P. Samui, S. Chakraborty, & D. Kim (Eds.), *Modeling and Simulation Techniques in Structural Engineering* (pp. 266–293). Hershey, PA: IGI Global. doi:10.4018/978-1-5225-0588-4.ch009

Samal, M., & Balakrishnan, K. (2017). Experiments on a Ring Tension Setup and FE Analysis to Evaluate Transverse Mechanical Properties of Tubular Components. In P. Samui, S. Chakraborty, & D. Kim (Eds.), *Modeling and Simulation Techniques in Structural Engineering* (pp. 91–115). Hershey, PA: IGI Global. doi:10.4018/978-1-5225-0588-4.ch004

Samarasinghe, D. A., & Wood, E. (2021). Innovative Digital Technologies. In J. Underwood & M. Shelbourn (Eds.), *Handbook of Research on Driving Transformational Change in the Digital Built Environment* (pp. 142–163). IGI Global. https://doi.org/10.4018/978-1-7998-6600-8.ch006

Sawant, S. (2018). Deep Learning and Biomedical Engineering. In U. Kose, G. Guraksin, & O. Deperlioglu (Eds.), *Nature-Inspired Intelligent Techniques for Solving Biomedical Engineering Problems* (pp. 283–296). Hershey, PA: IGI Global. doi:10.4018/978-1-5225-4769-3.ch014

Schulenberg, T. (2021). Energy Conversion Using the Supercritical Steam Cycle. In L. Chen (Ed.), *Handbook of Research on Advancements in Supercritical Fluids Applications for Sustainable Energy Systems* (pp. 659–681). IGI Global. doi:10.4018/978-1-7998-5796-9.ch018

Sezgin, H., & Berkalp, O. B. (2018). Textile-Reinforced Composites for the Automotive Industry. In K. Kumar & J. Davim (Eds.), *Composites and Advanced Materials for Industrial Applications* (pp. 129–156). Hershey, PA: IGI Global. doi:10.4018/978-1-5225-5216-1.ch007

Shaaban, A. A., & Shehata, O. M. (2021). Combining Response Surface Method and Metaheuristic Algorithms for Optimizing SPIF Process. *International Journal of Manufacturing, Materials, and Mechanical Engineering, 11*(4), 1–25. https://doi.org/10.4018/IJMMME.2021100101

Shafaati Shemami, M., & Sefid, M. (2022). Implementation and Demonstration of Electric Vehicle-to-Home (V2H) Application: A Case Study. In M. Alam, R. Pillai, & N. Murugesan (Eds.), *Developing Charging Infrastructure and Technologies for Electric Vehicles* (pp. 268–293). IGI Global. https://doi.org/10.4018/978-1-7998-6858-3.ch015

Shah, M. Z., Gazder, U., Bhatti, M. S., & Hussain, M. (2018). Comparative Performance Evaluation of Effects of Modifier in Asphaltic Concrete Mix. *International Journal of Strategic Engineering, 1*(2), 13–25. doi:10.4018/IJoSE.2018070102

Sharma, N., & Kumar, K. (2018). Fabrication of Porous NiTi Alloy Using Organic Binders. In K. Kumar & J. Davim (Eds.), *Composites and Advanced Materials for Industrial Applications* (pp. 38–62). Hershey, PA: IGI Global. doi:10.4018/978-1-5225-5216-1.ch003

Shivach, P., Nautiyal, L., & Ram, M. (2018). Applying Multi-Objective Optimization Algorithms to Mechanical Engineering. In M. Ram & J. Davim (Eds.), *Soft Computing Techniques and Applications in Mechanical Engineering* (pp. 287–301). Hershey, PA: IGI Global. doi:10.4018/978-1-5225-3035-0.ch014

Shmelova, T. (2018). Stochastic Methods for Estimation and Problem Solving in Engineering: Stochastic Methods of Decision Making in Aviation. In S. Kadry (Ed.), *Stochastic Methods for Estimation and Problem Solving in Engineering* (pp. 139–160). Hershey, PA: IGI Global. doi:10.4018/978-1-5225-5045-7.ch006

Siero González, L. R., & Romo Vázquez, A. (2017). Didactic Sequences Teaching Mathematics for Engineers With Focus on Differential Equations. In M. Ramírez-Montoya (Ed.), *Handbook of Research on Driving STEM Learning With Educational Technologies* (pp. 129–151). Hershey, PA: IGI Global. doi:10.4018/978-1-5225-2026-9.ch007

Sim, M. S., You, K. Y., Esa, F., & Chan, Y. L. (2021). Nanostructured Electromagnetic Metamaterials for Sensing Applications. In M. Bhat, I. Wani, & S. Ashraf (Eds.), *Applications of Nanomaterials in Agriculture, Food Science, and Medicine* (pp. 141–164). IGI Global. https://doi.org/10.4018/978-1-7998-5563-7.ch009

Singh, R., & Dutta, S. (2018). Visible Light Active Nanocomposites for Photocatalytic Applications. In K. Kumar & J. Davim (Eds.), *Composites and Advanced Materials for Industrial Applications* (pp. 270–296). Hershey, PA: IGI Global. doi:10.4018/978-1-5225-5216-1.ch012

Skripov, P. V., Yampol'skiy, A. D., & Rutin, S. B. (2021). High-Power Heat Transfer in Supercritical Fluids: Microscale Times and Sizes. In L. Chen (Ed.), *Handbook of Research on Advancements in Supercritical Fluids Applications for Sustainable Energy Systems* (pp. 424–450). IGI Global. https://doi.org/10.4018/978-1-7998-5796-9.ch012

Sözbilir, H., Özkaymak, Ç., Uzel, B., & Sümer, Ö. (2018). Criteria for Surface Rupture Microzonation of Active Faults for Earthquake Hazards in Urban Areas. In N. Ceryan (Ed.), *Handbook of Research on Trends and Digital Advances in Engineering Geology* (pp. 187–230). Hershey, PA: IGI Global. doi:10.4018/978-1-5225-2709-1.ch005

Stanciu, I. (2018). Stochastic Methods in Microsystems Engineering. In S. Kadry (Ed.), *Stochastic Methods for Estimation and Problem Solving in Engineering* (pp. 161–176). Hershey, PA: IGI Global. doi:10.4018/978-1-5225-5045-7.ch007

Strebkov, D., Nekrasov, A., Trubnikov, V., & Nekrasov, A. (2018). Single-Wire Resonant Electric Power Systems for Renewable-Based Electric Grid. In V. Kharchenko & P. Vasant (Eds.), *Handbook of Research on Renewable Energy and Electric Resources for Sustainable Rural Development* (pp. 449–474). Hershey, PA: IGI Global. doi:10.4018/978-1-5225-3867-7.ch019

Sukhyy, K., Belyanovskaya, E., & Sukhyy, M. (2021). *Basic Principles for Substantiation of Working Pair Choice*. IGI Global. doi:10.4018/978-1-7998-4432-7.ch002

Suri, M. S., & Kaliyaperumal, D. (2022). Extension of Aspiration Level Model for Optimal Planning of Fast Charging Stations. In A. Fekik & N. Benamrouche (Eds.), *Modeling and Control of Static Converters for Hybrid Storage Systems* (pp. 91–106). IGI Global. https://doi.org/10.4018/978-1-7998-7447-8.ch004

Tallet, E., Gledson, B., Rogage, K., Thompson, A., & Wiggett, D. (2021). Digitally-Enabled Design Management. In J. Underwood & M. Shelbourn (Eds.), *Handbook of Research on Driving Transformational Change in the Digital Built Environment* (pp. 63–89). IGI Global. https://doi.org/10.4018/978-1-7998-6600-8.ch003

Terki, A., & Boubertakh, H. (2021). A New Hybrid Binary-Real Coded Cuckoo Search and Tabu Search Algorithm for Solving the Unit-Commitment Problem. *International Journal of Energy Optimization and Engineering, 10*(2), 104–119. https://doi.org/10.4018/IJEOE.2021040105

Tüdeş, Ş., Kumlu, K. B., & Ceryan, S. (2018). Integration Between Urban Planning and Natural Hazards For Resilient City. In N. Ceryan (Ed.), *Handbook of Research on Trends and Digital Advances in Engineering Geology* (pp. 591–630). Hershey, PA: IGI Global. doi:10.4018/978-1-5225-2709-1.ch017

Ulamis, K. (2018). Soil Liquefaction Assessment by Anisotropic Cyclic Triaxial Test. In N. Ceryan (Ed.), *Handbook of Research on Trends and Digital Advances in Engineering Geology* (pp. 631–664). Hershey, PA: IGI Global. doi:10.4018/978-1-5225-2709-1.ch018

Valente, M., & Milani, G. (2017). Seismic Assessment and Retrofitting of an Under-Designed RC Frame Through a Displacement-Based Approach. In V. Plevris, G. Kremmyda, & Y. Fahjan (Eds.), *Performance-Based Seismic Design of Concrete Structures and Infrastructures* (pp. 36–58). Hershey, PA: IGI Global. doi:10.4018/978-1-5225-2089-4.ch002

Vargas-Bernal, R. (2021). Advances in Electromagnetic Environmental Shielding for Aeronautics and Space Applications. In C. Nikolopoulos (Ed.), *Recent Trends on Electromagnetic Environmental Effects for Aeronautics and Space Applications* (pp. 80–96). IGI Global. https://doi.org/10.4018/978-1-7998-4879-0.ch003

Vasant, P. (2018). A General Medical Diagnosis System Formed by Artificial Neural Networks and Swarm Intelligence Techniques. In U. Kose, G. Guraksin, & O. Deperlioglu (Eds.), *Nature-Inspired Intelligent Techniques for Solving Biomedical Engineering Problems* (pp. 130–145). Hershey, PA: IGI Global. doi:10.4018/978-1-5225-4769-3.ch006

Verner, C. M., & Sarwar, D. (2021). Avoiding Project Failure and Achieving Project Success in NHS IT System Projects in the United Kingdom. *International Journal of Strategic Engineering*, *4*(1), 33–54. https://doi.org/10.4018/IJoSE.2021010103

Verrollot, J., Tolonen, A., Harkonen, J., & Haapasalo, H. J. (2018). Challenges and Enablers for Rapid Product Development. *International Journal of Applied Industrial Engineering*, *5*(1), 25–49. doi:10.4018/IJAIE.2018010102

Wan, A. C., Zulu, S. L., & Khosrow-Shahi, F. (2021). Industry Views on BIM for Site Safety in Hong Kong. In J. Underwood & M. Shelbourn (Eds.), *Handbook of Research on Driving Transformational Change in the Digital Built Environment* (pp. 120–140). IGI Global. https://doi.org/10.4018/978-1-7998-6600-8.ch005

Yardimci, A. G., & Karpuz, C. (2018). Fuzzy Rock Mass Rating: Soft-Computing-Aided Preliminary Stability Analysis of Weak Rock Slopes. In N. Ceryan (Ed.), *Handbook of Research on Trends and Digital Advances in Engineering Geology* (pp. 97–131). Hershey, PA: IGI Global. doi:10.4018/978-1-5225-2709-1.ch003

You, K. Y. (2021). Development Electronic Design Automation for RF/Microwave Antenna Using MATLAB GUI. In A. Nwajana & I. Ihianle (Eds.), *Handbook of Research on 5G Networks and Advancements in Computing, Electronics, and Electrical Engineering* (pp. 70–148). IGI Global. https://doi.org/10.4018/978-1-7998-6992-4.ch004

Yousefi, Y., Gratton, P., & Sarwar, D. (2021). Investigating the Opportunities to Improve the Thermal Performance of a Case Study Building in London. *International Journal of Strategic Engineering*, *4*(1), 1–18. https://doi.org/10.4018/IJoSE.2021010101

Zindani, D., & Kumar, K. (2018). Industrial Applications of Polymer Composite Materials. In K. Kumar & J. Davim (Eds.), *Composites and Advanced Materials for Industrial Applications* (pp. 1–15). Hershey, PA: IGI Global. doi:10.4018/978-1-5225-5216-1.ch001

Zindani, D., Maity, S. R., & Bhowmik, S. (2018). A Decision-Making Approach for Material Selection of Polymeric Composite Bumper Beam. In K. Kumar & J. Davim (Eds.), *Composites and Advanced Materials for Industrial Applications* (pp. 112–128). Hershey, PA: IGI Global. doi:10.4018/978-1-5225-5216-1.ch006

About the Contributors

Tanupriya Choudhury completed his undergraduate studies in Computer Science and Engineering at the West Bengal University of Technology in Kolkata (2004-2008), India, followed by a Master's Degree in the same field from Dr. M.G.R University in Chennai, India (2008-2010). In 2016, he successfully obtained his PhD degree from Jagannath University Jaipur. With a total of 14 years of experience in both teaching and research, Dr. Choudhury holds the position of Professor at CSE Department,Symbiosis Institute of Technology, Symbiosis International University, Pune, Maharashtra, 412115, India and also he is holding Visiting Professor at Daffodil International University Bangladesh and Director Research (Honorary) at AI University, Montana US. Prior to this role, he served Graphic Era Hill University Dehradun (Research Professor), UPES Dehradun (Professor), Amity University Noida (Assistant Professor), and other prestigious academic institutions (Dronacharya College of Engineering Gurgaon,Lingaya's University Faridabad, Babu Banarsi Das Institute of Technology Ghaziabad, Syscon Solutions Pvt. Ltd. Kolkata etc.).Recently recognized for his outstanding contributions to education with the Global Outreach Education Award for Excellence in Best Young Researcher Award at GOECA 2018. His areas of expertise encompass Human Computing, Soft Computing, Cloud Computing, Data Mining among others. Notably accomplished within his field thus far is filing 25 patents and securing copyrights for 16 software programs from MHRD (Ministry of Human Resource Development). He has actively participated as an attendee or speaker at numerous National and International conferences across India and abroad. With over hundred plus quality research papers (Scopus) authored to date on record; Dr. Choudhury has also been invited as a guest lecturer or keynote speaker at esteemed institutions such as Jamia Millia Islamia University India, Maharaja Agersen College (Delhi University), Duy Tan University Vietnam etc.He has also contributed significantly to various National/ International conferences throughout India and abroad serving roles like TPC chair/ member and session chairperson. As an active professional within the technical community; Dr.Choudhury holds lifetime membership with IETA (International Engineering & Technology Association) along with being affiliated with IEEE (Institute of Electrical and Electronics Engineers), IET(UK) (Institution of Engineering & Technology UK),and other reputable technical societies.Additionally, he is associated with corporate entities and serves as a Technical Adviser for Deetya Soft Pvt. Ltd., Noida, IVRGURU, and Mydigital360.He is also serving a Editor's in reputed Journals. He currently serves as the Honorary Secretary in IETA (Indian Engineering Teacher's Association-India), alongside his role as the Senior Advisor Position in INDO-UK Confederation of Science, Technology and Research Ltd., London, UK and International Association of Professional and Fellow Engineers-Delaware-USA.

Anitha Mary X completed her B.E.Electonics and Instrumentation Engineering from Bharathiar University in the year 2001, M.E VLSI Design from Anna University in the year 2009, and Ph.D from Karunya University in the year 2015. She is working as an Associate Professor in the Department of Robotics Engineering, Karunya Institute of Technology and Sciences, Coimbatore. Her academic record holds fifteen years of teaching experience. She has 30 Scopus Indexed Publications to her credit. She works in Multidisciplinary research areas in the field of Embedded Systems and sensor interfacing with experts from countries like Israel and Canada. The areas of her expertise are Sensor design, embedded system control algorithms, and Machine Learning Techniques. She is the author of the book "A Beginners Guide for Machine Learning Models with Python Environment" published by LAP LAMBERT Academic Publishing

Subrata Chowdhury is working in the Department of the Computer Science of Engineering of Sreenivasa Institute of Technology And Management as a Associate Professor. He is been working in the IT Industry for more than 5 years in the R&D developments, he has handled many projects in the industry with much dedications and perfect time limits. He has been handling projects related to AI, Blockchains and the Cloud Computing for the companies from various National and Internationals Clients. He had published (4) books from 2014 - 2019 at the domestic market and Internationally Publishers CRC, River . And he been the editor for the 2 books for the CRC& River publisher. He has participated in the Organizing committee, Technical Programmed Committee and Guest Speaker for more than 10 conference and the webinars. He also Reviewed and evaluated more than 50 papers from the conferences and

Karthik Chandran (Member, ACM, Senior Member, IEEE) was born in Madurai, Tamil Nadu, India in 1986. He received the Bachelor of Engineering in Electronics and Instrumentation Engineering at Kamaraj College of Engineering and Technology, India in 2007, the Master's Degree and Ph.D. Degree in Control and Instrumentation Engineering from Kalasalingam Academy of Research and Education (KARE), in 2011 and 2017. In 2011, he joined the Department of Instrumentation and Control Engineering of KARE, India as Assistant Professor. After that, He served as a Lecturer in the Department of Electrical and Computer Engineering, University of Woldia, Ethiopia from 2016–2018. Presently, He was served as a Postdoctoral Researcher at Shanghai Jiaotong University, China. He is serving as Associate Professor in Mechatronics Engineering, at Jyothi Engineering College, Kerala. He is currently involved in research related to Time delay Control problems, Nonlinear system identification, Cascade Control system, and Unmanned vehicle.

C. Suganthi Evangeline currently working as Assistant Professor in the department of Electronics and Communication Engineering at Sri Eshwar College of Engineering, Coimbatore.

A. Jainulafdeen is a dedicated researcher currently pursuing a Ph.D. in Soft Robotics with a focus on Design, Development, and Control using Artificial Intelligence at Anna University, Chennai. His educational journey began with a Bachelor's in Electrical and Electronics Engineering from C.S.I College of Engineering, Ooty, and continued with a Master's in Power Electronics and Drives from Anna University's Regional Centre, Coimbatore. With a strong academic background Jainulafdeen embarked on a career in academia, starting as a Lecturer in EEE and progressing to become the Head of the Department at Sri Ramakrishna Mission Vidyalaya Polytechnic College, a government-aided autonomous institution in Coimbatore. This was followed by a position as Assistant Professor at VSB Engineering College, Karur. His current role is as an Assistant Professor at K. Ramakrishnan College of Engineering, Trichy, where he imparts knowledge and fosters the development of young engineering minds. His areas of interest encompass Power Electronics, Microprocessor and Microcontrollers, Electrical Circuits and Machines, and the exciting domain of Soft Robotics and Artificial Intelligence. His objective is to advance research in Soft Robotics while contributing to academia through teaching and research activities. Jainulafdeen is driven by a passion for innovation and the pursuit of knowledge, making him a promising researcher and educator in the field of engineering.

Sampath Boopathi is an accomplished individual with a strong academic background and extensive research experience. He completed his undergraduate studies in Mechanical Engineering and pursued his postgraduate studies in the field of Computer-Aided Design. Dr. Boopathi obtained his Ph.D. from Anna University, focusing his research on Manufacturing and optimization. Throughout his career, Dr. Boopathi has made significant contributions to the field of engineering. He has authored and published 200 more research articles in internationally peer-reviewed journals, highlighting his expertise and dedication to advancing knowledge in his area of specialization. His research output demonstrates his commitment to conducting rigorous and impactful research. This indicates his innovative thinking and ability to develop practical solutions. With 17 years of academic and research experience, Dr. Boopathi has enriched the engineering community through his teaching and mentorship roles.

Rvea Judie Dolly is an enthusiastic and dedicated professional educator, committed to fostering the social and academic growth of the students. Possessing a strong passion for research, her interests lie in image processing, video compression, machine learning, biomedical signal processing, and pattern recognition. She has successfully published articles in esteemed journals and secured two funded projects from the Indian Council of Medical Research (ICMR) to conduct impactful research benefiting society.

Prabhakar Gunasekaran is currently working as an Assistant Professor in the Department of Electronics and Communication Engineering, Thiagarajar College of Engineering, Madurai – 15 (A Govt. Aided Autonomous Institution Affiliated to Anna University) Tamilnadu, India. He obtained his B.E degree in Electronics & Communication Engineering from Arulmigu Kalasalingam College of Engineering, Krishnankoil, under Anna University, Chennai in the year of2009, and his M.Tech. Degree in the specialization of Embedded Systems from Hindustan University, Chennai in the year 2011. He obtained his Ph.D. degree in the year 2018 under the faculty of Electrical Engineering, at Anna University, Chennai. He is a recognized Ph. D Supervisor of Anna University, Chennai, under the Faculty of Electrical Engineering, and also guiding 5 Ph.D. scholars. He has published more than 35 research articles around the world including reputed journal transactions like IET, Springer, Taylor & Francis, and Elsevier.

J inesh Peter is currently working as Professor, Department of Computer Science and Engineering at Karunya Institute of Technology and Sciences. He holds the positions include heading the office of international affairs and IAESTE at Karunya from 2017. His recent activities include sending students abroad for semester abroad programs and international internships through IAESTE. Prior to this, he was a full-time research scholar at National Institute of Technology, Calicut, India, from where he received his PhD in computer science and engineering. His research focus includes Big-data, cloud computing, image processing, computer vision, artificial intelligence, machine learning and medicinal plants. He has several publications in various reputed international journals and conference paper which are widely referred to. He is a member of IEEE, CSI & IEI and has served as session chairs and delivered plenary speeches for various international conference and workshops. He has conducted many international conferences and been as editor for springer proceedings and many special issues in journals.

D J. Jagannath, a doctorate from the Faculty of Engineering, specialization in Signal Processing, is currently working as Assistant Professor, Department of Electronics and Communication Engineering at Karunya Institute of Technology and Sciences. He is an experienced researcher in the field of Vision or Pattern Recognition, Bio-inspired Systems, Signal Processing, Fuzzy Logic, Neural Networks, Evolutionary Computing, Machine learning, IoT & Cyber Physical Human Systems. He has worked in R & D research projects funded by DST and ICMR, Indian Government and other funding agencies. He has several publications in various reputed international journals, conference publications and book chapters. He has chaired in several internal conferences and has also delivered plenary speeches for various international conference and workshops. He has also conducted IEEE sponsored international conferences. He is a member of IAENG, ISTE, ISHNE and MISTE.

Anupama Jawale holds a Master's degree in Computer Applications (MCA), a Master's degree in Computer Management (MCM), and an MPhil. She started her career in 2004 as a lecturer. Currently she is Head of the Department and Academic Coordinator for B.Sc, IT program at Narsee Monjee College of Commerce & Economics, Vile Parle. Her research interests lie in cutting-edge areas such as the Internet of Things (IoT), Time Series Analysis, and Data Science.

Jsret Kaur is currently working as an Assistant Professor in University Business School,Chandigarh University,Mohali,Punjab.She is a post graduate (MBA-H.R) from Panjab University,Chandigarh.She has also qualified UGC NET JRF in Human Resource Management/Labour and Social Welfare and has pursued PhD in Business Management from Chandigarh University,Mohali. She has over 8 years of experience in academic and administrative assignments. She also received "Best Teacher of the Department Award " in the year 2019 and 2021 in the field of imparting quality education. Her research interests include Employee Engagement, Management of Organizational Change and Organization Development. She has published several research papers and articles in reputed international and national journals and has presented papers in various national and international conferences. She also contributed one edited book and 10 book chapters on various topics.

T adiarajan completed his UG degree in the field of Computer Science Engineering in the year 2007 and postgraduate in 2010 in field of computer science. He has more than 13 years of teaching experience. His area of interest includes Machine Learning and datamining, data science. He has published research papers in various international journals. Currently, he is pursuing his PhD in field of deep learning.

Anshit Mukherjee is pursuing his BTech degree in computer science from the Abacus Institute of Engineering and Management. He has published many papers in journals, conferences and book chapters to his own credit.

Rjalakshmi Murugesan was born in Madurai, Tamil Nadu, India in 1988. She graduated from Electronics and Instrumentation Engineering of Kamaraj College of Engineering and Technology in 2010. She completed post-graduation in Degree in the Faculty of Instrumentation and Control Engineering from Kalasalingam Academy of Research and Education (KARE), in 2012 and completed her Ph.D. in the faculty of Electrical Engineering from Anna University, Chennai, 2020. She is currently employed as an Assistant Professor in the Thiagarajar College of Engineering, Mechatronics Department, Madurai. She has academic background, with Ten years of teaching experience as an Assistant/Associate professor at various institutions. Besides a research background, she has published several international journals (SCI/Scopus) and conferences (Scopus) from 2012 to till date. Her professional interests focus on Machine learning, Artificial Intelligence, linear and nonlinear control systems, system identification, and her current projects include modeling and controlling nonlinear processes (machine learning algorithms for Biomedical & Robotics).

N. Ayyanar received the B.E. degree in electronics and communication from the Narasu's Sarathy Institute of Technology, Salem, India, in 2013, the M.E. degree in optical communication from the Alagappa Chettiar Government College of Engineering and Technology, Karaikudi, India, in 2015, and the Ph.D. degree in electronics and communication engineering from the National Institute of Technology, Tiruchirappalli, India, in 2020. He is currently working as an Assistant Professor with the Department of Electronics and Communication Engineering, Thiagarajar College of Engineering, Madurai, Tamil Nadu, India. He has published 35 articles in refereed international journals and over 21 papers in conferences. His research interests include PCF-based optical fiber sensor, few mode fibers, fiber laser, plasmonics, and few mode amplifier system designs.

RjrmVssudev Pai Kuchelkar did his graduation in Mechanical Engineering from Goa Engineering College-Goa University (India) in 1994. He began his career in 1994 and has been working in the Automotive Engineering domain in a private organisation. As a part of his interest in the field of computers and information technology, he pursued a PG Diploma in Software Technology from National Center for Software Technology, Mumbai (India) in 2001, PG Diploma in Software Development (Full Stack Development) from upGrad (India)-IIIT Bengaluru (India) in 2021 followed by MS in Computer Science from upGrad (India)-Liverpool John Moores University, UK in 2022.

S enakshi is currently doing Ph.D in the Department of Electronics and Communication Engineering, Thiagarajar College of Engineering, Madurai – 15 (A Govt. Aided Autonomous Institution Affiliated to Anna University) Tamilnadu, India. She obtained her B.E degree in Instrumentation and control Engineering from Sethu Institute of Technology, Kariapatti, under Anna University, Chennai in the year of2009, and her M.Tech. Degree in the specialization of Control and Instrumentation Engineering from Kalasalingam University, Krishnankoil in the year 2012.

S Vijayalakshmi is presently working as Assistant Professor of Computer Science and Technology department at SNS COLLEGE OF ENGINEERING, COIMBATORE since July 2023. Earlier she has worked in PTR COLLEGE OF ENGINEERING AND TECHNOLOGY,MADURAI. She obtained her Master's degree M.E.,CSE in the year 2010 from Anna University Tirunelveli. She possess a vast experience of 13 years of teaching undergraduate and postgraduate engineering students.Her major areas of interest including software testing and Artificial Intelligence and ML.She guided many undergraduate and postgraduate projects.

I**R. Sharmila** had completed her Doctoral Degree in Computer Applications from Manonmaniam Sundaranar University, Tirunelveli. She has the teaching experience of 19 years and presently she is working as Professor and Head in the Department of Computer Applications, Karpagam Academy of Higher Education, Coimbatore. Her area of interest is Data mining, Bigdata Analytics, and Networks. She had published more than 15 Research articles and two books on Opinion mining and Data mining.

Piyanga Subbiah received the B.Tech degree in Information Technology from New Prince Shri Bhavani College of Engineering and Technology, Chennai, India, in 2015 and the M.E. degree in Computer Science and Engineering from Thiagarajar College of Engineering, Madurai, in 2020. He is currently pursuing the Ph.D. degree in Computer Science and Engineering at SRM Institute of Science and Technology, Chengalpattu, Tamil Nadu, India. From 2022, she was a Research Scholar with the Computer Science and Engineering. His research interest includes the processing of images and detection of disease in the plant using Artificial Intelligence.

Priyanga Subbiah received the B.Tech degree in Information Technology from New Prince Shri Bhavani College of Engineering and Technology, Chennai, India, in 2015 and the M.E. degree in Computer Science and Engineering from Thiagarajar College of Engineering, Madurai, in 2020. He is currently pursuing the Ph.D. degree in Computer Science and Engineering at SRM Institute of Science and Technology, Chengalpattu, Tamil Nadu, India. From 2022, she was a Research Scholar with the Computer Science and Engineering. His research interest includes the processing of images and detection of disease in the plant using Artificial Intelligence.

Krthikeyan Subramanian working as a Senior Lecturer in the Automobile Engineering Department at Birmingham City University, Ras Al Khaimah, UAE. He previously worked as an Associate Professor in the Mechanical Engineering Department at Kalasalingam University, India. He has been in the teaching profession for more than 13 years. He started his academic journey with a Bachelor's degree in Automobile Engineering from Sriram Engineering College, Anna University, Chennai. He pursued a Master's degree in Production Engineering from PSG College of Technology, Coimbatore. This educational experience provided me with a solid foundation in automobile and production engineering principles and practices. He earned a Ph.D. in Mechanical Engineering from Kalasalingam University. Additionally, he got experience as a Post-Doctoral Researcher at Szent Istvan University, Godollo, Hungary, where he contributed to cutting-edge research in a dynamic international environment. His academic and research background reflects my dedication to the field of Mechanical and Automobile Engineering.

Preethiya Thandapani received her B.E. degree from Anna University,Chennai in the year 2007 and M.E. degree from Anna University,Thiruchirapalli, Tamil Nadu, India in the year 2009. She has completed Ph.D from Kalasalingam Academy of Research and Education in the field of Mobile Wireless Sensor Network in the year 2020. Her research primarily centered on improving the energy efficiency of mobile unmanned vehicles engaged in communication, with a specific focus on factors related to their speed while in motion. After completing her Ph.D., her curiosity extended to Data Science, and she seamlessly merged it with her WSN expertise, charting a new research trajectory. With over 11 years of teaching experience, she has authored numerous research papers in the field of WSN and Data Science, particularly in addressing environmental monitoring healthcare challenges.

Preethiya Thandapani received her B.E. degree from Anna University,Chennai in the year 2007 and M.E. degree from Anna University,Thiruchirapalli, Tamil Nadu, India in the year 2009. She has completed Ph.D from Kalasalingam Academy of Research and Education in the field of Mobile Wireless Sensor Network in the year 2020. Her research primarily centered on improving the energy efficiency of mobile unmanned vehicles engaged in communication, with a specific focus on factors related to their speed while in motion. After completing her Ph.D., her curiosity extended to Data Science, and she seamlessly merged it with her WSN expertise, charting a new research trajectory. With over 11 years of teaching experience, she has authored numerous research papers in the field of WSN and Data Science, particularly in addressing environmental monitoring healthcare challenges.

Index

A

Abaqus 100, 102-104, 108
Artificial Intelligence 2-14, 16-23, 30-31, 37-38, 52-53, 55-58, 60, 65, 71, 131-135, 143-146, 150, 152-153, 155, 157, 159-160, 163-167, 169-170, 175, 177-178, 181, 184, 215, 219, 227, 231, 233-234, 238, 240, 242
Authentication 109, 113, 122, 124, 142, 174, 189
Autonomous Driving 10, 12, 16-18, 20-21, 30-32, 35, 38, 54-55, 62, 70-72
Autonomous Systems 1-8, 11-12, 16-17, 19, 96, 109, 131-133, 143-150, 157-158, 195, 227
Autonomous Vehicles 3, 6-7, 13, 15-21, 29, 31, 33-34, 52-55, 60, 70, 152, 155-156, 163

B

Bio-Inspired 196-198, 201, 203-208, 210-211, 213, 221, 225, 229, 231
Biomarkers 196, 198, 205, 210
Block Chain 131
Breakbone Fever 232

C

Cabin Monitoring 33, 52
Chatbots 7, 10, 131, 134, 144-145, 153
Chemotherapy 196, 206-207
Chicken Swarm Optimization 33, 37, 42-43, 45, 52
Clinical Translation 179, 196, 211
Computer Vision 1, 6-8, 10, 18, 32, 71, 161, 193
Continuum Robot 213-217, 220-225, 231

D

Data Acquisition Device 73-75, 80, 82-83, 85, 87, 91
Dengue 232-244
Diagnostics 1, 7, 57-58, 166, 194
Diffie-Helman 109
Dijkstra 109, 112, 114-115, 118, 121-122, 127, 155

E

Edge Computing 54, 57, 60, 65
Explainable AI 15-20, 30
External Stimuli 196, 210

F

Fault Diagnosis of Battery Cells 54, 64
Finite Element Method 95

H

Healthcare 2-4, 7-10, 12, 50, 131-143, 145-146, 148-150, 152-161, 163, 166-169, 171-184, 186-195, 213, 215, 218, 220, 227, 231, 233, 236, 240
Healthcare Robotics 166, 194-195
Hospital Automation 180
Human Interaction 1, 54, 61, 144, 190, 230, 241
Human-Machine Interaction 152
Hyperelastic 95, 103

I

Image Segmentation 15, 40-41
Immunotherapy 196-197, 203, 206
Industry Automation 1
Intelligent Coordination 166
Intelligent Interactions 152

K

K Means Clustering 15

L

Laparoscope 218, 231

M

Machine Learning 1-3, 5-6, 8-9, 14, 18, 21, 25-26, 36-37, 40, 51-52, 54, 56-57, 59-60, 65, 113-114, 128-136, 141, 143-147, 152, 155-156, 160-161, 163, 167, 169, 171, 176-177, 181, 183, 190, 193, 214-215, 227-228, 233, 237-240
Medical Operations 166
Medical Robotics 2, 170, 179-180, 194
MIS 213, 216-220, 222, 225-227, 231

N

Nanomachines 196, 210
Nanorobots 196-198, 200-212
Nanoscale 197-198, 203, 207-208, 210
NS2 109

O

Object Detection 7-8, 15-21, 25, 30-33, 35-37, 39, 46, 50, 62, 70-71
Outbreak Prediction 232

P

Patient Care 7-9, 159, 166, 180, 191, 225
Patient Care Technology 180
Predictive Analytics 131, 134-135, 137, 144-145

R

Radiotherapy 196, 206
Random Forest 15, 25, 33, 37, 42, 44-45, 52
Robot Operating 59, 64
Robotic-Assisted Procedures 180
Robotics 2-4, 7, 9-10, 12-13, 52, 54, 60, 71, 96-97, 99, 107, 153-156, 158-170, 172-174, 176-195, 211-215, 218-221, 226-231, 240

S

Seaweed Farming 95-98, 104
Secure 34, 55, 109, 112, 114-115, 119, 121-125, 127-130, 138-139, 146, 161, 174, 188
Semantic Object 15-19, 31
Sensor Networks 53, 92, 113, 128-130, 166
Smart Fogger Robot 232-233, 240-242
Soft Robotic Gripper 95, 98-100, 103-104
Soft Tissue 227, 231
Surgery 9, 13, 159, 184, 192, 196-198, 203, 206, 208, 213-214, 216-231
Surgical Robot 231

T

Telemedicine 131, 138-139, 143, 149, 160, 172, 186

U

Under Water Vehicle 95

V

VANET 109-113, 115, 117-119, 128-130
Vehicle Body Fatigue 73-75, 80
Vehicle Camera 33
Vertical Acceleration 73-75, 80, 84-85
Virus 232-233, 235-238, 240, 242

W

Wearables 139-140
Weight Optimization 73

Printed in the United States
by Baker & Taylor Publisher Services

Printed in the United States
by Baker & Taylor Publisher Services